Industrial Safety
and Health for
Goods and Materials Services

Handbook of Safety and Health for the Service Industry

Industrial Safety and Health for Goods and Materials Services

Industrial Safety and Health for Infrastructure Services

Industrial Safety and Health for Administrative Services

Industrial Safety and Health for People-Oriented Services

HANDBOOK OF
SAFETY AND HEALTH FOR THE SERVICE INDUSTRY

Industrial Safety
and Health for
Goods and Materials Services

Charles D. Reese

CRC Press
Taylor & Francis Group
Boca Raton   London   New York

CRC Press is an imprint of the
Taylor & Francis Group, an **informa** business

CRC Press
Taylor & Francis Group
6000 Broken Sound Parkway NW, Suite 300
Boca Raton, FL 33487-2742

© 2009 by Taylor & Francis Group, LLC
CRC Press is an imprint of Taylor & Francis Group, an Informa business

No claim to original U.S. Government works
Printed in the United States of America on acid-free paper
10 9 8 7 6 5 4 3 2 1

International Standard Book Number-13: 978-1-4200-5378-4 (Hardcover)

Library of Congress Cataloging-in-Publication Data

Reese, Charles D.
   Industrial safety and health for goods and materials services / Charles D. Reese.
     p. cm.
   Includes bibliographical references and index.
   ISBN 978-1-4200-5378-4 (alk. paper)
    1. Service industries--United States--Safety measures. 2. Service industries--Employees--Health and hygiene--United States. 3. Retail trade--Employees--Health and hygiene--United States. 4. Wholesale trade--Employees--Health and hygiene--United States. 5. Materials handling--United States--Safety measures. I. Title.

HD7269.S452U6742 2009
658.3'82--dc22
                                              2008013268

Visit the Taylor & Francis Web site at
http://www.taylorandfrancis.com

and the CRC Press Web site at
http://www.crcpress.com

Contents

Preface

Industrial Safety and Health for Goods and Materials Services deals with goods and materials services, which include the wholesale trade, retail trade, and warehousing and storage sectors. These three sectors handle myriad goods and materials such as furniture, construction materials, machines, equipment, paper products, appliances, hardware items, drugs and sundries, apparel, groceries, chemicals, petroleum products, beverages, assorted parts, electronic devices and products, and farming products. These are all packed in both large and small boxes or containers that must be handled, unpacked, sorted, and placed on displays or shelves by the workforce.

The workers in these three sectors face similar occupationally related safety and health hazards. Some of these hazards are compressed gases, ergonomics-related issues, lifting, material handling, slips, trips, falls, fires, hand tools, chemicals, machines, and equipment.

Although the aforementioned hazards are not the only hazards faced by workers in these sectors, they are the most common ones. Because of the diversity in the wholesale trade, retail trade, and warehousing and storage services, other job-specific hazards may result.

This workforce is susceptible to several hazards because of the constant handling and moving of a variety of goods and materials. These hazards can be managed by applying the principles of identification, intervention, and prevention, all of which are proven techniques of occupational safety and health.

In today's work environment, workers must be trained for emergencies and security must be provided for both workers and patrons. By adhering to acceptable safe work practices and occupational safety and health regulations, the safety and health of the workforce can be ensured while running a productive business. This book serves as a guide in achieving this objective.

Charles D. Reese, PhD

Author

For 30 years, **Charles D. Reese, PhD,** has been involved with occupational safety and health as an educator, manager, and consultant. In his early career, Dr. Reese was an industrial hygienist at the National Mine Health and Safety Academy. He later became manager for the nation's occupational trauma research initiative at the National Institute for Occupational Safety and Health's Division of Safety Research. Dr. Reese has played an integral role in trying to ensure workplace safety and health. As the managing director for the Laborers' Health and Safety Fund of North America, he was responsible for the welfare of the 650,000 members of the laborers' union in the United States and Canada.

Dr. Reese has developed many occupational safety and health training programs, which range from radioactive waste remediation to confined space entry. He has also written numerous articles, pamphlets, and books on related issues.

Dr. Reese, professor emeritus, was a member of the graduate and undergraduate faculty at the University of Connecticut, where he taught courses on Occupational Safety and Health Administration regulations, safety and health management, accident-prevention techniques, industrial hygiene, and ergonomics. As professor of environmental/occupational safety and health, he was instrumental in coordinating the safety and health efforts at the University of Connecticut. He is often invited to consult with industry on safety and health issues and is asked for expert consultation in legal cases.

Dr. Reese is also the principal author of the *Handbook of OSHA Construction Safety and Health (Second Edition)*; *Material Handling Systems: Designing for Safety and Health*; *Annotated Dictionary of Construction Safety and Health*; *Occupational Health and Safety Management: A Practical Approach*; and *Office Building Safety and Health and Accident/Incident Prevention Techniques*.

1 Introduction to the Service Industry

Retail stores sell many of the goods and materials provided by the goods and materials services sector.

The service industry consists of many different places of work, called establishments. Establishments are physical locations in which people work, such as a branch office of a bank, a gasoline station, a school, a department store, or an electricity generation facility. Establishments can range from large retail operations with corporate office complexes employing thousands of workers to small community stores, restaurants, professional offices, and service businesses employing only a few workers. Establishments should not be confused with companies or corporations that are legal entities. Thus, a company or corporation may have a single establishment or more than one establishment.

Establishments that use and provide the same services are organized together into industries. Industries are in turn organized together into industry sectors. These are further organized into subsectors. Each of the industry groups requires workers with varying skills and employs unique service techniques. An example of this is found in utilities, which employs workers in establishments that provide electricity, natural gas, and water. The service industry is broken down into the following supersectors:

Trade
    Retail trade (44 and 45)
    Wholesale trade (42)

Transportation and utilities
    Transportation (48)
    Warehousing (49)
    Utilities (22)
Information services (51)
Financial activities
    Financial and insurance sector (52)
    Real estate and rental and leasing sector (53)
Professional and business services
    Professional, scientific, and technical services (54)
    Management of companies and enterprises (55)
    Administrative and support and waste management and remediation (56)
Education and health services
    Educational services sector (61)
    Health care and social assistance sector (62)
Leisure and hospitality
    Arts, entertainment, and recreation sector (71)
    Accommodation and food sector (72)
Other services (81)

The service industry is the fastest growing industrial sector in the United States and has seen growth in the international arena. The service industry accounts for approximately 70% of the total economic activity in the United States according to the U.S. Bureau of Census. This non-goods-producing industry, which includes retail trade, wholesale trade, and other service-related industries as previously mentioned, has a very diverse grouping.

1.1 NAICS

With the passage of the North American Free Trade Agreement (NAFTA), it became apparent that the long employed standard industrial classification (SIC) was no longer very useful when dealing with industries found in Canada and Mexico. Consequently, the Bureau of Labor Statistics has developed a new system entitled the North American Industrial Classification System (NAICS).

NAICS uses a six-digit hierarchical coding system to classify all economic activity into 20 industry sectors. Five sectors are mainly goods-producing sectors and 15 are entirely services-producing sectors.

1.2 EMPLOYMENT IN THE SERVICE INDUSTRY

To have some idea of the numbers of employees addressed when speaking of the service industry, the worker population in each of the service industry sector is provided. The fast growing service industry as well as the number and variety of

TABLE 1.1

Employment in the Service Industry (2004)

Service Industry Sectors	Employment	Percentage of Service Industry (84,896,300)	Percentage of All Private Industries (107,551,800)
Wholesale trade	5,642,500	6.6	5.2
Retail trade	15,060,700	17.7	14.0
Warehousing	555,800	0.65	0.52
Transportation	3,450,400	4.1	3.2
Utilities	583,900	0.69	0.54
Administrative and support and waste management and remediation	7,829,400	9.2	7.3
Information	3,099,600	3.7	2.9
Finance and insurance	5,813,300	6.8	5.4
Real estate	2,077,500	2.4	1.9
Management of companies and enterprises	1,696,500	2.0	1.6
Professional, scientific, and technical services	6,768,900	8.0	6.3
Education services	2,079,200	2.4	1.9
Health and social services	14,005,700	16.5	13.0
Art, entertainment, and recreation	1,852,900	2.2	1.7
Accommodation and food services	10,614,700	12.5	9.9
Other services	3,785,200	4.5	3.5

Source: From Bureau of Labor Statistics. United States Department of Labor. Available at http://www.bls.gov, Washington 2007.

occupations within each sector provides a window into the safety and health hazards that need to be addressed within each sector of the service industry's workforce (Table 1.1).

1.3 SAFETY

One of the most telling indicators of working condition is an industry's injury and illness rates. Overexertion, being struck by an object, and falls on the same level are among the most common incidents causing work-related injuries.

The service industry is a large umbrella that encompasses many types of businesses, each of which has its own safety and health issues. Some of the service industries' businesses are more hazardous than others. This book does not address each sector independently, but provides the tools and information needed to address the hazards and safety and health issues within each sector of the service industry.

The service industry is made up of a large number of widely dissimilar industry sectors. Each sector has its own unique functions that result in each sector having its own set of unique hazards that the individual workforces must face and that their employers must address.

The functioning of each sector results in different types of energy being released, and therefore the differences in the types of accidents, incidents, injuries, and illnesses that occur. Thus, the hazards and energy sources dictate the specific Occupational Safety and Health Administration (OSHA) regulations that each sector is compelled to be in compliance with.

For these reasons, it is not possible to describe or address the service industry in the same manner as construction, shipyard, or office building industries, where workforces perform similar tasks and thus face similar hazards.

Each hazard is discussed based upon the type of energy released, and its ability to cause specific accidents or incidents. In each section that describes hazards, the best practices for intervention and prevention of the release of the specific energy are emphasized. This approach will allow for the identification and prevention of hazards, and for framing regulations by any service industry sector. It will allow for a similar approach to address areas where the service industry sectors are alike and can be addressed as a collective industry rather than as individual sectors, for example, the same sectors will need to address areas such as compliance with OSHA regulations, conducting training, and effective management of safety and health.

The intent of this book is to provide a source for the identification and prevention of most of the injuries and illnesses occurring in the service industry. Also, it summarizes applicable safety standards that impact the service industry as well as address how to work with and around OSHA to comply with its regulations. The book covers safety hazards involved with confined spaces, electrical equipment, falls, forklifts, highway vehicles, preventive maintenance activities, handling chemicals, radiation, welding, etc. The content describes the safety hazard as applied to the type of energy released or to the unique event that occurs from exposure to the hazard.

The question that we should be asking, "is the kind of safety being practiced preventing the destruction that we see in the American workforce?" Maybe we ought to ask how we define safety for a start. Here are some definitions of safety that may be useful:

1. Doing things in a manner so that no one will get hurt and so the equipment and product will not get damaged
2. Implementation of good engineering design, personnel training, and the common sense to avoid bodily harm or material damage
3. Systematic planning and execution of all tasks so as to produce safe products and services with relative safety to people and property
4. Protection of persons and equipment from hazards that exceed normal risk
5. Application of techniques and controls to minimize the hazards involved in a particular event or operation, considering both potential personal injury and property damage
6. Employing processes to prevent accidents both by conditioning the environment as well as conditioning the person toward safe behavior
7. Function with minimum risk to personal well-being and to property
8. Controlling exposure to hazards that could cause personal injury and property damage

FIGURE 1.1   Safety should be an integral part of goods and material handling.

9. Controlling people, machines, and the environment that could cause personal injury or property damage (Figure 1.1)
10. Performing your daily tasks in the manner that they should be done, or when you do not know, seek the necessary knowledge
11. Elimination of foreseen hazards and the necessary training to prevent accidents or to provide limited acceptable risk to personnel and facilities

Do any of these definitions match the safety guidelines practiced in workplaces where you have responsibility? If so, have you had any deaths or injuries to any of your workforce? When injuries and illnesses are not occurring anymore then the ultimate goals have been reached. Now comes the task of maintaining what has been gained.

As one can see, safety can be defined in many different ways. Nearly all of these definitions include property damage as well as personal injury. It shows that the thinking is in the right direction and that safety consists of a total loss-control activity. The book's content contains only one facet of a comprehensive safety effort, which is a never ending journey.

1.4   WHY TRAUMA PREVENTION?

There are very real advantages when addressing trauma prevention, which do not exist when addressing illness prevention. The advantages are

- Trauma occurs in real time with no latency period (an immediate sequence of events).
- Accident/incident outcomes are readily observable (only have to reconstruct a few minutes to a few hours).
- Root or basic causes are more clearly identified.
- It is easy to detect cause and effect relationships.
- Traumas are not difficult to diagnose.
- Trauma is highly preventable.

1.5   ACCIDENTS OR INCIDENTS

The debate over the use of the term "accidents" versus "incidents" has been long and continual. Although these terms are used virtually interchangeably in the context of this book, you should be aware of the distinction between the two. Accidents are usually defined as an unexpected, unplanned, or uncontrollable event or mishap. This undesired event results in personal injury and property damage or both and may also lead to equipment failure. An incident is all of the above as well as the adverse effects on production.

This definition for an accident underlies the basic foundation of this book. The philosophy behind this book is that we can control these types of events or mishaps by addressing the existence of hazards and taking steps to remove or mitigate them as part of the safety effort. This is why we spend time identifying hazards and determining risk. Thus, the striving for a safe workplace, where the associated risks are judged to be acceptable, is the goal of safety. This will result in freedom from those circumstances that can cause injury or death to workers, and damage to or loss of equipment or property.

The essence of this book's approach is that we can control those factors, which are the causing agents of accidents. Hazard prevention described in this book is addressed both from practical and regulatory approaches.

1.6   COMPREHENSIVE ACCIDENT PREVENTION

Accident prevention is very complex because of interactions that transpire within the workplace. These interactions are between

- Workers
- Management
- Equipment/machines
- Environment

The interaction between workers, management, equipment/machinery, and the workplace environment have enough complexity themselves as they try to blend together in the physical workplace environment. However, this physical environment is not the only environment that has an impact upon the accident prevention effort in companies. The social environment is also an interactive factor that encompasses our lives at work and beyond. Government entities that establish rules and regulations leave their mark upon the workplace. But others in the social arena such as unions, family, peer pressure, friends, and associates also exert pressure on the workplace environment. The extent of the interactions that must be attended to for having a successful accident prevention effort is paramount.

Many workplaces have high accident incidence rates because they are hazardous. Hazards are dangerous situations or conditions that can lead to accidents. The more hazards present, the greater the chance of accidents. Unless safety procedures are followed, there will be a direct relationship between the number of hazards in the workplace and the number of accidents that will occur there.

In most industries, people work together with machines in an environment that causes employees to face hazards that can lead to injury, disability, or even death. To prevent industrial accidents, the people, machines, and other factors, which can cause accidents, including the energies associated with them, must be controlled. This can be done through education and training, good safety engineering, and enforcement.

Many accidents can be prevented. One study showed that 88% were caused by human failure (unsafe acts), 10% by mechanical failure (unsafe conditions), and only 2% were beyond human control (acts of God).

If workers are aware of what hazards are, and what can be done to eradicate them, many accidents can be prevented. For a situation to be called an accident, it must have certain characteristics. The personal injury may be considered minor when it requires no treatment or only first aid. Personal injury is considered serious if it results in a fatality or in a permanent, partial, or temporary total disability (lost-time injuries). Property damage may also be minor or serious.

1.7 FATALITY AND INJURY PROFILE FOR THE SERVICE INDUSTRY

In 2005, there were 5702 occupationally related deaths in all of private industry, while the service industry had 2736 (48%) of these fatalities the goods-producing industry had 42% fatalities. In Table 1.2, the major contributors to these fatalities are depicted.

TABLE 1.2
Occupational Death Cause in Percent

Cause	Service Industry (%)	All Private Industries (%)
Highway	34	25
Homicides	16	10
Falls	9	13
Struck-by	7	11

Source: From Bureau of Labor Statistics, United States Department of Labor. *National Census of Fatal Occupational Injuries in 2005.* Available at http://bls.gov.

Injuries are examined somewhat differently and the statistical data are presented usually in four different ways. These are as follows:

1. Nature of injury or illness names the principal physical characteristic of a disabling condition, such as sprain/strain, cut/laceration, or carpal tunnel syndrome.
2. Part of body affected is directly linked to the nature of injury or illness cited, for example, back sprain, finger cut, or wrist and carpal tunnel syndrome.
3. Source of injury or illness is the object, substance, exposure, or bodily motion that directly produced or inflicted the disabling condition cited. Examples are a heavy box, a toxic substance, fire/flame, and bodily motion of injured/ill worker.
4. Event or exposure (type of accident) signifies the manner in which the injury or illness was produced or inflicted, for example, overexertion while lifting or fall from a ladder (see Appendix A).

Tables 1.3 through 1.6 allow us to start identifying the most common facets of an injury profile. The total employment for the service industry in 2004 was 84,896,300 and the total number of injuries was 850,930. The data in the tables denote the most frequently occurring factor resulting in the injury/incident or resulting from the injury/incident.

It would appear from a rough observation of Tables 1.3 through 1.6 that a service industry employee would suffer a sprain or strain to the trunk and in most cases the back or possibly the lower or upper extremities because of one of the three causes: worker motion/position; floors, walkways, or ground surfaces; or containers that

---

TABLE 1.3
Nature of Injury by Number and Percent for the Service Industry

Nature of Injury	Number	Percent
Sprains/strains[a]	377,760	44
Fractures[a]	55,450	6.5
Cuts/punctures[a]	63,220	7
Bruises[a]	82,610	10
Heat burns	12,780	1.5
Chemical burns	4,330	0.5
Amputations	2,710	0.3
Carpal tunnel syndrome	10,810	1.3
Tendonitis	3,950	0.4
Multiple trauma[a]	34,450	4
Back pain (only)	28,600	3

Source: From Bureau of Labor Statistics, United States Department of Labor. *Workplace Injuries and Illnesses in 2004.* Available at http://bls.gov.

[a] Five most frequently occurring conditions.

TABLE 1.4
Body Part Injured by Number and Percent
for the Service Industry

Body Part Injured	Number	Percent
Head	51,500	6
Eyes	19,070	2
Neck	15,960	1.8
Trunk[a]	314,190	37
Back[a]	204,240	24
Shoulder	56,350	7
Upper extremities[a]	173,260	20
Finger	58,080	6.8
Hand	30,810	3.6
Wrist	38,000	4.5
Lower extremities[a]	183,780	22
Knee	69,250	8
Foot and toe	39,050	4.6
Body systems	10,940	1.3
Multiple body parts[a]	95,490	11

Source: From Bureau of Labor Statistics, United States Department of Labor. *Workplace Injuries and Illnesses in 2004.* Available at http://bls.gov.

[a] Five most frequently injured body parts.

TABLE 1.5
Source of Injury by Number and Percent
for the Service Industry

Sources of Injuries	Number	Percent
Parts and materials	51,680	6
Worker motion/position[a]	119,340	14
Floor, walkways, or ground surfaces[a]	168,620	20
Hand tools	29,420	3.5
Vehicles[a]	88,830	10
Health care patient[a]	57,220	6.7
Chemicals and chemical products	11,070	1.3
Containers[a]	124,700	15
Furniture and fixtures	36,700	4
Machinery	40,940	4.8

Source: From Bureau of Labor Statistics, United States Department of Labor. *Workplace Injuries and Illnesses in 2004.* Available at http://bls.gov.

[a] Five most frequent sources of injury.

TABLE 1.6
Exposure/Accident Type by Number and Percent for the Service Industry

Type of Accidents	Number	Percent
Struck by an object[a]	101,390	12
Struck against an object	51,670	6
Caught in or compressed or crushed	25,290	3
Fall to lower level	46,820	5.5
Fall on same level[a]	130,260	15
Slips or trips without a fall	27,400	3
Overexertion[a]	227,350	27
Lifting[a]	126,380	15
Repetitive motion	27,180	3.2
Exposure to harmful substance or environment	36,070	4
Transportation accidents[a]	51,070	6
Fires and explosions	1,100	0.1
Assaults/violent acts	22,790	2.7

Source: From Bureau of Labor Statistics, United States Department of Labor. *Workplace Injuries and Illnesses in 2004.* Available at http://bls.gov.

[a] Five most frequent exposures or type of accidents that led to an injury.

resulted in an overexertion/lifting or fall on the same level. As it can be seen, these data give us some information to start our search for the hazards that contributed to these injuries.

1.8   OCCUPATIONAL ILLNESSES IN THE SERVICE INDUSTRY

Occupational illnesses have always been underreported. For this reason, they do not seem to get the same attention as injuries since their numbers or causes are not of epidemic proportions. The reasons why illnesses are not reported include the following:

- Not occurring in real time and usually having a latency period before signs and symptoms occur.
- Not readily observable and have been linked to personal habits and exposure from hobbies. There is the question of multiple exposures and synergistic effects on-the-job and off-the-job.
- Not always easy to detect cause and effect relationships.
- Often difficult to diagnose since many exhibit flu or cold symptoms.

These are not excuses for not pursuing preventive strategies, but an explanation of why occupational illnesses are more difficult to accurately describe and identify their

TABLE 1.7

Occupational Illnesses by Number of Cases and Percent for the Service Industry

Illness Type	Number	Percent
Skin diseases and disorders	24,900	19
Respiratory conditions	13,000	10
Poisoning	2,000	1.5
Hearing loss	4,000	3
All others	87,400	66.5

Source: From Bureau of Labor Statistics, United States Department of Labor. *Workplace Injuries and Illnesses in 2004.* Available at http://bls.gov.

root cause. The 84,896,300 employees experienced 131,500 (53%) cases of illness during 2004 according to the Bureau of Labor Statistics. This compares to the total number of illnesses reported by all of industry that equaled 249,000 of which 53% was attributed to the service industry. The most common reported types of occupational illnesses for this period are found in Table 1.7.

The remainder of this book is directed toward managing, preventing, and controlling hazards that occur within the goods and material service sector of the service industry. This includes the wholesale trade, retail trade, and warehousing sectors.

It is important to keep in mind that because of the complexity and diversity within the industry sectors of the service industry, no cookie cutter approach could be used nor is a one-size-fits-all approach possible. There has to be a mixing of information and data from diverse sources such as the NAICS and the outdated SIC, since all agencies have not changed to the new system. Also, even within the supersectors and the sectors themselves there is not a common approach to the management of safety and health, identification of hazards compatible, or the same approach for each varied sector, nor should we expect there to be. This is the reason that by consulting the table of contents of this book and *Industrial Safety and Health for Infrastructure Services, Industrial Safety and Health for Administrative Services,* and *Industrial Safety* and *Health for People-Oriented Services,* decisions can be made regarding which book would be most useful to your particular business. In some cases, one book will fulfill a company's safety and health needs while in other cases all the four books will be most beneficial.

REFERENCES

Bureau of Labor Statistics, United States Department of Labor. Available at http://www.bls. gov, Washington, 2007.

Bureau of Labor Statistics, United States Department of Labor. *National Census of Fatal Occupational Injuries in 2005.* Available at http://bls.gov.

Bureau of Labor Statistics, United States Department of Labor. *Workplace Injuries and Illnesses in 2004.* Available at http://bls.gov.

2 Wholesale Trade

Goods and materials are delivered by the wholesaler to the retailer.

The wholesale trade (42) sector comprises establishments engaged in wholesaling merchandise, generally without transformation, and rendering services incidental to the sale of merchandise. The wholesaling process is an intermediate step in the distribution of merchandise. Wholesalers are organized to sell or arrange the purchase or sale of (1) goods for resale (i.e., goods sold to other wholesalers or retailers), (2) capital or durable nonconsumer goods, and (3) raw and intermediate materials and supplies used in production. Wholesalers sell merchandise to other businesses and normally operate from a warehouse or office. BLS (Bureau of Labor Statistics) data show that wholesale and retail trades make up a large part of the nation's employment and business establishments. In the economy as a whole, wholesale trade represents about 4.4% of all employment and 7.1% of all establishments. BLS estimates show that wholesale trade employment averaged 5,749,500 during 2005.

The composition of the wholesale trade in the North American Industry Classification System (NAICS) breakdown is as follows:

Wholesale trade (42)
  Merchant wholesaler, durable goods (423000)
    Motor vehicle and motor vehicle parts and supplies merchant wholesalers (423100)
    Furniture and home furnishing merchant (423200)
    Lumber and other construction materials merchant wholesalers (423300)

Professional and commercial equipment and supplies merchant wholesalers (423400)

Metal and mineral (except petroleum) merchant wholesalers (423500)

Electrical and electronic goods merchant wholesalers (423600)

Hardware and plumbing and heating equipment and supplies merchant wholesalers (423700)

Machinery, equipment, and supplies merchant wholesalers (423800)

    Farm and garden machinery and equipment merchant wholesalers (423820)

Miscellaneous durable goods merchant wholesalers (423900)

Merchant wholesalers, nondurable goods (424000)

Paper and paper product merchant wholesalers (424100)

Drugs and druggist' sundries merchant wholesalers (424200)

Apparel, piece goods, and notions merchant wholesalers (424300)

Grocery and related product wholesalers (424400)

Farm product raw material merchant wholesalers (424500)

Chemical and allied products merchant wholesalers (424600)

Petroleum and petroleum products merchant wholesalers (424700)

Beer, wine, and distilled alcoholic beverage merchant wholesalers (424800)

Miscellaneous nondurable goods merchant wholesalers (424900)

Wholesale electronic markets and agents and brokers (425000)

Wholesale electronic markets and agents and brokers (425100)

The wholesale sector is characterized by most workplaces being small, employing fewer than 50 workers. About 7 in 10 work in office and administrative support, sales, or transportation and material-moving occupations. While some jobs require a college degree, a high school education is sufficient for most jobs.

When consumers purchase goods, they usually buy them from a retail establishment, such as a supermarket, department store, gas station, or Internet site. When retail establishments, other businesses, governments, or institutions—such as universities or hospitals—need to purchase goods for their own use—such as equipment, motor vehicles, office supplies, or any other items—or for resale to consumers, they normally buy them from wholesale trade establishments (Figure 2.1).

The size and scope of firms in the wholesale trade industry vary greatly. Wholesale trade firms sell any and every type of goods. Customers of wholesale trade firms buy goods for making other products, as in the case of a bicycle manufacturer who purchases steel tubing, wire cables, and paint. Customers may also purchase items for daily use, as when a corporation buys office furniture, paper clips, or computers, or for resale to the public, as does a department store that purchases socks, flatware, or televisions. Wholesalers may offer only a few items for sale, perhaps all made by one manufacturer, or they may offer thousands of items produced by hundreds of different manufacturers. Some wholesalers sell only a narrow range of goods, such as very specialized machine tools; while others sell a broad range of goods, such as all the supplies necessary to open a new store, including shelving, light fixtures, wallpaper, floor coverings, signs, cash registers, accounting ledgers, and perhaps even some merchandise for resale.

FIGURE 2.1 Wholesalers process and deliver goods utilizing smaller trucks in most instances.

Wholesale trade firms are essential to the economy. They simplify product, payment, and information flows by acting as intermediaries between the manufacturer and the customer. They store goods that neither manufacturers nor retailers can store until consumers require them. In so doing, they fulfill several roles in the economy. They provide businesses with a nearby source of goods made by many different manufacturers; they provide manufacturers with a manageable number of customers, while allowing their products to reach a large number of users; and they allow manufacturers, businesses, institutions, and governments to devote minimal time and resources to transactions by taking on some sales and marketing functions—such as customer service, sales contact, order processing, and technical support—that manufacturers otherwise would have to perform.

There are two main types of wholesalers: merchant wholesalers and wholesale electronic markets, agents, and brokers. Merchant wholesalers generally take title to the goods that they sell; in other words, they buy and sell goods on their own account. They deal in either durable or nondurable goods. Durable goods are new or used items that generally have a normal life expectancy of 3 years or more. Establishments in this sector of wholesale trade are engaged in wholesaling goods, such as motor vehicles, furniture, construction materials, machinery and equipment (including household appliances), metals and minerals (except petroleum), sporting goods, toys and hobby goods, recyclable materials, and parts. Nondurable goods are items that generally have a normal life expectancy of less than 3 years. Establishments in this sector of wholesale trade are engaged in wholesaling goods, such as paper and paper products, chemicals and chemical products, drugs, textiles and textile products, apparel, footwear, groceries, farm products, petroleum and petroleum products, alcoholic beverages, books, magazines, newspapers, flowers and nursery stock, and tobacco products. The merchant wholesale sector also includes the individual sales offices and sales branches (but not retail stores) of manufacturing and mining enterprises that are specifically set up to perform the sales and marketing of their products.

Firms in the wholesale electronic markets, agents, and brokers subsector arrange for the sale of goods owned by others, generally on a fee or commission basis. They act on behalf of the buyers and sellers of goods, but generally do not take ownership of the goods. This sector includes agents and brokers as well as business-to-business electronic markets that use electronic means, such as the Internet or electronic data interchange (EDI), to facilitate wholesale trade.

Only firms that sell their wares to businesses, institutions, and governments are considered part of wholesale trade. As a marketing ploy, many retailers that sell mostly to the general public present themselves as wholesalers. For example, wholesale price clubs, factory outlets, and other organizations are retail establishments, even though they sell their goods to the public at wholesale prices.

Besides selling and moving goods to their customers, merchant wholesalers may provide other services to clients, such as the financing of purchases, customer service and technical support, marketing services such as advertising and promotion, technical or logistical advice, and installation and repair services. After customers buy equipment, such as cash registers, copiers, computer workstations, or various types of industrial machinery, assistance may be needed to integrate the products into the customer's workplace. Wholesale trade firms often employ workers to visit customers, install or repair equipment, train users, troubleshoot problems, or advise on how to use the equipment most efficiently.

Working conditions and physical demands of wholesale trade jobs vary greatly. Moving stock and heavy equipment can be strenuous, but freight, stock, and material movers may make use of forklifts in large warehouses. Workers in some automated warehouses use computer-controlled storage and retrieval systems that further reduce labor requirements. Employees in refrigerated meat warehouses work in a cold environment and those in chemical warehouses often wear protective clothing to avoid harm from toxic chemicals. Outside sales workers are away from the office for much of the workday and may spend a considerable amount of time traveling. On the other hand, most management, administrative support, and marketing staff work in offices.

Overall, working conditions are relatively safe in wholesale trade. In 2003, there were 4.7 work-related injuries or illnesses per 100 full-time workers, as against 5.0 per 100 for the entire private sector. However, not all wholesale trade sectors are equally safe. Occupational injury and illness rates were considerably higher than the national average for wholesale trade workers who dealt with lumber and construction materials (7.1 per 100 workers); motor vehicle and motor vehicle parts and supplies (6.2 per 100 workers); groceries (7.5 per 100 workers); and beer, wine, and distilled beverages (10.9 per 100 workers).

Most workers put in long shifts, particularly during peak times, and others, such as produce wholesalers who start work before dawn to receive shipments of vegetables and fruits, work unusual hours.

2.1 PROFILE OF RETAIL WORKERS' DEATHS, INJURIES, AND ILLNESSES

2.1.1 DEATHS

There were 204 occupationally related deaths to retail workers in 2005. The wholesale sector accounted for 7% of the service industry deaths (2736). Table 2.1 shows the percent of those deaths from each major category.

2.1.2 INJURIES

There were 178,760 reported injuries for retail workers in 2004; this was 21% of the total injuries (850,930) for the service industry. The distributions for the nature, body part, source, and exposure (accident type) for the 178,760 injuries are presented in Tables 2.2, 2.3, 2.4, and 2.5, respectively.

2.1.3 ILLNESSES

In the wholesales sector, 7100 cases of occupational illness were reported; this is 5% of the total reported illnesses for the service industry. Table 2.6 provides the breakdown of the illnesses.

2.2 HAZARDS FACED BY WHOLESALE WORKERS

The hazards of working with all types of products and the handling of these products as well as the assurance that the products are delivered accurately and in good condition contribute to the hazards faced by wholesale workers.

The hazards covered in this book are the primary ones that affect wholesale workers in processing and warehousing facilities. In most cases, the most frequent hazards faced by wholesale workers are as follows:

- Walking and working surfaces
- Electrocutions
- Overexertion
- Material handling/lifting of containers
- Slips, trips, and falls
- Strains/sprains
- Trauma injuries
- Vehicle accidents
- Forklifts
- Power tools
- Office hazards

2.3 OCCUPATIONS

Many occupations are involved in wholesale trade, but not all are represented in every type of wholesale trade firm. Merchant wholesalers, by far, make up the largest

TABLE 2.1
Occupational Death Cause by Percent
for Wholesale Sector

Cause	Wholesale Sector (%)
Highway	44
Homicides	2
Falls	8
Struck-by	13

Source: From Bureau of Labor Statistics, U.S. Department of Labor. *National Census of Fatal Occupational Injuries in 2005.* Available at http://bls.gov.

part of the industry. The activities of these wholesale trade firms commonly center on storing, selling, and transporting goods. As a result, the three largest occupational groups in the industry are office and administrative support workers, many of whom work in inventory management; sales and related workers; and workers in transportation and material-moving occupations, most of whom are truck drivers and material movers. In 2004, 70% of wholesale trade workers were concentrated in these three groups. Common occupations in the wholesale sector are bookkeeping, accounting, and auditing clerks; computer, automated teller, and office-machine repairers; order clerks; purchasing managers, buyers, and purchasing agents; sales

TABLE 2.2
Nature of Injury by Number and Percent
for the Wholesale Sector

Nature of Injury	Number	Percent
Sprains/strains[a]	35,700	44
Fractures[a]	6,310	7.8
Cuts/punctures[a]	6,320	7.8
Bruises[a]	7,100	8.8
Heat burns	490	0.6
Chemical burns	310	0.3
Amputations	470	0.6
Carpal tunnel syndrome	800	1.1
Tendonitis	130	0.2
Multiple trauma[a]	2,950	3.6
Back pain only	3,300	4

Source: From Bureau of Labor Statistics, U.S. Department of Labor. *Workplace Injuries and Illnesses in 2004.* Available at http://bls.gov.

[a] Five most frequently occurring conditions.

TABLE 2.3
Body Part Injured by Number and Percent
for the Wholesale Sector

Body Part Injured	Number	Percent
Head	4,580	5.7
Eyes	1,810	2
Neck	1,330	1.6
Trunk[a]	32,370	40
Back[a]	22,260	27
Shoulder	4,570	5.6
Upper extremities[a]	14,290	17.6
Finger	5,530	6.8
Hand	2,850	3.5
Wrist	2,740	3
Lower extremities[a]	20,410	25
Knee	6,340	7.8
Foot and toe	5,660	7
Body systems	460	0.6
Multiple body parts[a]	7,240	8.9

Source: From Bureau of Labor Statistics, U.S. Department of Labor. *Workplace Injuries and Illnesses in 2004.* Available at http://bls.gov.

[a] Five most frequently injured body parts.

TABLE 2.4
Source of Injury by Number and Percent for the Wholesale
Sector

Sources of Injuries	Number	Percent
Parts and materials[a]	9,730	12
Worker motion/position[a]	12,940	16
Floor, walkways, or ground surfaces[a]	13,100	16
Hand tools	2,590	3
Vehicles[a]	12,040	15
Health care patient	0	0
Chemicals and chemical products	690	0.9
Containers[a]	15,260	19
Furniture and fixtures	1,980	2.4
Machinery	5,080	6

Source: From Bureau of Labor Statistics, U.S. Department of Labor. *Workplace Injuries and Illnesses in 2004.* Available at http://bls.gov.

[a] Five most frequent sources of injury.

TABLE 2.5

Exposure/Accident Type by Number and Percent for the Wholesale Sector

Type of Accidents	Number	Percent
Struck by object[a]	11,750	14
Struck against object	5,080	6
Caught in or compressed or crushed	4,100	5
Fall to lower level	5,800	7
Fall on same level[a]	7,690	9
Slips or trips without a fall	3,280	4
Overexertion[a]	21,350	39
Lifting[a]	13,290	16
Repetitive motion	2,190	2.7
Exposure to harmful substance or environment	1,950	2.4
Transportation accident[a]	6,270	7.7
Fires and explosions	160	0.2
Assaults/violent acts	390	0.4

Source: From Bureau of Labor Statistics, U.S. Department of Labor. *Workplace Injuries and Illnesses in 2004.* Available at http://bls.gov.

[a] Five most frequent exposures or type of accidents that led to an injury.

engineers; sales representatives, wholesale and manufacturing; shipping, receiving, and traffic clerks; stock clerks and order fillers; and truck drivers and driver/sales workers.

Most office and administrative support workers need to have at least a high school diploma, and some related experience or additional schooling is an asset. As in most industries, many secretaries and administrative assistants; bookkeeping, accounting, and auditing clerks; and general office clerks are employed in wholesale trade. Most of the other administrative support workers are needed to control inventory. Shipping,

TABLE 2.6

Occupational Illnesses by Number of Cases and Percent for the Wholesale Sector

Illness Type	Number	Percent
Skin diseases and disorders	1500	21
Respiratory conditions	700	10
Poisoning	0	0
Hearing loss	600	8
All others	4300	61

Source: From Bureau of Labor Statistics, U.S. Department of Labor. *Workplace Injuries and Illnesses in 2004.* Available at http://bls.gov.

receiving, and traffic clerks check the contents of all shipments, and verify condition, quantity, and sometimes shipping costs. They use computer terminals or barcode scanners and, in small firms, pack and unpack goods. Order clerks handle order requests from customers, or from the firm's regional branch offices in the case of a large, decentralized wholesaler. These workers take and process orders, and route them to the warehouse for packing and shipment. Often, they must be able to answer customer inquiries about products and monitor inventory levels or record sales for the accounting department. Stock clerks and order fillers code or price goods and store them in the appropriate warehouse sections. They also retrieve from stock the appropriate type and quantity of goods ordered by customers. In some cases, they also perform tasks similar to those performed by shipping and receiving clerks.

Like office and administrative support workers, sales and related workers typically do not need postsecondary training, but many employers seek applicants with prior sales experience. Generally, workers in marketing and sales occupations try to interest customers in purchasing a wholesale firm's goods and assist them in buying the goods. There are three primary types of salespeople in wholesale firms: inside sales workers, outside sales workers, and sales worker supervisors.

Inside sales workers generally work in sales offices taking sales orders from customers. They are also increasingly performing duties such as problem solving, solicitation of new and existing customers, and handling complaints. Outside sales workers, also called sales representatives or sales engineers, are the most skilled workers and one of the largest occupations in wholesale trade. They travel to places of business—whether manufacturers, retailers, or institutions—to maintain contact with current customers or to attract new ones. They make presentations to buyers and management or demonstrate items to production supervisors. In the case of complex equipment, sales engineers often need a great deal of highly technical knowledge, often obtained through postsecondary training. As more customers gather information and complete orders through the Internet, outside sales workers are devoting more time to developing prospective clients and offering services to existing clients such as installation, maintenance, and advising on the most efficient use of purchases. Sales representatives and sales engineers also may be known as manufacturers' representatives or agents in some wholesale trade firms. Sales worker supervisors monitor and coordinate the work of the sales staff and often do outside sales work themselves. Counter clerks wait on customers who come to the firm to make a purchase.

Transportation and material-moving workers move goods around the warehouse, pack and load goods for shipment, and transport goods to buyers. Laborers and freight, stock, and material movers manually move goods to or from storage and help to load delivery trucks. Hand packers and packagers also prepare items for shipment. Industrial truck and tractor operators use forklifts and tractors with trailers to transport goods within the warehouse, to outdoor storage facilities, or to trucks for loading. Truck drivers transport goods between the wholesaler and the purchaser or between distant warehouses. Drivers of medium and heavy trucks need a state commercial driver's license (CDL). Driver/sales workers deliver goods to customers, unload goods, set up retail displays, and take orders for future deliveries. They are responsible for maintaining customer confidence and keeping clients well stocked. Sometimes these workers visit prospective clients, hoping to generate new business.

Management and business and financial operations workers direct the operations of wholesale trade firms. General and operations managers and chief executives supervise workers and ensure that operations meet standards and goals set by the top management. Managers with ownership interest in smaller firms often also have some sales responsibilities. First-line supervisors oversee warehouse workers—such as clerks, material movers, and truck drivers—and see that standards of efficiency are maintained.

To provide manufactured goods to businesses, governments, or institutional customers, merchant wholesalers employ large numbers of wholesale buyers and purchasing managers. Wholesale buyers purchase goods from manufacturers for resale, based on price and what they think customers want. Purchasing managers coordinate the activities of buyers and determine when to purchase what types and quantities of goods.

Many wholesalers do not just sell goods to other businesses, they also install and service these goods. Installation, maintenance, and repair workers set up, service, and repair these goods. Others maintain vehicles and other equipment. For these jobs, firms usually hire workers with maintenance and repair experience or mechanically inclined individuals who can be trained on the job.

2.4   APPLICABLE OSHA REGULATIONS

Another way to gather an understanding of the hazards faced by retail workers is to see the types of violations that the Occupational Safety and Health Administration (OSHA) has found during their inspections of retail establishments. These violations provide another way of targeting hazards that have the potential to cause injury, illness, and death of workers. As can be seen from the 50 most frequently cited violations, OSHA cites this industry under the general industry standard (29 CFR 1910) and the recordkeeping standard (29 CFR 1904) (Table 2.7).

With the hazards faced by this sector, it is imperative that safety and health be an integral part of doing business and with the specific purpose of protecting its employees.

TABLE 2.7
Fifty Most Frequent OSHA Violations for the Wholesale Sector

CFR Standard	Number Cited	Description
1910.178	379	Powered industrial trucks (forklifts)
1910.1200	277	Hazard communication
1910.147	189	The control of hazardous energy, lockout/tagout
1910.305	181	Electrical, wiring methods, components and equipment
1910.303	141	Electrical systems design, general requirements
1910.23	126	Guarding floor and wall openings and holes
1910.134	114	Respiratory protection
1910.157	114	Portable fire extinguishers

TABLE 2.7 (continued)
Fifty Most Frequent OSHA Violations for Retail Sector

CFR Standard	Number Cited	Description
1910.212	113	Machines, general requirements
1910.37	112	Maintenance, safeguards, and operational features for exit routes
1910.132	112	Personal protective equipment, general requirements
5A1	91	General duty clause (section of OSHA Act)
1910.22	81	Walking–working surfaces, general requirements
1910.151	73	Medical services and first aid
1910.219	73	Mechanical power-transmission apparatus
1910.215	67	Abrasive wheel machinery
1910.95	65	Occupational noise exposure
1904.29	63	Forms
1910.304	57	Electrical, wiring design and protection
1910.1025	57	Lead
1910.253	54	Oxygen-fuel gas welding and cutting
1910.176	53	Materials handling, general
1910.272	51	Grain handling facilities
1904.32	43	Annual summary
1910.36	41	Design and construction requirements for exit routes
1910.213	34	Woodworking machinery requirements
1910.146	33	Permit-required confined spaces
1910.110	31	Storage and handling of liquefied petroleum gases
1910.106	28	Flammable and combustible liquids
1910.133	27	Eye and face protection
1910.184	25	Slings
1910.101	23	Compressed gases, general requirements
1910.141	20	Sanitation
1910.24	19	Fixed industrial stairs
1910.38	19	Emergency action plans
1910.107	19	Spray finishing using flammable/combustible materials
1910.179	18	Overhead and gantry cranes
1910.27	15	Fixed ladders
1910.242	15	Hand and portable powered tools and equipment, general
1910.334	14	Electrical, use of equipment
1904.2	13	Partial exemption for establishments in certain industries
1904.41	12	Annual OSHA injury and illness survey of 10 or more employees
1910.252	11	Welding, cutting and brazing, general requirements
1910.1000	11	Air contaminants
1910.180	10	Crawler locomotive and truck cranes
1904.4	9	Recording criteria
1910.26	9	Portable metal ladders
1910.333	8	Selection and use of work practices
1910.138	7	Hand protection
1910.1001	7	Asbestos

Note: Standards cited by the Federal OSHA for the retail service sector from October 2005 to September 2006 are included here.

REFERENCES

Bureau of Labor Statistics, U.S. Department of Labor. *National Census of Fatal Occupational Injuries in 2005.* Available at http://bls.gov.

Bureau of Labor Statistics, U.S. Department of Labor. *Workplace Injuries and Illnesses in 2004.* Available at http://bls.gov.

3 Retail Trade

Retailers provide the public access to the goods and materials they need for everyday living.

The retail trade (44 and 45) sector comprises establishments engaged in retailing merchandise, generally without transformation, and rendering services incidental to the sale of merchandise. The retailing process is the final step in the distribution of merchandise; retailers are, therefore, organized to sell merchandise in small quantities to the general public. This sector comprises two main types of retailers: store and nonstore retailers. The North American Industry Classification System (NAICS) includes the following industries under the retail trade sector:

Retail trade (44 and 45)
    Motor vehicle and part dealers (441000)
        Automobile dealers (441100)
        Automotive parts, accessories, and tire stores (441300)
    Furniture and home furnishing stores (442000)
        Furniture stores (442100)
        Home furnishing stores (442200)
    Electronic and appliance stores (443000)
        Electronic and appliance stores (443100)
        Computer and software stores (443120)
    Building material and garden equipment and supplies dealers (444000)
        Building material and supplies dealers (444100)
        Lawn and garden equipment and supplies stores (444200)
    Food and beverage stores (445000)

Grocery stores (445100)
Specialty food stores (445200)
Beer, wine, and liquor stores (445300)
Health and personal care stores (446000)
    Health and personal care stores (446100)
       Pharmacies and drug stores (446110)
Gasoline stations (447000)
    Gasoline stations (447100)
Clothing and clothing accessories stores (448000)
    Clothing stores (448100)
    Shoe stores (448200)
    Jewelry, luggage, and leather goods stores (448300)
Sports goods, hobby, book, musical instrument stores (451100)
    Sporting goods stores (451110)
    Book, periodical, and music stores (451120)
General merchandise stores (452000)
    Department stores (452100)
    Other general merchandise stores (452900)
Miscellaneous store retailers (453000)
    Florist (453100)
    Office supplies, stationery, and gift stores (453200)
    Used merchandise stores (453300)
    Other miscellaneous store retailers (453900)
Nonstore retailers (454000)
    Electronic shopping and mail order houses (454100)
    Vending machine operators (454200)
    Direct selling establishments (454300)

3.1 RETAIL TRADE

Wholesale and retail trades make up a large part of the nation's employment and business establishments. In the economy as a whole, retail trade is about 11.6% of all employment and 12.4% of all establishments. Retail trade employment averaged 15,254,900 in 2005.

In the retail sector, the three most representative subsectors are motor vehicle and part dealers; clothing, accessory, and general merchandise stores; and grocery stores. These three are used to describe the general nature and working conditions of the retail trade sector.

3.1.1 MOTOR VEHICLE AND PART DEALERS

It is interesting to note that about half of all workers in this industry have no formal education beyond high school. Employment is expected to grow, but will remain sensitive to downturns in the economy. Opportunities should be plentiful in vehicle maintenance and repair occupations, especially for persons who complete formal automotive service technician training.

FIGURE 3.1 A typical automotive dealership.

Automobile dealers are the bridge between automobile manufacturers and the U.S. consumers. New car dealers are primarily engaged in retailing new cars, sport utility vehicles (SUVs), and passenger and cargo vans. New car dealers employ 9 out of 10 workers in the industry. Most new car dealers combine vehicle sales with other activities, such as providing repair services, retailing used cars, and selling replacement parts and accessories. These dealers offer one-stop shopping for customers who wish to buy, finance, and service their next vehicle. On the other hand, stand-alone used car dealers specialize in used vehicle sales and account for only 1 out of 10 jobs in the industry. By putting new vehicles on the road, dealers can count on aftermarket additions, new repair and service customers, and future trade-ins of used vehicles (Figure 3.1).

The aftermarket sales department in a new car dealer sells additional services and merchandise after the vehicle salesperson has closed a deal. Aftermarket sales workers sell service contracts and insurance to buyers of new and used cars and arrange financing for their purchases. Representatives offer extended warranties and additional services, such as undercoat sealant and environmental paint protection packages, to increase the revenue generated for each vehicle sold.

3.1.2 Clothing, Accessory, and General Merchandise Stores

Clothing, accessory, and general merchandise stores are represented by sales and administrative support jobs that account for 83% of employment in this subsector.

Most jobs do not require formal education; many people get their first jobs in this industry. Clothing, accessory, and general merchandise stores offer many part-time jobs, but earnings are relatively low. Despite relatively slow employment growth, turnover will produce numerous job openings in this large industry.

Clothing, accessory, and general merchandise stores are some of the most visited establishments in the country. Whether shopping for an item of clothing, a piece of jewelry, a household appliance, or even food, you will likely go to one of these stores to make your purchase or compare selections with other retail outlets. Composed of department stores (including discount department stores), supercenters, and warehouse club stores, general merchandise stores in particular sell a large assortment of items. Also included among general merchandise stores are dollar stores that sell a wide variety of inexpensive merchandise.

Department stores sell an extensive selection of merchandise, with no one line predominating. As the name suggests, these stores generally are arranged into departments, each headed by a manager. The various departments can sell apparel, furniture, appliances, home furnishings, cosmetics, jewelry, paint and hardware, electronics, and sporting goods. They also may sell services such as optical, photography, and pharmacy services. Discount department stores typically have fewer sales workers, relying more on self-service features, and have centrally located cashiers. Department stores that sell bulk items, like major appliances, usually provide delivery and installation services. Upscale department stores may offer tailoring for their clothing lines and more personal service.

Warehouse club stores and supercenters, the fastest growing segment of this industry, sell an even more eclectic mix of products and services, in fixed quantities and at low prices. These stores typically include an assortment of food items, often sold in bulk, along with an array of household and automotive goods, clothing, and services that may vary over time. Often, such stores require that shoppers purchase a membership that entitles them to shop there. They offer very little service and usually require the customer to take home the item.

Compared with department stores, clothing and accessory stores sell a much narrower group of items that include apparel for all members of the family, as well as shoes, luggage, leather goods, lingerie, jewelry, uniforms, and bridal gowns. Stores in this sector may sell a relatively broad range of these items or concentrate on a few. They often are staffed with knowledgeable salespersons who can help in the selection of sizes, styles, and accessories. Many of these stores are located in shopping malls across the country and have significantly fewer workers than department stores.

3.1.3 GROCERY STORES

Grocery stores have numerous job openings, many of them part time and relatively low paying, and are usually available because of the industry's large size and high turnovers. Many grocery store workers are young (16–24 years) and hold 32% of the jobs. Cashier, stock clerks, and order fillers account for 49% of all jobs. College graduates fill most new management positions.

Grocery stores, also known as supermarkets, are familiar to everyone. They sell an array of fresh and preserved foods, primarily for preparation and consumption at home. They also often sell prepared food, such as hot entrées or salads, for takeout meals. Stores range in size from supercenters—which may employ hundreds of workers, provide a variety of consumer services, and sell numerous food and nonfood items—to traditional supermarkets to convenience stores with small staffs and limited selections.

Convenience stores, however, also often sell fuel, including gasoline, diesel, kerosene, and propane. Recently, many convenience stores have expanded their scope of services by providing ATMs, money orders, and a more comprehensive selection of products, including food for immediate consumption and an assortment of nonfood items.

Specialty grocery stores—meat and fish markets; fruit and vegetables markets; candy, nut, and confectionery stores; dairy products stores; retail bakeries; and health and dietetic food stores, for example—are not covered in this section. Food services and drinking places that sell food and beverages for consumption on the premises are also excluded.

Grocery stores are found everywhere, although the size of the establishment and the range of goods and services offered vary. Traditionally, inner-city stores are small and offer a limited selection, although larger stores, including specialty grocers and a few supercenters, are now being built in many urban areas; suburban stores are predominantly large supermarkets and supercenters with a more diverse stock. Most supermarkets include several specialty departments that offer the products and services of seafood stores, bakeries, delicatessens, pharmacies, or florist shops. Household goods, health and beauty care items, automotive supplies, pet products, greeting cards, and clothing also are among the nonfood items that can be found at large supermarkets. Some of the largest supermarkets, including wholesale clubs, even have cafeterias or food courts, and a few feature convenience stores, automotive services, and full-service banks. In addition, most grocery stores offer basic banking services and ATMs, postal services, on-site film processing, dry cleaning, video rentals, and catering services.

Working conditions in most grocery stores are pleasant, with clean, well-lighted, climate-controlled surroundings. Work can be hectic, and dealing with customers can be stressful.

Grocery stores are open more hours and days than most work establishments, so workers are needed for early morning, late night, weekend, and holiday work. With employees working 30.8 h a week, on average, these jobs are particularly attractive to workers who have family or school responsibilities or another job.

Most grocery store workers wear some sort of uniform, such as a jacket or an apron that identifies them as store employees and keep their personal clothing clean. Health and safety regulations require some workers, such as those who work in the delicatessen or meat department, to wear head coverings, safety glasses, or gloves.

3.2 PROFILE OF RETAIL WORKERS' DEATHS, INJURIES, AND ILLNESSES

3.2.1 DEATHS

There were 397 occupationally related deaths to retail workers in 2005. The retail sector accounted for 15% of the service industry deaths (2736). Table 3.1 shows the percent of those deaths from each major category.

TABLE 3.1

Occupational Death Cause by Percent for Retail Sector

Cause	Retail Industry (%)
Highway	21
Homicides	46
Falls	9
Struck-by	5

Source: From Bureau of Labor Statistics, U.S. Department of Labor. *National Census of Fatal Occupational Injuries in 2005.* Available at http://bls.gov.

3.2.2 INJURIES

There were 178,760 reported injuries for retail workers in 2004; this was 21% of the total injuries (850,930) for the service industry. The distributions for the nature, body part, source, and exposure (accident type) for the 178,760 injuries are presented in Tables 3.2, 3.3, 3.4, and 3.5, respectively.

3.2.3 ILLNESSES

In the retail sector, there were 17,200 cases of occupationally related illnesses; this is 13% of the total for the service industry (Table 3.6).

TABLE 3.2

Nature of Injury by Number and Percent for the Retail Sector

Nature of Injury	Number	Percent
Sprains/strains[a]	79,700	45
Fractures[a]	11,830	6.6
Cuts/punctures[a]	17,640	10
Bruises[a]	17,940	10
Heat burns	1,700	1
Chemical burns	1,120	0.6
Amputations	1,000	0.5
Carpal tunnel syndrome	2,110	1.2
Tendonitis	930	0.5
Multiple trauma[a]	6,250	3
Back pain only	4,950	2.8

Source: From Bureau of Labor Statistics, U.S. Department of Labor. *Workplace Injuries and Illnesses in 2004.* Available at http://bls.gov.

[a] Five most frequently occurring conditions.

TABLE 3.3
Body Part Injured by Number and Percent for the Retail Sector

Body Part Injured	Number	Percent
Head	10,760	6
Eyes	4,020	2
Neck	3,320	1.8
Trunk[a]	66,970	37
Back[a]	42,780	24
Shoulder	12,010	7
Upper extremities[a]	38,950	22
Finger	15,260	8.5
Hand	6,280	3.5
Wrist	7,740	4
Lower extremities[a]	39,720	22
Knee	14,080	8
Foot and toe	11,040	6
Body systems	1,950	1
Multiple body parts[a]	15,030	9

Source: From Bureau of Labor Statistics, U.S. Department of Labor. *Workplace Injuries and Illnesses in 2004.* Available at http://bls.gov.

[a] Five most frequently injured body parts.

TABLE 3.4
Source of Injury by Number and Percent for the Retail Sector

Sources of Injuries	Number	Percent
Parts and materials[a]	15,100	8
Worker motion/position[a]	23,160	13
Floor, walkways, or ground surfaces[a]	32,720	18
Hand tools	8,250	4.6
Vehicles[a]	14,680	8
Health care patient	0	0
Chemicals and chemical products	2,230	1
Containers[a]	38,750	22
Furniture and fixtures	1,136	0.6
Machinery	12,390	7

Source: From Bureau of Labor Statistics, U.S. Department of Labor. *Workplace Injuries and Illnesses in 2004.* Available at http://bls.gov.

[a] Five most frequent sources of injury.

TABLE 3.5

**Exposure/Accident Type by Number and Percent
for the Retail Sector**

Type of Accidents	Number	Percent
Struck by object[a]	29,610	17
Struck against object[a]	12,100	6.8
Caught in or compressed or crushed	5,830	3
Fall to lower level	9,640	5
Fall on same level[a]	24,760	14
Slips or trips without a fall	4,860	2.7
Overexertion[a]	51,950	29
Lifting[a]	32,770	18
Repetitive motion	5,150	2.8
Exposure to harmful substance or environment	5,530	3
Transportation accident	5,950	3
Fires and explosions	210	0.1
Assaults/violent acts	1,970	1

Source: From Bureau of Labor Statistics, U.S. Department of Labor.
Workplace Injuries and Illnesses in 2004. Available at http://
bls.gov.

[a] Five most frequent exposures or type of accidents that led to an
injury.

3.3 HAZARDS FACED BY RETAIL WORKERS

The hazards of working with all types of products and the handling of these products
as well as interacting with the general public contribute to the hazards faced by retail
workers.

TABLE 3.6

**Occupational Illnesses by Number of Cases and Percent
for the Retail Sector**

Illness Type	Number	Percent
Skin diseases and disorders	2,900	17
Respiratory conditions	1,300	8
Poisoning	300	2
Hearing loss	500	3
All others	12,200	71

Source: From Bureau of Labor Statistics, U.S. Department of Labor.
Workplace Injuries and Illnesses in 2004. Available at http://bls.gov.

The hazards covered in this book are the primary ones that affect retail workers in facility and stores operations. In most cases, the most frequent hazards faced by retail workers are as follows:

- Walking and working surfaces
- Electrocutions
- Material handling/lifting of containers
- Slips, trips, and falls
- Strains/sprains
- Trauma injuries
- Vehicle accidents
- Fires
- Power tools
- Office hazards
- Cutting, slicing, or other power equipment
- Repetitive/cumulative trauma
- Violence and security

3.4 OCCUPATIONS

A variety of occupations provide service to buyers and customers' needs and expectations, for example, motor vehicle and part dealers, and employees in clothing and grocery stores. Employees are the interface for the retail sector.

3.4.1 MOTOR VEHICLE AND PART DEALERS

Employees in automobile dealers work longer hours than those in most other industries. An overwhelming 84% of automobile dealer employees worked full time in 2004, and 38% worked more than 40 h a week. To satisfy customer service needs, many dealers provide evening and weekend service. The 5 day, 40 h week usually is the exception, rather than the rule, in this industry. Some of the most common occupations are advertising, marketing, promotions, public relations, and sales managers; automotive body and related repairers; automotive service technicians and mechanics; retail sales personnel; and sales worker supervisors.

Because most automobile salespersons and administrative workers spend their time in dealer showrooms, individual offices are a rarity. Multiple users share limited office space that may be cramped and sparsely equipped. The competitive nature of selling is stressful to automotive salespersons, as they try to meet company sales quotas and personal earning goals. Compared with all other occupations in general, the proportion of workers who transfer from automotive sales jobs to other occupations is relatively high.

Service technicians and automotive body repairers generally work indoors in well-ventilated and well-lighted repair shops. However, some shops are drafty and noisy. Technicians and repairers frequently work with dirty and greasy parts, and in awkward positions. They often lift heavy parts and tools. Minor cuts, burns, and bruises are common, but serious accidents are avoided when shops are kept clean and orderly and when safety practices are observed. Despite hazards, precautions taken by dealers to prevent injuries have kept the workplace relatively safe. In 2003, there were 5.1 cases of work-related injuries and illnesses per 100 full-time workers in the automobile dealers industry, close to the national average of 5.0 per 100.

Sales and related occupations are among the most important occupations in automobile dealers and account for 36% of industry employment. Sales workers' success in selling vehicles and services determines the success of the dealer. Automotive retail salespersons usually are the first to greet customers and determine their interests through a series of questions.

Installation, maintenance, and repair occupations are another integral part of automobile dealers, constituting 27% of industry employment. Automotive service technicians and mechanics service, diagnose, adjust, and repair automobiles and light trucks, such as vans, pickups, and SUVs. Automotive body and related repairers repair and finish vehicle bodies, straighten bent body parts, remove dents, and replace crumpled parts that are beyond repair. Shop managers usually are among the most experienced service technicians.

Service advisors handle the administrative and customer relations part of the service department. They greet customers, listen to their description of problems or service desired, write repair orders, and estimate the cost and time needed to do the repair. They also handle customer complaints, contact customers when technicians discover new problems while doing the work, and explain to customers the work performed and the charges associated with the repairs.

In support of the service and repair departments, parts salespersons supply vehicle parts to technicians and repairers. They also sell replacement parts and accessories to the public. Parts managers run the parts department and keep the automotive parts inventory. They display and promote sales of parts and accessories and deal with garages and other repair shops seeking to purchase parts.

Office and administrative support workers handle the paperwork of automobile dealers and make up about 15% of employment in the industry. Bookkeeping, accounting, and auditing clerks; general office clerks; and secretaries and administrative assistants prepare reports on daily operations, inventory, and accounts receivable. They gather, process, and record information and perform other administrative support and clerical duties. Office managers organize, supervise, and coordinate administrative operations. Many office managers also are responsible for collecting and analyzing information on each department's financial performance.

Transportation and material moving occupations account for about 12% of jobs in automobile dealers. Cleaners of vehicles and equipment prepare new and used cars for display in the showroom or parking lot and for delivery to customers. They may wash and wax vehicles by hand and perform simple services such as changing a tire or a battery. Truck drivers typically operate light delivery trucks to pick up and deliver automotive parts; some drive tow trucks that bring damaged vehicles to the dealer for repair.

Management jobs often are filled by promoting workers with years of related experience. For example, most sales managers start as automotive salespersons. Sales managers hire, train, and supervise the dealer's sales force. They are the lead negotiators in all transactions between sales workers and customers. Most advance to their positions after success as salespersons. They review market analyses to determine consumer needs, estimate volume potential for various models, and develop sales campaigns.

General and operations managers are in charge of all dealer operations. They need extensive business and management skills, usually acquired through experience as a manager in one or more of the dealer departments. Dealer performance and profitability ultimately are up to them. General managers sometimes have an ownership interest in the dealer.

Requirements for many jobs vary from dealer to dealer. To find out exactly how to qualify for a specific job, ask the dealer or manager in charge. Many jobs require no postsecondary education; about half of all workers in the industry have no formal education beyond high school. In today's competitive job market, however, nearly all dealers demand a high school diploma.

3.4.2 CLOTHING, ACCESSORY, AND GENERAL MERCHANDISE STORES

Most employees in clothing, accessory, and general merchandise stores work under clean, well-lighted conditions. Many jobs are part time with most employees working during peak selling times, including nights, weekends, and holidays. Because weekends are busy days in retailing, almost all employees work at least one of these days and have a weekday off. During busy periods, such as holidays and the back-to-school season, longer than normal hours may be scheduled, and vacation time is limited for most workers, including buyers and managers. Some of the most common occupations in this subsector are advertising, marketing, promotions, public relations, and sales managers; cashiers; customer service representatives; purchasing managers, buyers, and purchasing agents; retail salespersons; sales worker supervisors; security guards and gaming surveillance officers; and stock clerks and order fillers.

Retail salespersons and cashiers often stand for long periods, and stock clerks may perform strenuous tasks, such as moving heavy, cumbersome boxes. Sales representatives and buyers often travel to visit clients and may be away from home for several days or weeks at a time. Those who work for large manufacturers and retailers may travel outside of the country (Figure 3.2).

The incidence of work-related illnesses and injuries varies greatly among segments of the industry. In 2003, workers in clothing and accessory stores had 2.8 cases of injury and illness per 100 full-time workers, while those in general merchandise stores had 7.2 cases per 100 full-time workers. These figures compare with an average of 5.0 throughout private industry.

It is of interest to note that sales and related occupations accounted for 65% of workers in this industry in 2004. Retail salespersons, which make up 43% of employment in the industry, help customers select and purchase merchandise. A salesperson's primary job is to interest customers in the merchandise and to answer any questions the customers may have. To do this, the worker describes the product's various models, styles, and colors or demonstrates its use. To sell expensive and complex items, workers need extensive knowledge of the products.

In addition to selling, most retail salespersons register the sale electronically on a cash register or terminal; receive cash, checks, and charge payments; and

FIGURE 3.2   Carrying clothing and standing for long periods are tiring for salespersons.

give change and receipts. Depending on the hours they work, they may open or close their cash registers or terminals. Either of these operations may include counting the money in the cash register; separating charge slips, coupons, and exchange vouchers; and making deposits at the cash office. Salespersons are held responsible for the contents of their register, and repeated shortages often are cause for dismissal.

Salespersons may be responsible for handling returns and exchanges of merchandise, wrapping gifts, and keeping their work areas neat. In addition, they may help stock shelves or racks, arrange for mailing or delivery of a purchase, mark price tags, take inventory, and prepare displays. They also must be familiar with the store's security practices to help prevent theft of merchandise. Cashiers total bills, receive money, make change, fill out charge forms, and give receipts. Retail salespersons and cashiers often have similar duties.

Office and administrative support occupations make up the next largest group of employees, accounting for 19% of the total employment in the industry. Stock clerks and order fillers bring merchandise to the sales floor and stock shelves and racks. They also mark items with identifying codes or prices so that they can be recognized quickly and easily, although many items today arrive pre-ticket. Customer service representatives investigate and resolve customers' complaints about merchandise, service, billing, or credit ratings. The industry also employs administrative occupations found in most industries, such as general office clerks and bookkeepers.

Management and business and financial operations occupations accounted for 2% of the industry employment. This does not include corporate managers.

Department managers oversee sales workers in a department or section of the store. They set the work schedule, supervise employee performance, and are responsible for the overall sales and profitability of their departments. They also may be called upon to settle a dispute between a customer and a salesperson.

Buyers purchase merchandise for resale from wholesalers or manufacturers. Using historical records, market analysis, and their sense of consumer demand, they buy merchandise, keeping in mind their customer's demand for style, quality, and low price. Wrong decisions mean that the store will mark down slow-selling merchandise, thus losing profits. Buyers for larger stores or chains usually buy one classification of merchandise, such as casual menswear or home furnishings; those working for smaller stores may buy all the merchandise sold in the store. They also plan and implement sales promotion plans for their merchandise, such as arranging for advertising and ensuring that the merchandise is displayed properly.

Merchandise managers are in charge of a group of buyers and department managers; they plan and supervise the purchase and marketing of merchandise in a broad area, such as women's apparel or appliances. In department store chains, with numerous stores, many of the buying and merchandising functions are centralized in one location. Some local managers might decide which merchandise, among that bought centrally, would be best for their own stores.

Department store managers direct and coordinate the activities in these stores. They set pricing policies to maintain profitability and notify senior management of concerns or problems. Department store managers usually directly supervise department managers and indirectly oversee other department store workers.

Clothing and accessory store managers—often the only managers in smaller stores—combine many of the duties of department managers, department store managers, and buyers. Retail chain store area managers or district managers oversee the activities of clothing and accessory store managers in an area. They hire managers, ensure that company policies are carried out, and coordinate sales and promotional activities.

Various other store-level occupations in this diversified industry include pharmacists, hairdressers, material moving workers, food preparation and serving workers, and security guards.

There are no formal educational requirements for most sales and administrative support jobs; in fact, many people get their first jobs in this industry. A high school education is preferred, especially by larger employers. Because many of the new workers in the industry are recent immigrants, employers may require proficiency in English and may even offer language training to employees.

3.4.3 Grocery Stores

In 2003, cases of work-related injury and illness averaged 7.2 per 100 full-time workers in grocery stores, compared with 5.0 per 100 full-time workers in the entire private sector. Some injuries occur while workers transport or stock goods. Persons in food-processing occupations, such as butchers and meat cutters, as well as cashiers working with computer scanners or traditional cash registers, may be vulnerable to cumulative trauma and other repetitive motion injuries.

Grocery store workers stock shelves on the sales floor; prepare food and other goods; assist customers in locating, purchasing, and understanding the content and uses of various items; and provide support services to the establishment. However, 49% of all grocery store employees are cashiers or stock clerks and order fillers. The most common occupations in grocery stores are advertising, marketing, promotions, public relations, and sales managers; building cleaning workers; cashiers; chefs, cooks, and food preparation workers; demonstrators, product promoters, and models; food and beverage serving and related workers; food-processing occupations; food service managers; human resources, training, and labor relations managers and specialists; material moving occupations; pharmacists; pharmacy aides; pharmacy technicians; purchasing managers, buyers, and purchasing agents; retail salespersons; sales worker supervisors; and stock clerks and order fillers.

Cashiers make up the largest occupation in grocery stores, accounting for 34% of all workers. They scan the items being purchased by customers, total the amount due, accept payment, make change, fill out charge forms, and produce a cash register receipt that shows the quantity and price of the items. In most supermarkets, the cashier passes the Universal Product Code (UPC) on the item's label across a computer scanner that identifies the item and its price, which is automatically relayed to the cash register. In some grocery stores, customers themselves scan and bag their purchases, and pay using an automatic payment terminal, a system known as self-checkout. Cashiers verify that the items have been paid for before the customer leaves, and if needed, assist the customer in completing the transaction. In other grocery stores, the cashier reads a hand-stamped price on each item and keys that price directly into the cash register. Cashiers then place items in bags for customers; accept cash, personal checks, credit cards, or electronic debit card payments; and make change. When cashiers are not needed to check out customers, they sometimes assist other workers (Figure 3.3).

Stock clerks and order fillers are the second largest occupation in grocery stores, accounting for 15% of workers. They fill the shelves with merchandise and arrange displays to attract customers. In stores without computer-scanning equipment, stock clerks and order fillers have to manually mark prices on individual items and count stock for inventory control.

Many office clerical workers—such as secretaries and administrative assistants; general office clerks; and bookkeeping, accounting, and auditing clerks—prepare and maintain the records necessary to run grocery stores smoothly.

Butchers and other meat-, poultry-, and fish-processing workers prepare meat, poultry, and fish for purchase by cutting up and trimming carcasses and large sections into smaller pieces, which they package, weigh, price, and place on display. They also prepare ground meat from other cuts and fill customers' special orders. These workers also prepare ready-to-heat foods by filleting or cutting meat, poultry, or fish into bite-sized pieces, preparing and adding vegetables, or applying sauces or breading. While most butchers work in the meat section of grocery stores, many other meat-, poultry-, and fish-processing workers are employed at central processing facilities, from which smaller packages are sent to area stores.

Some specialty workers prepare food for sale in the grocery store but work in kitchens that may not be located in the store. Bakers produce breads, rolls, cakes,

FIGURE 3.3   Cashiers must stand, lift groceries, and use the scanner that creates ergonomic issues.

cookies, and other baked goods. Chefs and head cooks direct the preparation, seasoning, and cooking of salads, soups, fish, meats, vegetables, desserts, or other foods. Some plan and price menu items, order supplies, and keep records and accounts. Cooks and food preparation workers make salads—such as coleslaw or potato, macaroni, or chicken salad—and other entrées, and prepare ready-to-heat foods—such as burritos, marinated chicken breasts, or chicken stir-fry—for sale in the delicatessen or in the gourmet food or meat department. Other food preparation workers arrange party platters or prepare various vegetables and fruits that are sold at the salad bar.

Demonstrators and product promoters offer samples of various products to entice customers to purchase them. In supermarkets that serve food and beverages for consumption on the premises, food and beverage serving workers take orders and serve customers at counters. They prepare short-order items, such as salads or sandwiches, to be taken out and consumed elsewhere. Building cleaning workers keep the stores clean and orderly.

In the warehouses and stockrooms of large supermarkets, hand laborers and freight, stock, and material movers move stock and goods in storage and deliver them to the sales floor; they also help load and unload delivery trucks. Hand packers and packagers, also known as courtesy clerks or baggers, perform a variety of simple tasks, such as bagging groceries, loading parcels in customers' cars, and returning unpurchased merchandise from the checkout counter to shelves.

First-line managers of retail sales workers supervise mostly entry-level employees in the grocery, produce, meat, and other specialty departments. These managers train employees and schedule their hours; oversee ordering, inspection, pricing, and inventory of goods; monitor sales activity; and make reports to store managers. General and operations managers are responsible for the efficient and profitable operation of grocery stores. Working through their department managers, general and operations managers may set store policy, hire and train employees, develop merchandising plans, maintain good customer and community relations, address customer complaints, and monitor the store's profits or losses.

Purchasing managers plan and direct the task of purchasing goods for resale to consumers. They must thoroughly understand grocery store foods, other items, and each store's customers and must select the best suppliers and maintain a good relationship with them. Purchasing managers evaluate their store's sales reports to determine what products are in demand and plan purchases according to their budget.

Because of the expansion of the industry to meet the consumers' desire for "one-stop shopping," grocery stores have begun to employ an array of workers to help meet that need. For example, marketing and sales managers forecast sales and develop a marketing plan based on demographic trends, sales data, community needs, and consumer feedback. Pharmacists fill customers' drug prescriptions and advise them on over-the-counter medicines. Inspectors, testers, sorters, samplers, and weighers assess whether products and facilities meet quality, health, and safety standards. Human resources, training, and labor relations specialists are responsible for making sure that employees maintain and, if necessary, improve their skill levels.

3.5   APPLICABLE OSHA REGULATIONS

Another way to gather an understanding of the hazards faced by retail workers is to see the types of violations that the Occupational Safety and Health Administration (OSHA) has found during their inspections of retail establishments. These violations provide another way of targeting hazards that have the potential to cause injury, illness, and death of workers. As can be seen from the 50 most frequently cited violations, OSHA cites this industry under the general industry standard (29 CFR 1910) and the recordkeeping standard (29 CFR 1904) (Table 3.7).

Although the violations cited in Table 3.7 are the 50 most frequently issued violations, OSHA has cited other hazards with less frequency. Some of these are as follows:

- Hand protection
- Slings
- Occupational foot protection
- Hazardous locations

- Hydrogen
- Automatic sprinklers
- Ladders
- Fixed ladders
- Air receivers
- Safeguard for personnel protection
- Bakery equipment

With the hazards faced by this sector, it is imperative that safety and health be an integral part of doing business and with the specific purpose of protecting its employees.

TABLE 3.7
Fifty Most Frequent OSHA Violations for the Retail Sector

CFR Standard	Number Cited	Description
1910.1200	2538	Hazard communication
1910.134	422	Respiratory protection
1910.178	180	Powered industrial trucks (forklifts)
1910.305	172	Electrical, wiring methods, components and equipment
1910.132	131	Personal protective equipment, general requirements
1910.303	129	Electrical systems design, general requirements
1910.157	121	Portable fire extinguishers
1910.37	115	Maintenance, safeguards, and operational features for exit routes
1910.22	104	Walking–working surfaces, general requirements
1910.23	81	Guarding floor and wall openings and holes
1910.151	54	Medical services and first aid
1910.212	54	Machines, general requirements
1910.213	52	Woodworking machinery requirements
1904.29	44	Forms
1910.36	43	Design and construction requirements for exit routes
1910.133	40	Eye and face protection
1910.304	39	Electrical, wiring design and protection
1910.147	36	The control of hazardous energy, lockout/tagout
1910.176	35	Handling materials, general
5A1	32	General duty clause (section of OSHA Act)
1910.1000	30	Air contaminants
1910.141	28	Sanitation
1910.215	28	Abrasive wheel machinery
1910.106	27	Flammable and combustible liquids
1910.138	24	Hand protection
1910.107	21	Spray finishing using flammable/combustible materials
1910.1030	21	Bloodborne pathogens
1910.253	18	Oxygen-fuel gas welding and cutting
1910.38	16	Emergency action plans

(continued)

TABLE 3.7 (continued)
Fifty Most Frequent OSHA Violations for the Retail Sector

CFR Standard	Number Cited	Description
1910.334	16	Electrical, use of equipment
1910.24	14	Fixed industrial stairs
1910.219	14	Mechanical power-transmission apparatus
1904.32	13	Annual summary
1904.41	13	Annual OSHA injury and illness survey of 10 or more employees
1910.26	13	Portable metal ladders
1910.110	11	Storage and handling of liquefied petroleum gases
1910.101	10	Compressed gases, general requirements
1910.242	10	Hand and portable powered tools and equipment
1904.39	8	Reporting fatalities and multiple hospitalization incidents to OSHA
1910.1025	8	Lead
1926.451	8	Scaffolds, general requirements
1904.40	7	Providing records to government representatives
1910.25	7	Portable wood ladders
1910.29	7	Manually propelled mobile ladder stands and scaffolds (towers)
1910.145	7	Specifications, accident prevention signs and tags
1910.1052	7	Methylene chloride
1926.501	7	Duty to have fall protection
1910.95	6	Occupational noise exposure
1926.1101	6	Asbestos
1910.146	5	Permit-required confined spaces

Source: From Occupational Safety and Health Administration, U.S. Department of Labor. Available at http://www.osha.gov.

Note: Standards cited by the Federal OSHA for the retail service sector from October 2005 to September 2006 are included here.

REFERENCES

Bureau of Labor Statistics, U.S. Department of Labor. *National Census of Fatal Occupational Injuries in 2005.* Available at http://bls.gov.

Bureau of Labor Statistics, U.S. Department of Labor. *Workplace Injuries and Illnesses in 2004.* Available at http://bls.gov.

Occupational Safety and Health Administration, U.S. Department of Labor. Available at http://www.osha.gov.

4 Warehousing

Warehouses are containers for the transfer and storage of goods and materials.

Warehousing was selected to be in this book because its function was to store materials and goods that made the exposures of the workforce very similar to those faced by workers in wholesale and retail trades. It is usually housed in the same grouping as the transportation sector. The North American Industrial Classification System (NAICS) lists it as follows:

Warehousing and storage (493000)
   Warehousing and storage (493100)

Warehousing and storage facilities comprised 13,000 establishments in 2004. These firms are engaged primarily in operating warehousing and storage facilities for general merchandise and refrigerated goods (Figure 4.1). They provide facilities to store goods; self-storage mini-warehouses that rent to the general public are also included in this segment of the industry.

The deregulation of interstate trucking in 1980 encouraged many firms to add a wide range of customer-oriented services to complement trucking and warehousing services and led to innovations in the distribution process. Increasingly, trucking and warehousing firms are providing logistical services encompassing the entire transportation process. Firms that offer these services are called third-party logistics providers. Logistical services manage all aspects of the movement of goods between producers and consumers. Among their value-added services are sorting bulk goods into customized lots, packaging and repackaging goods, controlling and managing

43

FIGURE 4.1   Warehousing provides for the orderly staging and storage of goods and materials.

inventory, order entering and fulfillment, labeling, performing light assembly, and marking prices. Some full-service companies even perform warranty repair work and serve as local parts distributors for manufacturers. Some of these services, such as maintaining and retrieving computerized inventory information on the location, age, and quantity of goods available, have helped to improve the efficiency of relationships between manufacturers and customers.

Many firms are relying on new technologies and the coordination of processes to expedite the distribution of goods. Voice control software allows a computer to coordinate workers through audible commands—telling workers what items to pack for which orders—helping to reduce errors and increase efficiency. Voice control software can also be used to perform inventory checks and reordering. Some firms use radio frequency identification devices (RFIDs) to track and manage incoming and outgoing shipments. RFID simplifies the receiving process by allowing entire shipments to be scanned without unpacking a load to manually compare it against a bill of lading. Just-in-time shipping is a process whereby goods arrive just before they are needed, saving recipients money by reducing their need to carry large inventories. These technologies and processes reflect two major trends in warehousing: supply chain integration, whereby firms involved in production, transportation, and storage all move in concert so as to act with the greatest possible efficiency; and ongoing attempts to reduce inventory levels and increase inventory accuracy.

The average annual employment for warehousing is 555,800 workers. This is approximately 0.4% of the U.S. workforce and 0.7% of the service industry workforce.

4.1 PROFILE OF WAREHOUSING WORKERS' DEATH, INJURIES, AND ILLNESSES

4.1.1 DEATHS

There were 27 deaths in the warehousing sector. These deaths account for 0.9% or the total deaths (2736) in the service industry. As can be noted in Table 4.1, fall was the most frequent cause of warehousing deaths.

4.1.2 INJURIES

There were 14,620 reported injuries for warehousing workers in 2004. The injury rate for warehousing was 9.3 per 100 full-time workers, while it was 4.2 and 4.8 for the service industry and all of private industry, respectively. This alarming injury rate indicates the danger involved in the warehousing sector. The distribution for the nature, body part, source, and exposure (accident type) for the 14,620 injuries are presented in Tables 4.2, 4.3, 4.4, and 4.5, respectively.

4.1.3 ILLNESSES

In the warehousing sector, 1900 cases of occupational illness were reported; this is 1% of the total reported illnesses for the service industry. Table 4.6 provides the breakdown of the illnesses.

4.2 HAZARDS FACED BY WAREHOUSING WORKERS

The fatal injury rate for warehousing is higher than the national average for all industries. Warehousing is often viewed as not being a very complex operation, but traffic patterns, fast pace, constant motion, and a myriad of various materials being handled and stored lead to exposure to many hazards. Some of the most common hazards faced by warehouse workers are as follows:

- Unsafe forklifts
- Unsafe docks
- Improper stacking or storage of materials
- Failure to wear proper personal protective equipment
- Conveyor hazards
- Hazardous substances, materials, and chemicals
- Manual lifting and handling task
- Improper fueling and charging batteries
- Poor ergonomic design
- Potential fires

4.3 OCCUPATIONS

The handling and storage of materials require a workforce that is not the most highly paid or skillfully qualified. The perception is that anyone can do warehousing type of

TABLE 4.1

Occupational Death Cause by Percent for the Warehousing Sector

Cause	Percent
Highway	0
Homicides	0
Falls	15
Struck-by	0

Source: From Bureau of Labor Statistics, U.S. Department of Labor. *National Census of Fatal Occupational Injuries in 2005.* Available at http://bls.gov.

work. However, the more experienced, trained, and skilled the workforce, the less the chances of potential injury and death. With the use of computers and automated equipment in warehouse, there is more of a presence of administrative types of workers who spend most of their time in an office environment. (Safety and health for office workers is covered in *Industrial Safety and Health for Administrative Services.*)

Laborers and hand freight, stock, and material movers help load and unload freight and move it around warehouses and terminals. Often, these unskilled employees work together in groups of three or four. They may use conveyor belts, hand trucks, pallet jacks, or forklifts to move freight. They may place heavy or bulky items

TABLE 4.2

Nature of Injury by Number and Percent for the Warehousing Sector

Nature of Injury	Number	Percent
Sprains/strains[a]	6300	43
Fractures[a]	940	6
Cuts/punctures[a]	630	4
Bruises	1170	12
Heat burns	—	—
Chemical burns	—	—
Amputations	—	—
Carpal tunnel syndrome	230	1.5
Tendonitis	290	1.9
Multiple trauma[a]	470	3
Back pain (only)[a]	360	2

Source: From Bureau of Labor Statistics, U.S. Department of Labor. *Workplace Injuries and Illnesses in 2004.* Available at http://bls.gov.

[a] Five most frequently occurring conditions.

TABLE 4.3
Body Part Injured by Number and Percent
for the Warehousing Sector

Body Part Injured	Number	Percent
Head	640	4
Eyes	280	1.9
Neck	260	1.7
Trunk[a]	6410	44
Back[a]	3820	26
Shoulder[a]	1250	8.5
Upper extremities[a]	3070	21
Finger	620	4
Hand	760	5
Wrist	660	4.5
Lower extremities[a]	3390	23
Knee	810	5.5
Foot and toe	1050	7
Body systems	50	0.3
Multiple body parts	800	5

Source: From Bureau of Labor Statistics, U.S. Department of Labor.
Workplace Injuries and Illnesses in 2004. Available at http://bls.gov.
[a] Five most frequently injured body parts.

TABLE 4.4
Source of Injury by Number and Percent
for the Warehousing Sector

Sources of Injuries	Number	Percent
Parts and materials[a]	720	5
Worker motion/position[a]	1810	12
Floor, walkways, or ground surfaces[a]	1410	10
Hand tools	230	1.5
Vehicles[a]	2390	16
Health care patient	—	—
Chemicals and chemical products	—	—
Containers[a]	5030	34
Furniture and fixtures	730	5
Machinery	520	4

Source: From Bureau of Labor Statistics, U.S. Department of Labor.
Workplace Injuries and Illnesses in 2004. Available at http://bls.gov.
[a] Five most frequent sources of injury.

TABLE 4.5

Exposure/Accident Type by Number and Percent for the Warehousing Sector

Type of Accidents	Number	Percent
Struck by object[a]	1570	11
Struck against object	900	6
Caught in or compressed or crushed	590	4
Fall to lower level	450	3
Fall on same level[a]	1110	7.6
Slips or trips without a fall	820	2
Overexertion[a]	5720	39
Lifting[a]	3700	25
Repetitive motion	530	3
Exposure to harmful substance or environment	80	0.5
Transportation accident[a]	1170	12
Fires and explosions	—	—
Assaults/violent acts	80	0.5

Source: From Bureau of Labor Statistics, U.S. Department of Labor. *Workplace Injuries and Illnesses in 2004.* Available at http://bls.gov.

[a] Five most frequent exposures or type of accidents that led to an injury.

on wooden skids or pallets and have industrial truck and tractor operators move them (Figure 4.2).

Office and administrative support workers perform the daily recordkeeping operations for the truck transportation and warehousing industry. Dispatchers coordinate the movement of freight and trucks, and provide the main communication link that informs the truck drivers of their assignments, schedules, and routes.

TABLE 4.6

Occupational Illnesses by Number of Cases and Percent for the Warehousing Sector

Illness Type	Number	Percent
Skin diseases and disorders	100	5
Respiratory conditions	0	0
Poisoning	0	0
Hearing loss	100	5
All others	1700	90

Source: From Bureau of Labor Statistics, U.S. Department of Labor. *Workplace Injuries and Illnesses in 2004.* Available at http://bls.gov.

FIGURE 4.2 Although forklifts are useful, they pose danger when misused.

Dispatchers frequently receive new shipping orders on short notice and must juggle drivers' assignments and schedules to accommodate a client. Shipping, receiving, and traffic clerks keep records of shipments arriving and leaving. They verify the contents of trucks' cargo against shipping records. They also pack and move stock. Billing and posting clerks and machine operators maintain company records of the shipping rates negotiated with customers and shipping charges incurred; they also prepare customer invoices.

Sales and related workers sell warehousing services to shippers of goods. They meet with prospective buyers, discuss the customers' needs, and suggest appropriate services. Travel may be required, and many analyze sales statistics, prepare reports, and handle some administrative duties.

Managerial staff provides general direction to the firm. They staff, supervise, and provide safety and other training to workers in the various occupations. They also resolve logistical problems such as forecasting the demand for transportation, mapping out the most efficient traffic routes, ordering parts and equipment service support, and scheduling the transportation of goods.

Many jobs in the warehousing industry require only a high school education, although an increasing number of workers have at least some college education. College education is most important for those seeking positions in management. Increasing emphasis on formal education stems from the increasing use of technology in the industry. Nearly all operations involve computers and information management systems. Many occupations—especially those involved in scheduling, ordering, and receiving—require detail-oriented people with computer skills. A growing number of employers recommend some form of formal training. Some companies provide such training in-house. Other sources of training include trade

---

TABLE 4.7

Twenty Nine Most Frequent OSHA Violations for the Warehousing Sector

Standard	Number Cited	Description
1910.178	258	Powered industrial trucks
1910.1200	141	Hazard communications
1910.305	112	Wiring methods, components, and equipment for general use
1910.37	97	Maintenance, safeguards, and operation features of exit routes
1910.303	96	General requirements
1910.157	68	Portable fire extinguishers
1910.132	67	Guarding floor and wall opening and holes
1910.147	59	The control of hazardous energy (lockout/tagout)
1904.29	49	Forms
5A1	45	General duty clause
1910.134	44	Respiratory protection
1910.151	44	Medical services and first aid
1910.1030	44	Bloodborne pathogens
1904.41	39	Annual OSHA injury and illness survey of 10 or more employees
1910.215	39	Abrasive wheel machinery
1910.22	37	General requirements
1910.212	35	General requirements for all machines
1910.176	34	Handling materials, general
1910.304	29	Wiring design and protection
1910.36	26	Design and construction requirements for exit routes
1910.146	24	Permit-required confined spaces
1904.32	20	Annual summary
1910.253	20	Oxygen-fuel welding and cutting
1910.119	19	Process safety management of highly hazardous chemical
1910.106	18	Flammable and combustible liquids
1910.110	17	Storage and handling of liquefied petroleum gases
1910.219	16	Mechanical power-transmission apparatus
1910.133	15	Eye and face protection
1910.141	14	Sanitation

Source: From Occupational Safety and Health Administration, U.S. Department of Labor. Available at http://www.osha.gov.

associations, unions, and vocational schools. Many companies have specific curricula on safety and procedural issues, as well as on occupational duties.

4.4 APPLICABLE OSHA REGULATIONS

One of the best ways to understand the hazards faced by utility workers is to see the types of violations that the Occupational Safety and Health Administration (OSHA) has found during their inspections of the workplace. These violations provide another way of targeting hazards that have the potential to cause injury, illness, and death of workers. As can be seen from the 30 most frequently cited violations, OSHA cites this industry under the general industry standard (29 CFR 1910) (Table 4.7).

Although the violations cited in Table 4.7 were the 30 most frequently issued violations, OSHA has cited other hazards with less frequency. Some of these are as follows:

- Woodworking machinery
- Overhead cranes and gantry cranes
- Hazardous waste operations and emergency response
- Hand protection
- Asbestos
- Portable metal ladders
- Emergency action plans
- Fixed industrial stairs
- Hand and portable power tools
- Arc welding and cutting
- Occupational noise exposure
- Compressed gases
- Foot protection

With the hazards faced by this sector, it is imperative that safety and health be an integral part of doing business and with the specific purpose of protecting its employees.

REFERENCES

Bureau of Labor Statistics, U.S. Department of Labor. *National Census of Fatal Occupational Injuries in 2005*. Available at http://bls.gov.

Bureau of Labor Statistics, U.S. Department of Labor. *Workplace Injuries and Illnesses in 2004*. Available at http://bls.gov.

Occupational Safety and Health Administration, U.S. Department of Labor. Available at http://www.osha.gov.

5 OSHA and Its Regulations

Hazards abound from the handling of goods and materials and workers should be protected. An example is that of a worker moving heavy furniture that could result in injury.

A worker in the goods and materials services should be able to go to work each day and expect to return home uninjured and in good health. There is no logical reason that a worker should be part of workplace carnage. Workers do not have to become one of the yearly workplace statistics.

Workers who know the occupational safety and health rules and safe work procedures and follow them are less likely to become one of the 5700 occupational trauma deaths, one of the 90,000 occupational illness deaths, or even one of the 6.8 million nonfatal occupational injuries and illnesses.

The essence of workplace safety and health should not completely depend on the Occupational Safety and Health Administration (OSHA) and its regulations, since they are not the driving forces behind workplace safety and health. OSHA has limited resources for inspection and limited inspectors. Enforcement is usually based on serious complaints, catastrophic events, and workplace deaths. An employer with a good safety and health program and safety record has a better opportunity to get a contract/order tied to his/her workplace safety and health record and reap the benefits of low insurance premiums for workers compensation and liability. Usually, safety and health is linked to the bottom line, which is seldom perceived as humanitarian.

This chapter tries to answer many of the questions regarding OSHA compliance, workplace safety and health, and coordination between workers and employers to have a safe and healthy workplace.

5.1   FEDERAL LAWS

Congress establishes federal laws (legislation or acts) and the president signs them into law. These laws often require that regulations (standards) be developed by the federal agencies that are responsible for the intent of the law.

5.2   OSHACT

The Occupational Safety and Health Act (OSHACT) of 1970, also known as the Williams–Steiger Act, is such a law. It was signed by President Richard Nixon on December 29, 1970 and became effective from April 29, 1971. (The OSHACT was not amended until November 5, 1990 by Public Law 101-552.) The OSHACT assigned the responsibility of implementing and enforcing the law to a newly created agency, the OSHA, located in the U.S. Department of Labor (DOL).

5.3   CONTENT OF THE OSHACT

Before the OSHACT, there were some state laws, a few pieces of federal regulations, and a small number of voluntary programs by employers. Most of the state programs were limited in scope and the federal laws only partially covered workers.

Another important reason for the OSHACT was the increasing number of injuries and illnesses within the workplace. Thus, the OSHACT was passed with the express purpose of assuring that every working man and woman in the nation would be provided safe and healthful work conditions while preserving this national human resource, the American worker. The OSHACT is divided into 34 sections with each having a specific purpose. The full text of the OSHACT is approximately 31 pages and a copy can be obtained from your local OSHA office.

5.4   REGULATION PROCESS

The OSHA was mandated to develop, implement, and enforce regulations relevant to workplace safety and health and the protection of workers. Time constraints prevented the newly formed OSHA from developing brand new regulations. Therefore, OSHA adopted previously existing regulations from other government regulations, consensus standards, proprietary standards, professional groups' standards, and accepted industry standards. This is why the hazardous chemical exposure levels today, with a few exceptions, are the same as the existing threshold limit values (TLVs) published by the American Conference of Governmental Industrial Hygienist in 1968. Once these TLVs were adopted, it became very difficult to revise them. Even though research and knowledge have fostered newer and safer TLVs in the past 30 years, they have not been adopted by the OSHA.

As stated previously, the original OSHA standards and regulations have come from three main sources: consensus standards, proprietary standards, and federal laws that existed when the OSHACT became law.

Consensus standards are developed by industry-wide standard developing organizations and are discussed and substantially agreed upon through industry consensus. OSHA has incorporated into its standards the standards of two primary groups: the American National Standards Institute (ANSI) and the National Fire Protection Association (NFPA).

Proprietary standards are prepared by professional experts within specific industries, professional societies, and associations. The proprietary standards are determined by a straight membership vote, not by consensus.

Some of the preexisting federal laws that are enforced by OSHA include the Federal Supply Contracts Act (Walsh–Healy), the Federal Service Contracts Act (McNamara–O'Hara), the Contract Work Hours and Safety Standard Act (Construction Safety Act), and the National Foundation on the Arts and Humanities Act. Standards issued under these acts are now enforced in all industries where they apply. When OSHA needs to develop a new regulation or even revise an existing one, it becomes a lengthy and arduous process, often taking more than a decade.

Standards are sometimes referred to as being either horizontal or vertical in their application. Most standards are horizontal or general. This means they apply to any employer in any industry. Fire protection, working surfaces, and first aid standards are examples of horizontal standards. Some standards are only relevant to a particular industry and are called vertical or particular standards.

It seems, through newspapers and conversations, as though OSHA is producing new standards every day that will impact the workplace. This simply is not true. The regulatory process is very slow. The steps are as follows:

1. Agency (OSHA) opens a regulatory development docket for a new or revised regulation.
2. This indicates that OSHA believes a need for a regulation exists.
3. Advanced notice of proposed rulemaking (ANPRM) is published in the *Federal Register* and written comments are requested to be submitted within 30–60 days.
4. Comments are analyzed.
5. Notice of proposed rulemaking (NPRM) is published in the *Federal Register* with a copy of the proposed regulation.
6. Another public comment period transpires usually for 30–60 days.
7. If no additional major issues are raised by the comments, the process continues to step 10.
8. If someone raises some serious issues, the process goes back to step 4 for review and possible revision of the NPRM.
9. Once the concerns have been addressed, it continues forward to steps 5 and 6 again.
10. If no major issues are raised, a final rule (FR) will be published in the *Federal Register*, along with the date when the regulation will be effective (usually 30–120 days).
11. There can still be a petition of reconsideration of the FR. There are times when an individual or industry may take legal action to bar the regulations' promulgation.

12. If the agency does not follow the correct procedures or acts arbitrarily or capriciously, the court may void the regulation and the whole process will need to be repeated.

Comments for or against a regulation during the development process are welcome. This is an opportunity to speak up. No comments indicate a lack of concern. You must be specific. Give examples, be precise, give alternatives, and provide any data that can back up your opinion. Federal agencies always welcome good data, which substantiates your case. Cost/benefit data is always important in the regulatory process and any valid cost data that you are able to provide may be very beneficial. But, make sure that your comments are based upon what is published in the *Federal Register* and not based upon hearsay information. Remember that the agency proposing the regulation may be working under specific restraints. Make sure you understand these constraints. Because of restrictions the agency may not have the power to do what you think ought to be done.

5.5 FEDERAL REGISTER

The *Federal Register* is the official publication of the U.S. government. If you are involved in regulatory compliance, you should obtain a subscription to the *Federal Register*. The reasons for obtaining this publication are clear. It is official, comprehensive, and not a summary done by someone else. It provides immediate accurate information and early notices of forthcoming regulations, informs you of comment periods, and gives the preamble and responses to questions that are raised about a final regulation. It provides notices of meetings, gives information on obtaining guidance documents, and supplies guidance on findings, on cross references, and gives the yearly regulatory development agenda. It is the "Bible" for regulatory development. It is published daily and is recognizable by brown paper and newsprint quality printing (Figure 5.1).

5.6 PURPOSE OF OSHA

The OSHA's purpose is to ensure, as much as possible, a healthy and safe workplace and conditions for workers in the United States. The OSHA was created by the act to

- Encourage employers and employees to reduce workplace hazards and to implement new and improve existing safety and health programs.
- Provide for research in occupational safety and health to develop innovative ways of dealing with occupational safety and health problems.
- Establish separate but dependent responsibilities and rights for employers and employees for the achievement of better safety and health conditions.
- Maintain a reporting and recordkeeping system to monitor job-related injuries and illnesses.
- Establish training programs to increase the number and competence of occupational safety and health personnel.

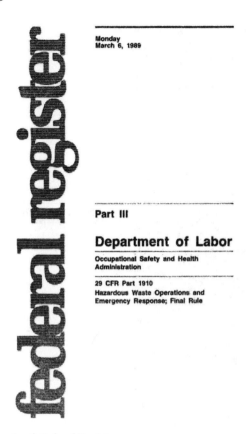

Monday
March 6, 1989

Part III

Department of Labor

Occupational Safety and Health
Administration

29 CFR Part 1910
Hazardous Waste Operations and
Emergency Response; Final Rule

FIGURE 5.1 Example of *Federal Register* cover.

- Develop mandatory job safety and health standards and enforce them effectively.
- Provide for the development, analysis, evaluation, and approval of state occupational safety and health programs.

5.7 CODE OF FEDERAL REGULATIONS

Probably, one of the most common complaints from people who use the U.S. Code of Federal Regulations (CFR) is, "How do you wade through hundreds of pages of standards and make sense out of them?" You may have experienced this frustration and been tempted to throw the standards in the round file.

The CFR is a codification of the general and permanent rules published in the *Federal Register* by the executive departments and agencies of the federal government. The code is divided into 50 titles, which represent broad areas that are subject to federal regulations. Each title is divided into chapters that usually bear the name of the issuing agency. Each chapter is further subdivided into parts covering specific regulatory areas. Based on this breakdown, the OSHA is designated Title 29—Labor, Chapter XVII

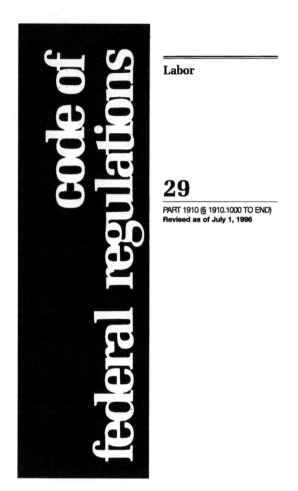

FIGURE 5.2   Example of Code of Federal Regulations cover.

(Occupational Safety and Health Administration) and Part 1910 for the General Industry Sector (Figure 5.2).

Each volume of the CFR is revised at least once each calendar year and issued on a quarterly basis. OSHA issues regulations at the beginning of the fourth quarter, or July 1 of each year (the approximate revision date is printed on the cover of each volume).

The CFR is kept up-to-date by individual revisions issued in the *Federal Register*. These two publications (the CFR and the *Federal Register*) must be used together to determine the latest version of any given rule.

To determine whether there have been any amendments since the revision date of the U.S. code volume in which you are interested, the following two lists must be consulted: the Cumulative List of CFR Sections Affected, issued monthly; and the Cumulative List of Parts Affected, appearing daily in the *Federal Register*. These two lists refer you to the *Federal Register* page where you may find the latest amendment of any given rule. The pages of the *Federal Register* are numbered sequentially from January 1 to January 1 of the next year.

As stated previously, Title 29, Chapter XVII has been set aside for the OSHA. Chapter XVII is broken down into parts. Part 1910 contains the general industry standards. The general industry standards are further broken down into subparts, sections, and paragraphs.

5.8   CFR NUMBERING SYSTEM

To use the CFR, you need an understanding of the hierarchy of the paragraph numbering system. The numbering system is a mixture of letters and numbers. Before 1979, italicized small case letters and small case Roman numerals were used. A change was made after 1979.

The CFR numbering hierarchy is a follows:

Before 1979	1980
(a)	(a)
(1)	(1)
(i)	(i)
Italicized (*a*)	(A)
Italicized (*1*)	{1}
Italicized (*i*)	(i)

When trying to make use of the regulations, prior knowledge will be beneficial. This should make comprehension easier and more user friendly. The following illustrates and explains the numbering system using an example:

29 CFR 1910.110 (b)(13)(ii)(*b*)(*7*)(*iii*)

Portable containers shall not be taken into buildings except as provided in Paragraph (b)(6)(i) of this section.

Title	Code of Federal Regulations	Part	Subpart	Section	Paragraph
29	CFR	1910	H	.110	

As can be seen from this example, the first number (29) stands for the title. CFR stands for the Code of Federal Regulations, followed by the Part 1910 and the Subpart H. Finally, there is a period which is followed by an Arabic number. This will always be the section number. In this case, Section .110 is the storage and handling of liquefied petroleum gases standard. If the number had been .146, the section would pertain to permit-required confined spaces.

29 CFR 1910.110 (b)(13)(ii)(*b*)(*7*)(*iii*)

Portable containers shall not be taken into buildings except as provided in paragraph (b)(6)(i) of this section.

Title	Code of Federal Regulations	Part	Subpart	Section	Paragraph
29	CFR	1910	H	.110	(b)

This means that the next breakdown of paragraphs will be sequenced by using small case letters in parentheses (a), (b), (c), etc. If you had three major paragraphs of information under a section, they would be lettered .110(a), .110(b), and .110(c).

29 CFR 1910.110 (b)(13)(ii)(*b*)(*7*)(*iii*)

Portable containers shall not be taken into buildings except as provided in paragraph (b)(6)(i) of this section.

Title	Code of Federal Regulations	Part	Subpart	Section	Paragraph and Subparagraph
29	CFR	1910	H	.110	(b)(13)

The next level of sequencing involves the use of Arabic numbers. As illustrated, if there were three paragraphs of information between subheadings (a) and (b), they would be numbered (a)(1), (a)(2), and (a)(3).

29 CFR 1910.110 (b)(13)(ii)(*b*)(*7*)(*iii*)

Portable containers shall not be taken into buildings except as provided in paragraph (b)(6)(i) of this section.

Title	Code of Federal Regulations	Part	Subpart	Section	Paragraph and Subparagraph
29	CFR	1910	H	.110	(b)(13)(ii)

The next level uses the lower case Roman numerals. An example would be between Paragraphs (2) and (3). If there were five paragraphs of information pertaining to Arabic number 2, they would be numbered (2)(i), (2)(ii), (2)(iii), (2)(iv), and (2)(v).

Since 29 CFR 1910.110 was promulgated before 1979, all subparagraph numbering beyond this are italicized letters and numbers (e.g., (*b*)(*7*)(*iii*)).

If after 1979 there are subparagraphs to the lower case Roman numerals, then a capital or upper case letter is used such as (A), (B), . . . , (F). Any other subparagraph falling under an upper case letter is numbered using brackets, for example, {*1*}, {*2*}, . . . , {*23*}, and any subparagraph to the bracketed numbers would be denoted by an italicized Roman numeral, such as (*i*), (*ii*), . . . , (*ix*).

5.9 OSHA STANDARDS COVER

The OSHA standards cover construction and the general industry, which includes manufacturing, transportation and public utilities, wholesale and retail trades, finance and insurance services, as well as other industrial sector (i.e., longshoring). Some of the specific areas covered by regulations are as follows:

Lockout/Tagout	Electrical Safety	Housekeeping
Training requirements	Noise exposure	Fire prevention
Hazard communication	Confined spaces	Personal protection
Ventilation	Equipment requirements	Sanitation
Medical and first aid	Fall protection	Working with hazardous chemical
Emergency planning	Hazardous substances	Recordkeeping
Guarding	Use of hand tools	Machine safety
Ladders and scaffolds	Equipment safety	Radiation
Explosives/blasting	Hazardous chemicals	Lead

5.10 COPIES OF THE OSHA STANDARDS

The standards for Occupational Safety and Health are found in Title 29 of the CFR. The standards for specific industries are found in Title 29 of the CFR with separate parts:

- General Industry—29 CFR Part 1910
- Ship Yard Employment—29 CFR Part 1915
- Marine Terminals—29 CFR Part 1917
- Longshoring—29 CFR Part 1918
- Gear Certification—29 CFR Part 1919
- Construction—29 CFR Part 1926
- Agriculture—29 CFR Part 1928

5.11 RELIEF (VARIANCE) FROM AN OSHA STANDARD

Variance can be obtained for OSHA standards for the following reasons:

- Employer may not be able to comply with the standard by its effective date.
- Employer may not be able to obtain the materials, equipment, or professional or technical assistance.
- Employer has in place processes or methods that provide protection to workers, which is at least as effective as the standard's requirements.

A temporary variance meeting the above criteria may be issued until compliance is achieved or for 1 year, whichever is shorter. It can also be extended or renewed for 6 months (twice).

Employers can obtain a permanent variance if the employer can document with a preponderance of evidence that existing or proposed methods, conditions, processes, procedures, or practices provide workers with protections equivalent to or better than the OSHA standard.

Employers are required to post a copy of the variance in a visible area in the workplace as well as make workers aware of request for a variance.

5.12   OSHAcт PROTECTS

Usually, all employers and their employees are considered to be under the OSHAcт. But this statement is not entirely true since the following are not covered under the OSHAcт:

- Self-employed persons
- Farms at which only immediate family members are employees
- Workplaces already protected by other federal agencies under federal statues such as the U.S. Department of Energy and the Mine Safety and Health Administration
- Also, federal employees and state and local employees

5.13   NATIONAL INSTITUTE FOR OCCUPATIONAL SAFETY AND HEALTH'S ROLE

The National Institute for Occupational Safety and Health (NIOSH) is one of the centers for disease control under the Department of Health and Human Services with headquarters in Atlanta, Georgia, and it is not part of OSHA. Its functions are as follows:

- Recommend new safety and health standards to OSHA.
- Conduct research on various safety and health problems.
- Conduct health hazard evaluations (HHEs) of the workplace when called upon.
- Publish an annual listing of all known toxic substances and recommended exposure limits.
- Conduct training which will provide qualified personnel under the OSHAcт.

5.14   OCCUPATIONAL SAFETY AND HEALTH REVIEW COMMISSION'S ROLE

The Occupational Safety and Health Review Commission (OSHRC) was established, when the OSHAcт was passed, to conduct hearings when OSHA citations and penalties are contested by employers or by their employees.

5.15   EMPLOYERS ARE RESPONSIBLE FOR WORKERS' SAFETY AND HEALTH

The employer is held accountable and responsible under the OSHAcт. The General Duty Clause, Section 5(a)(1) of the OSHAcт states that employers are obligated to provide a workplace that is free of recognized hazards that are likely to cause death or serious physical harm to employees. Employers are responsible for the following:

- Must abide and comply with the OSHA standards.
- Maintain records of all occupational injuries and illnesses.
- Maintain records of workers exposure to toxic materials and harmful physical agents.
- Make workers aware of their rights under the OSHACT.
- Provide free medical examinations to workers at a convenient location when the OSHA standards require them.
- Report to the nearest OSHA office all occupational fatalities and/or a catastrophe where three or more employees are hospitalized within 8 h.
- Abate cited violations of the OSHA standards within the prescribed time period.
- Provide training on hazardous materials and make material safety data sheets (MSDSs) available to workers upon request.
- Post information required by OSHA such as citations, hazard warnings, and injury/illness records.

5.16 WORKERS' RIGHTS

Workers have many rights under the OSHACT. These rights include the following:

- Right to review copies of appropriate standards, rules, regulations, and requirements that the employer should have available at the workplace
- Right to request information from the employer on safety and health hazards in the workplace, precautions that may be taken, and procedures to be followed if an employee is involved in an accident or is exposed to toxic substances
- Right to access relevant worker exposure and medical records
- Right to be provided personal protective equipment (PPE)
- Right to file a complaint with OSHA regarding unsafe or unhealthy workplace conditions and request an inspection
- Right to not be identified to the employer as the source of the complaint
- Right to not be discharged or discriminated against in any manner for exercising rights under the OSHACT related to safety and health
- Rights to have an authorized employee representative accompany the OSHA inspector and point out hazards
- Right to observe the monitoring and measuring of hazardous materials and see the resulting of the sampling, as specified under the OSHACT and as required by OSHA standards
- Right to review the occupational injury and illness records (OSHA No. 300) at a reasonable time and in a reasonable manner
- Right to have safety and health standards established and enforced by law
- Right to submit to NIOSH a request for an HHE of the workplace
- Right to be advised of OSHA actions regarding a complaint and request an informal review of any decision not to inspect or issue a citation
- Right to participate in standard development
- Right of a worker to talk with the OSHA inspector related to hazards and violations during the inspection

- Right of the worker filing a complaint to receive a copy of any citations and the time for abatement
- Right to be notified by the employer if the employer applies for a variance from an OSHA standard and testify at a variance hearing and appeal the final decision
- Right to be notified if the employer intends to contest a citation, abatement period, or penalty
- Right to file a notice of contest with OSHA if the time period granted to the company for correcting the violation is unreasonable, provided it is contested within 15 working days of employers notice
- Right to participate at any hearing before the OSHRC or at any informal meeting with OSHA when the employer or a worker has contested an abatement date
- Right to appeal the OSHRC's decisions in the U.S. Court of Appeals
- Right to obtain a copy of the OSHA file regarding a facility or workplace

5.17 WORKERS' RESPONSIBILITIES UNDER THE LAW

Workers have certain responsibilities to which they must adhere, but the employer is ultimately responsible for the workers' safety and health. Workers should do the following:

- Comply with the OSHA regulations and standards.
- Are not to remove, displace, or interfere with the use of any safeguards.
- Comply with the employer's safety and health rules and regulations.
- Report any hazardous condition to the supervisor or employer.
- Report any job-related injuries and illness to the supervisor or employer.
- Cooperate with the OSHA inspector during inspections when requested to do so.

5.18 RIGHT TO NOT BE DISCRIMINATED AGAINST

Workers have the right to expect safety and health on the job without fear of punishment. This is spelled out in Section 11(c) of the OSHAct. The law states that employers shall not punish or discriminate against workers for exercising rights such as the following:

- Complaining to an employer, union, or OSHA (or other government agency) about job safety and health
- Filing a safety and health grievance
- Participating in an OSHA inspection, conferences, hearing, or OSHA-related safety and health activity

5.19 RIGHT TO KNOW

This means that the employer must establish a written, comprehensive hazard communication program that includes provisions for container labeling, materials safety data sheets, and an employee training program. The program must include the following:

1. List of the hazardous chemicals in the workplace
2. Means the employer uses to inform employees of the hazards of non-routine tasks
3. Way the employer will inform other employers of the hazards to which their employees may be exposed

Workers have the right to information regarding the hazards to which they are or will be exposed. They have the right to review plans such as the hazard communication program. They have a right to see a copy of an MSDS during their shift and receive a copy of an MSDS when requested. Also, information on hazards that may be brought to the workplace by another employer should be available to workers. Other forms of information such as exposure records, medical records, etc., are to be made available to workers upon request.

5.20   ENVIRONMENTAL MONITORING RESULTS

Workers have the right to receive the results of any OSHA test for vapors, noise, dusts, fumes, or radiation. This includes observation of any measurement of hazardous materials in the workplace.

5.21   PERSONAL PROTECTIVE CLOTHING

Workers are to be provided at no cost, the proper and well-maintained personal protective clothing, when appropriate for the job. A new regulation does not require employers to pay for safety toed footwear, prescription safety eyewear, everyday clothing, weather-related gear, logging boots, uniforms, items worn to keep clean, and for other items that are not PPE.

5.22   OSHA INSPECTIONS

OSHA can routinely initiate an unannounced inspection of a business. Inspections may occur due to routine inspections or by complaints. These occur during normal working hours. These inspections will include checking company records, review of compliance with hazard communication standard, fire protection, PPE, and review of the company's health and safety plan. This inspection will include conditions, structures, equipment, machinery, materials, chemicals, procedures, and processes. OSHA's priorities for scheduling an inspection are as follows:

1. Situations involving imminent danger
2. Catastrophes or fatal accidents
3. Complaint by workers or their representatives
4. Regular inspections targeted at high-hazard industries
5. Reinspections

OSHA can give an employer advance notice of a pending inspection at certain times:

- In case of an imminent danger
- When it would be effective to conduct an inspection after normal working hours
- When it is necessary to assure the presence of the employer, specific employer representative, or employee representatives
- When the area director determines that an advance notice would enhance the probability of a more thorough and effective inspection

No inspection will occur during a strike, work stoppage, or picketing unless the area director approves such action and usually this would occur due to extenuating circumstances (e.g., an occupational death inside the facility). The steps of an OSHA inspection include the following:

1. Inspector becomes familiar with the operation including previous citation, accident history, and business demographics. The inspector gains entry. OSHA is forbidden to make a warrantless inspection without the employer's consent.
2. Thus, the inspector may have to obtain a search warrant if reasonable grounds for an inspection exist.
3. Inspector will hold an opening conference with the employer or a representative of the company. It is required that a representative of the company be with the inspector during the inspection and a representative of the workers be given the opportunity to accompany the inspector.
4. Inspection tour may take from hours to days depending on the size of the operation. The inspector will usually cover every area within the operation while assuring compliance with OSHA regulations.
5. Closing conference will be conducted for the employer to review what the inspector has found. The inspector will request an abatement time for the violations from the employer.
6. Area director will issue the written citations with proposed penalties along with the abatement dates to the employer. This document is called Notification of Proposed Penalty.

5.23 OSHA RECEIVES A COMPLAINT

OSHA gathers information and decides whether to send a compliance officer (inspector) or inform the employee about a decision not to inspect. The time period for response is based upon the seriousness of the complaint. The usual times are as follows:

- Within 24 h if the complaint alleges an imminent danger
- Within 3 days if the complaint is serious
- Within 20 days for all other complaints

5.24 CITATIONS

If violations of OSHA standards are detected then the citations will include the following information:

- Violation
- Workplace affected by the violation
- Denote specific control measures
- Abatement period or the time to correct the hazard

Copies of the citation should be posted near the violations location for at least 3 days or until the violation is abated, whichever is longer.

5.25 TYPES OF VIOLATIONS

Violations are categorized in the following manner:

De Minimis	No Penalty
Nonserious	$1,000–$7,000/violation
Serious	$1,500–$7,000/violation
Willful, no death	Up to $70,000 and $7,000/day for each day it remains
Willful, repeat violations	Same as willful, no death
Willful, death results	Up to $250,000 or $500,000 for a corporation and 6 months in jail
Willful, death results, second violation	$2,500,000 and 1 year in jail
Failure to correct a cited violation	$7,000/day
Failure to post official documents	$1,000/poster
Falsification of documents	$10,000 and 6 months in jail

5.26 CHALLENGING CITATIONS, PENALTIES, AND OTHER ENFORCEMENT MEASURES

Upon receipt of penalty notification, the employer has 15 days to submit a notice of contest to OSHA, which must be given to the workers' authorized representative or if no such representative exists it must be posted in a prominent location at the workplace. An employer who has filed a notice of contest may withdraw it before the hearing date by

- Showing that the alleged violation has been abated or will be abated
- Informing the affected employees or their designated representative of the withdrawal of the contest
- Paying the fine that had been assessed for the violation

Employers can request an informal hearing with the area director to discuss these issues and the area director can enter into a settlement agreement if the situation merits it. But, if a settlement cannot be reached then the employer must notify the area director in writing a notice of contest of the citation, penalties, or abatement period within 15 days of receipt of the citation.

5.27 WORKERS GET THE RESULTS OF AN INSPECTION

The workers or their representative can request the inspector to conduct a closing conference for labor and all citations are to be posted by the employer for the workers information.

Workers can contest the length of time period for abatement of a citation and the employer's petition for modification for abatement (PMA), which requests a time extension for correcting the hazard (workers must do this within 10 working days of posting). Workers cannot contest the following:

- Employer's citations
- Employer's amendments to citations
- Penalties for the employer's citations
- Lack of penalties

Two items can be challenged by workers. They are as follows:

- Time element in the citation for abatement of the hazard
- Employer's PMA; workers have 10 days to contest the PMA

5.28 DETERMINING PENALTIES

Penalties are usually based upon four criteria:

- Seriousness or gravity of the alleged violation
- Size of the business
- Employer's good faith in genuinely and effectively trying to comply with the OSHAct before the inspection and effort to abate and comply with the law during and after the inspection
- Employer's history of previous violations

5.29 STATE PROGRAMS

Some states elect to enforce occupational safety and health in their states. They must develop a program that OSHA will review and approve. Approximately, 23 states have such programs. If a state has a state plan (program) that is approved, the following conditions must exist:

- State must create an agency to carry out the plan.
- State's plan must include safety and health standards and regulations.
- Enforcement of these standards must be at least as effective as the federal plan.

- State plan must include provisions for the right of entry and inspection of the workplace including a prohibition on advance notice of inspections.
- State's plan must also cover state and local government employees.

If a state has specific standards or regulations, they must be at least as stringent as the federal standards and regulations. Some states have standards and regulations that go beyond the requirements of the existing federal standards and regulations.

5.30 WORKERS' TRAINING

Many standards promulgated by OSHA specifically require the employer to train employees in the safety and health aspects of their jobs. Other OSHA standards make it the employer's responsibility to limit certain job assignments to employees, who are certified, competent, or qualified—meaning that they have had special previous training, in or out of the workplace. OSHA's regulations imply that an employer has assured that a worker has been trained before being designated his task.

To completely address this issue, one would have to go directly to the regulation that applies to the specific type of activity. The regulation may mandate hazard training, task training, length of training, as well as specifics to be covered by the training.

It is always a good idea for the employer to keep records of training. These may be used by a compliance inspector during an inspection, after an accident resulting in injury or illness, as a proof or good intention to comply by an employer, or when a worker goes to a new job.

5.31 OCCUPATIONAL INJURIES AND ILLNESSES

Any occupational illness that has resulted in an abnormal condition or disorder caused by exposure to environmental factors that may be acute or chronic and because of inhalation, absorption, ingestion, or direct contact with toxic substances or harmful agents, and any repetitive motion injury is classified as an illness. All illnesses are recordable, regardless of severity. Injuries are recordable when

- On-the-job death occurs (regardless of length of time between injury and death)
- One or more lost workdays occur
- Restriction of work or motion transpires
- Loss of consciousness occurs
- Worker is transferred to another job
- Worker receives medical treatment beyond first aid

Employers with more than 10 employees are required to complete the OSHA 301 and retain it for 5 years as well as complete the OSHA 300A log and post it yearly from February 1 to April 30.

5.32   MEDICAL AND EXPOSURE RECORDS

Medical examinations of certain specialized workers are a requirement of some OSHA regulations. This work includes the following:

- Asbestos abatement
- Lead abatement
- Hazardous waste remediation
- Physicians may require medical examinations of a worker before wearing a respirator

Exposure records (monitoring records) are to be maintained by the employer for 30 years. Medical records are to be maintained by the employer for the length of employment plus 30 years.

A worker must make a written request to obtain a copy of their medical record or make them available to their representative or physician. Workers' medical records are considered confidential and require a request in writing from the worker to the physician for the records to be released.

5.33   POSTING

Employers are required to post in a prominent location the following:

- Job safety and health protection workplace poster (OSHA Form 2203) or state equivalent (use new OSHA Form 3165)
- Copies of any OSHA citations of violations of the OSHA standards are to be at or near the location of the violation for at least 3 days or until the violation is abated, whichever is longer
- Copies of summaries of petitions for variances from any standard and this include recordkeeping procedures
- Summary portion of the Log and Summary of Occupational Injuries and Illnesses (OSHA Form 300A) annually from February 1 to April 30

5.34   SUMMARY

Although the employer is ultimately responsible for a safe and healthy workplace, the adherence to guidelines set by OSHA is the basis for a good safety and health program and assures the well-being of the employer, the managers, the supervisors, the workers, and their fellow workers.

Although many employers would bulk at the idea that OSHA is an asset to them, employers would not have a foundation upon which to enforce their safety and health program. After all, the employer can blame OSHA when questioned about a safety and health regulation or rule, and absolve himself/herself. "OSHA made me do it." Without OSHA's threats of enforcement, employers would have no leverage upon which to enforce the safety and health policies.

With proper cooperation and coordination, all parties can ensure a safe and healthy workplace by following good safety and health practices. A safe and healthy home away from home is and should be the ultimate goal.

6 Safety and Health Management

The safety and health program needs to be professionally managed. (Courtesy of U.S. Environmental Protection Agency.)

6.1 SAFETY AND HEALTH MANAGEMENT

Management of safety and health is accomplished through a strong leadership that provides the resources, motivation, priorities, and accountability for ensuring the safety and health of the workforce. This leadership involves setting up systems to ensure continuous improvement and maintaining a health and safety focus while attending to production concerns. Enlightened managers understand the value in creating and fostering a strong safety culture within their organization. Safety should be a priority so that it is a value of the organization as opposed to a mundane duty. Integrating safety and health concerns into the everyday management of the organization, just like production, quality control, and marketing allows for a proactive approach to accident prevention and demonstrates the importance of working safety in the entire organization.

You can increase worker protection, cut business costs, enhance productivity, and improve employee morale. Worksites participating in OSHA's voluntary protection programs (VPP) have reported OSHA-verified lost workday cases at rates 60%–80% lower than their industry averages. For every $1 saved on medical or insurance compensation costs (direct costs), an additional $5–$50 are saved on indirect costs, such as repair to equipment or materials, retraining new workers, or production delays.

71

During 3 years in the VPP, a Ford plant noted a 13% increase in productivity, and a 16% decrease in scrapped product that had to be reworked. Bottom line, safety does pay off. Losses prevented go straight to the bottom line profit of an organization. With today's competitive markets and narrow profit margins, loss control should be every manager's concern. Management actions include the following:

- Establishing a safety and health policy
- Establishing goals and objectives
- Providing visible top management leadership and involvement
- Ensuring employee involvement
- Ensuring assignment of responsibility
- Providing adequate authority and responsibility
- Ensuring accountability for management, supervisors, and rank and file employees
- Providing a program evaluation

6.1.1 Safety and Health Policy

By developing a clear statement of management policy, you help everyone involved with the worksite understand the importance of safety and health protection in relation to other organizational values (e.g., production vs. safety and health). A safety and health policy provides an overall direction or vision while setting a framework from which specific goals and objectives can be developed.

6.1.2 Goals and Objectives

Companies should make general safety and health policy specific by establishing clear goals and objectives, and make objectives realistic and attainable by aiming at specific areas of performance that can be measured or verified. Some examples are as follows: have weekly inspections and correct hazards found within 24 h, or train all employees about hazards of their jobs, and specific safe behaviors (use of job safety analysis sheets) before beginning work.

6.1.3 Visible Top Management Leadership

Values and goals of top management in an organization tend to get emulated and accomplished. If employees see the emphasis that the top management puts on safety and health, they are more likely to emphasize it in their own activities. Besides following set safety rules themselves, managers can also participate in plant-wide safety and health inspections, personally stopping activities or conditions that are hazardous until the hazards can be corrected, assigning specific responsibilities, participating in or helping to provide training, and tracking safety and health performance.

6.1.4 Assignment of Responsibility

Everyone in the workplace should have some responsibility for safety and health. Clear assignment helps avoid overlaps or gaps in accomplishing activities.

Safety and health is not the sole responsibility of the safety and health professional. Rather, it is everyone's responsibility, while the safety and health professional is a resource.

6.1.5 PROVISION OF AUTHORITY

Any realistic assignment of responsibility must be accompanied by the needed authority and by having adequate resources. This includes appropriately trained and equipped personnel as well as sufficient operational and capital funding.

6.1.6 ACCOUNTABILITY

Accountability is crucial to helping managers, supervisors, and employees understand that they are responsible for their own performance. Reward progress and punish when appropriate. Supervisors are motivated to do their best when management measures their performance, "what gets measured is what gets done." Take care to ensure that measures accurately depict accomplishments and do not encourage negative behaviors such as not reporting accidents or near misses. Accountability can be established in safety through a variety of methods:

- Charge backs—Charge accident costs back to the department or job, or prorate insurance premiums.
- Safety goals—Set safety goals for management and supervision (e.g., accident rates, accident costs, and loss ratios).
- Safety activities—Conduct safety activities to achieve goals (e.g., hazard hunts, training sessions, safety fairs, etc., activities that are typically developed from needs identified based on accident history and safety program deficiencies).

6.1.7 PROGRAM EVALUATION

Once your safety and health program is up and running, you will want to assure its quality, just like any other aspect of your company's operation. Each program goal and objective should be evaluated in addition to each of the program elements, for example, management leadership, employee involvement, worksite analysis (accident reporting, investigations, surveys, pre-use analysis, hazard analysis, etc.), hazard prevention and control, and training. The evaluation should not only identify accomplishments and the strong points of the safety and health program, but also identify weaknesses and areas where improvements can be made. Be honest and identify the true weaknesses. The audit can then become a blueprint for improvements and a starting point for the next year's goals and objectives (Figure 6.1).

6.2 SAFETY AND HEALTH PROGRAMS

The need for health and safety programs in the workplace has been an area of controversy for some time. Many companies feel that written safety and health programs are just more paperwork, a deterrent to productivity, and nothing more

FIGURE 6.1   Monitoring and evaluation are keys to assuring effectiveness. (Courtesy of U.S. Environmental Protection Agency.)

than another bureaucratic way of mandating safety and health on the job. But over a period of years, data and information have been mounting in support of the need to develop and implement written safety and health programs.

To effectively manage safety and health, a company must pay attention to some critical factors that were mentioned in Section 6.1. These factors are essential to manage safety and health on worksites. The written safety and health program is of primary importance in addressing these critical factors. Have you ever wondered how your company is doing in comparison with a company without a safety and health professional and a viable safety and health program? Well, wonder no more.

In research conducted by the Lincoln Nebraska Safety Council in 1981, the following conclusions were based on a comparison of responses from a survey of 143 national companies. All conclusions have a 95% confidence level or more. Table 6.1 is an abstraction of results from that study.

It seems apparent from the previous research that in order to have an effective safety program, at a minimum, an employer must

- Have a demonstrated commitment to job safety and health
- Commit budgetary resources
- Train new personnel
- Insure that supervisors are trained
- Have a written safety and health program
- Hold supervisors accountable for safety and health
- Respond to safety complaints and investigate accidents
- Conduct safety audits

Other refinements can always be part of the safety and health program, which will help in reducing those workplace injuries and illnesses. They are as follows: more worker involvement (e.g., joint labor/management committees), incentive or recognition programs, getting outside help from a consultant or safety association, and setting safety and health goals.

TABLE 6.1

Effectiveness of Safety and Health Program Findings

Fact	Statement	Findings
1	Do not have separate budget for safety	43% more accidents
2	No training for new hires	52% more accidents
3	No outside sources for safety training	59% more accidents
4	No specific training for supervisors	62% more accidents
5	Do not conduct safety inspections	40% more accidents
6	No written safety program compared with companies that have written programs	106% more accidents
7	Those using canned programs are not self-generated	43% more accidents
8	No written safety program	130% more accidents
9	No employee safety committees	74% more accidents
10	No membership in professional safety organizations	64% more accidents
11	No established system to recognize safety accomplishments	81% more accidents
12	Did not document/review accident reports and reviewers did not have safety as part of their job responsibility	122% more accidents
13	Did not hold supervisor accountable for safety through merit salary reviews	39% more accidents
14	Top management did not actively promote safety awareness	470% more accidents

A decrease in occupational incidents that result in injury, illness, or damage to property is enough reason to develop and implement a written safety and health program.

6.3 REASONS FOR A COMPREHENSIVE SAFETY PROGRAM

The three major considerations involved in the development of a safety program are as follows:

1. Humanitarian—Safe operation of workplaces is a moral obligation imposed by modern society. This obligation includes consideration for loss of life, human pain and suffering, family suffering, and hardships.
2. Legal obligation—Federal and state governments have laws charging the employer with the responsibility for safe working conditions and adequate supervision of work practices. Employers are also responsible for paying the costs incurred for injuries suffered by their employees during their work activities.
3. Economic—Prevention costs less than accidents. This fact is proven consistently by the experience of thousands of industrial operations. The direct cost is represented by medical care, compensation, etc. The indirect cost of 4–10 times the direct cost must be calculated, as well as the loss of wages to employees and the reflection of these losses on the entire community.

These three factors are reason enough to have a health and safety program. It is also important that these programs be formalized in writing, since a written program sets the foundation and provides a consistent approach to occupational health and safety for the company. There are other logical reasons for a written safety and health program. Some of them are as follows:

- It provides standard directions, policies, and procedures for all company personnel.
- It states specifics regarding safety and health and clarifies misconceptions.
- It delineates the goals and objectives regarding workplace safety and health.
- It forces the company to actually define its view of safety and health.
- It sets out in black and white the rules and procedures for safety and health that everyone in the company must follow.
- It is a plan that shows how all aspects of the company's safety and health initiative work together.
- It is a primary tool of communications of the standards set by the company regarding safety and health.

6.4   BUILDING A SAFETY AND HEALTH PROGRAM

The length of such a written plan is not as important as the content. It should be tailored to the company's needs and the health and safety of its workforce. It could be a few pages or a multiple page document. However, it is advisable to follow the KISS principle (Keep It Simple, Stupid). To ensure a successful safety program, three conditions must exist: management leadership, safe working conditions, and safe work habits by all employees. The employer must

- Let the employees know that he/she is interested in safety on the job by consistently enforcing and reinforcing safety regulations.
- Provide a safe working place for all employees; it pays dividends.
- Be familiar with federal and state laws applying to your operation.
- Investigate and report all OSHA recordable accidents and injuries. This information may be useful in determining areas where more work is needed to prevent such accidents in the future.
- Make training and information available to the employees, especially in such areas as first aid, equipment operation, and common safety policies.
- Develop a prescribed set of safety rules to follow, and see that all employees are aware of these rules.

The basic premise of this chapter is that all employers should establish a workplace safety and health program to assist them in compliance with OSHA standards and the General Duty Clause of the Occupational Safety and Health Act (OSHA<small>CT</small>) of 1970 (Section 5(a)(1)). Each employer should set up a safety and health program to manage workplace safety and health to reduce injuries, illnesses, and fatalities by a systematic approach to safety and health. The program should be appropriate to conditions in the workplace, such as the hazards to which employees are exposed and the number of employees there. The primary guideline for employers to develop an organized safety and health program are as follows:

- Employers are advised and encouraged to institute and maintain in their establishments a program, which provides systematic policies, procedures, and practices that are adequate to recognize and protect their employees from occupational safety and health hazards.
- Effective program includes provisions for the systematic identification, evaluation, and prevention or control of general workplace hazards, specific job hazards, and potential hazards that may arise from foreseeable conditions.
- Although compliance with the law, including specific OSHA standards, is an important objective, an effective program looks beyond specific requirements of law to address all hazards. This effectively will seek to prevent injuries and illnesses, whether or not compliance is at issue.
- Extent to which the program is described in writing is less important than how effective it is in practice. As the size of a worksite or the complexity of a hazardous operation increases, however, the need for written guidance also increases to ensure clear communications of policies and priorities and consistent and fair application of rules.

The primary elements that should be addressed within this program are management leadership and employee participation, hazard identification and assessment, hazard prevention and control, information and training, and evaluation of program effectiveness.

6.4.1 MANAGEMENT COMMITMENT AND EMPLOYEE INVOLVEMENT

Management commitment and employee involvement are complementary. Management commitment provides the motivating force and the resources for organizing and controlling activities within an organization. In an effective program, management regards workers' safety and health as a fundamental value of the organization and assigns as much importance to it as other organizational issues. Employee involvement provides the means through which workers develop and/or express their own commitment to safety and health protection, for themselves and for their fellow workers.

Management must state clearly a worksite policy on safe and healthful work and working conditions, so that all personnel with responsibility at the site and personnel at other locations with responsibility for the site understand the priority of safety and health protection in relation to other organizational values.

Management must establish and communicate a clear goal for the safety and health program and objectives for meeting that goal, so that all members of the organization understand the results desired and the measures planned for achieving them. There needs to be visible top management involvement in implementing the program, so that the management's commitments are taken seriously.

Employees must be encouraged to be involved in the structure and operation of the program and in decisions that affect their safety and health, so that their insight and energy help to achieve the safety and health program's goals and objectives.

Management should assign and communicate responsibility for all aspects of the program so that managers, supervisors, and employees in all parts of the organization know what performance is expected of them. Adequate authority and resources must be provided to responsible parties, so that assigned responsibilities can be met. Managers, supervisors, and employees must be held accountable for meeting their responsibilities, so that essential tasks will be performed. Ensure that managers understand their safety and health responsibilities, as described previously, so that the managers will effectively carry out those responsibilities.

Review program operations at least annually to evaluate their success in meeting the goals and objectives, so that deficiencies can be identified and the program and/or the objectives can be revised when they do not meet the goal of effective safety and health protection.

Management commitment and leadership provides a policy statement that should be signed by the top person in your company. Safety and health goals and objectives are also included to assist you with establishing workplace goals and objectives that demonstrate your company's commitment to safety. An enforcement policy is provided to outline disciplinary procedures for violations of your company's safety and health program. This enforcement policy should be communicated to everyone at the company.

Establish the program responsibilities of managers, supervisors, and employees for safety and health in the workplace and hold them accountable for carrying out those responsibilities; provide managers, supervisors, and employees with the authority, access to relevant information, training, and resources they need to carry out their safety and health responsibilities; and identify at least one manager, supervisor, or employee to receive and respond to reports about workplace safety and health conditions and, where appropriate, to initiate corrective action.

The safety and health program should contain the following to demonstrate management commitment and leadership:

- Policy statement: goals established, issued, and communicated to employees
- Program revised annually
- Participation in safety meetings and inspections; agenda item in meetings
- Commitment of resources is adequate
- Safety rules and procedures incorporated into jobsite operations
- Management observes safety rules

Assignment of responsibility identifies the responsibilities of management officials, supervisors, and employees. Emphasis on responsibility to safety and health is more creditable if everyone is held accountable for their safety and health performance as related to established safety and health goals. The assignment of responsibility should include the following aspects:

- Safety designee on site should be knowledgeable and accountable.
- Supervisors' (including foremen) safety and health responsibilities should be understood.
- Employees should be aware of and adhere to safety rules.

The employer must allow employees to establish, implement, and evaluate the program. The employer must regularly communicate with employees about workplace safety and health matters; provide employees with access to information relevant to the program; provide means for employees to become involved in hazard identification and assessment, prioritizing hazards, training, and program evaluation; establish means for employees to report job-related fatalities, injuries, illnesses, incidents, and hazards promptly and to make recommendations about appropriate ways to control those hazards; and provide prompt responses to such reports and recommendations.

The employer must not discourage employees from making reports and recommendations about fatalities, injuries, illnesses, incidents, or hazards in the workplace, or from otherwise participating in the workplace safety and health program.

6.4.2 Hazard Identification and Assessment

The employer must systematically identify and assess hazards to which employees are exposed and assess compliance with the General Duty Clause and OSHA standards. The employer must conduct inspections of the workplace; review safety and health information; evaluate new equipment, materials, and processes for hazards before they are introduced into the workplace; and assess the severity of identified hazards and rank those hazards that cannot be corrected immediately according to their severity.

Identification of hazards includes those items that can assist you with identifying workplace hazards and determining what corrective action is necessary to control them. These items include jobsite safety inspections, accident investigations, safety and health committees, and project safety meetings. To accomplish the identification of hazards, the following items should be addressed:

- Periodic site safety inspection program involves supervisors
- Preventative controls in place [personal protective equipment (PPE), maintenance, engineering controls]
- Action taken to address hazards
- Safety committee, where appropriate
- Technical references available
- Enforcement procedures implemented by management

The employer must carry out an initial assessment, and then as often thereafter as necessary ensure compliance, usually, at least once every 2 years. When safety and health information or a change in workplace conditions indicates that a new or increased hazard may be present, then the employer should conduct a reassessment. The employer should investigate each work-related death, serious injury or illness, or incident (near miss) having the potential to cause death or serious physical harm. The employer should keep records of the hazards identified and their assessment and the actions the employer has taken or plans to take to control those hazards. These will be positives if OSHA were to inspect the workplace. It shows good faith effort and commitment to safety and health.

Worksite analysis involves a variety of worksite examinations, to identify not only existing hazards but also potential hazards. Unawareness of a hazard that stems from failure to examine the worksite is a sure sign that safety and health policies and/or practices are ineffective. Effective management actively analyzes the work and worksite, to anticipate and prevent harmful occurrences. Worksite analysis is to assure all hazards are identified. This can be accomplished by the following:

- Conducting comprehensive baseline worksite surveys for safety and health and periodic comprehensive update surveys
- Analyzing planned and new facilities, processes, materials, and equipment
- Performing routine job hazard analyses

Providing for regular site safety and health inspection, so that new or previously missed hazards and failures in hazard controls are identified, is critical to worksite analysis.

So that employee insight and experience in safety and health protection may be utilized and employee concerns may be addressed, a reliable system for employees is to be provided, without fear of reprisal, to notify management personnel about conditions that appear hazardous and to receive timely and appropriate responses; and encourage employees to use the system.

All accidents and near miss incidents should be investigated, so that their causes and means for their prevention are identified. Analysis of injury and illness trends over time should be undertaken, so that patterns with common causes can be identified and prevented.

6.4.3   HAZARD PREVENTION AND CONTROL

The requirements of the General Duty Clause and OSHA standards are to be met. If immediate compliance is not possible, the employer must devise a plan for prompt compliance, which includes setting priorities and deadlines and tracking progress in controlling hazards. Note: Any hazard identified by the employer's hazard identification and assessment process that is covered by an OSHA standard or the General Duty Clause must be controlled as required by that standard or that clause, as appropriate. Control means to reduce exposure to hazards in accordance with the General Duty Clause or OSHA standards, including providing appropriate supplemental and/or interim protection, as necessary, to exposed employees. Prevention and elimination are the best forms of control.

Hazard prevention and controls are triggered by a determination that a hazard or potential hazard exists. Where feasible, hazards are prevented by effective design of the jobsite or job. Where it is not feasible to eliminate them, they are controlled to prevent unsafe and unhealthful exposure. Elimination or controls should be done in a timely manner, once a hazard or potential hazard is identified.

Procedures are to be established for the purpose, using the following measures, so that all current and potential hazards, however detected, are corrected or controlled in a timely manner:

- Engineering techniques where feasible and appropriate
- Procedures for safe work which are understood and followed by all affected parties, as a result of training, positive reinforcement, correction of unsafe performance, and, if necessary, enforcement through a clearly communicated disciplinary system
- Provision of PPE
- Administrative controls, such as reducing the duration of exposure

Facility and equipment maintenance is to be provided, so that hazardous breakdown is prevented. Plan and prepare for emergencies, and conduct training and drills as needed, so that the response of all parties to emergencies will be second nature. Establish a medical program that includes availability of first aid on site and of physician and emergency medical care nearby, so that harm will be minimized if any injury or illness does occur.

6.4.4 INFORMATION AND TRAINING

The employer must ensure that each employee is provided with information and training in the safety and health program, and each employee exposed to a hazard is provided with information and training in that hazard. Note: Some OSHA standards impose additional, more specific requirements for information and training. This rule does not displace those requirements.

Safety and health information means the establishment's fatality, injury, and illness experience; OSHA 300 logs; workers' compensation claims; nurses' logs; the results of any medical screening/surveillance; employee safety and health complaints and reports; environmental and biological exposure data; information from prior workplace safety and health inspections; MSDSs; the results of employee symptom surveys; safety manuals and health and safety warnings provided to the employer by equipment manufacturers and chemical suppliers; information about occupational safety and health provided to the employer by trade associations or professional safety or health organizations; and the results of prior accident and incident investigations at the workplace. The employer must provide information and training in the following subjects:

- Nature of the hazards to which the employee is exposed and how to recognize them
- What is being done to control these hazards
- What protective measures the employee must follow to prevent or minimize exposure to these hazards
- Provisions of applicable standards (Figure 6.2)

FIGURE 6.2   Make sure all workers are trained in appropriate safety and health aspects of their job. (Courtesy of U.S. Environmental Protection Agency.)

The employer must provide initial information and training as follows:

- For new employees, before initial assignment to a job involving exposure to a hazard.
- Employer is not required to provide initial information and training for which the employer can demonstrate that the employee has already been adequately trained.
- Employer must provide periodic information and training as often as necessary to ensure that employees are adequately informed and trained, and when safety and health information or a change in workplace conditions indicates that a new or increased hazard exists.

Safety and health training addresses the safety and health responsibilities of all personnel concerned with the site, whether salaried or hourly. It is often most effective when incorporated into other training about performance requirements and job practices. Its complexity depends on the size and complexity of the worksite, and the nature of the hazards and potential hazards at the site.

It must be ensured that all employees understand the hazards to which they may be exposed and how to protect themselves and others from exposure to these hazards, so that employees accept and follow established safety and health procedures.

Ensure that they understand those responsibilities and the reasons for them, including the following, so that supervisors will carry out their safety and health responsibilities effectively:

- Analyzing the work under their supervision to identify unrecognized potential hazards
- Maintaining physical protections in their work areas
- Reinforcing employee training on the nature of potential hazards in their work and on needed protective measures, through continual performance feedback and, if necessary, through enforcement of safe work practices

The employer must provide all employees who have program responsibilities with the information and training necessary for them to carry out their safety and health responsibilities.

6.4.5 EVALUATION OF PROGRAM EFFECTIVENESS

The employer's basic obligation is to evaluate the safety and health program to ensure that it is effective and appropriate to workplace conditions. The employer must evaluate the effectiveness of the program as often as necessary to ensure program effectiveness or at least once every 2 years. The employer must revise the program in a timely manner to correct deficiencies identified by the program evaluation.

6.4.6 MULTIEMPLOYER WORKPLACES

Multiemployer worksite means a workplace where there is a host employer and at least one contract employer. Host employer means an employer who controls conditions at a multiemployer worksite. The host employer's responsibilities are to

- Provide information about hazards, controls, safety and health rules, and emergency procedures to all employers at the workplace.
- Ensure that safety and health responsibilities are assigned as appropriate to other employers at the workplace.

The responsibilities of a contract employer are to

- Ensure that the host employer is aware of the hazards associated with the contract employer's work and what the contract employer is doing to address them.
- Advise the host employer of any previously unidentified hazards that the contract employer identifies at the workplace.

Contract employer is an employer who performs work for a host employer at the host employer's workplace. A contract employer does not include an employer who provides incidental services that do not influence the workplace safety and health program, whose employees are only incidentally exposed to hazards at the host employer's workplace (e.g., food and drink services, delivery services, or other supply services).

6.5 CHARACTERISTICS OF AN OCCUPATIONAL SAFETY AND HEALTH PROGRAM

A review of research on successful safety and health programs reveals a number of factors, which comprise these programs. Strong management commitment to health and safety and frequent, close contacts between workers, supervisors, and management on health and safety are the two most dominant factors in good health and safety programs. Other relevant factors include workforce stability, stringent housekeeping, training emphasizing early indoctrination and follow-up instruction, and special adaptation of conventional health and safety practices to enhance their suitability to the workplace.

6.5.1 FACTORS AFFECTING SAFETY AND HEALTH

The factors affecting safety and health are as follows:

1. Management factors
   a. Management commitment as reflected by management involvement in aspects of the health and safety program in a formal way and employers' resources committed to employers' health and safety program
   b. Management adherence to principles of good management in the utilization of resources (people, machinery, and materials), supervision of employees, and production planning and monitoring
   c. Designated health and safety personnel reporting directly to the top management
2. Motivational factors
   a. Humanistic approach to interacting with employees
   b. High levels of employee/supervisor contact
   c. Efficient production planning
3. Hazard control factors
   a. Effort to improve the workplace
   b. Continuing development of the employees
   c. Clean working environment
   d. Regular, frequent inspections
4. Illness and injury investigations and recordkeeping factors
   a. Investigation of all incidents of illness and injury as well as non-lost-time accidents
   b. Recording of all first aid cases

6.6   SUMMARY

As can be seen, it is critical to have an organized approach to occupational safety and health. The outcomes to effectively manage a company's safety and health initiative results in many positives, which include less carnage and suffering, but also a better bottom line because of reduced accidents, better productivity, better morale, and a decrease in the cost of doing business.

A listing of the components that comprise a successful health and safety program are as follows:

- Health and safety program management
- Inspections and job observations
- Illness and injury investigations
- Task analysis
- Training
- Personal protection
- Communication/promotion of health and safety
- Personal perception
- Off-the-job health and safety

This is only a representative list that could be either expanded or consolidated depending upon the unique needs of your company. Health and safety programs should be tailored to meet individual requirements. A sample written safety and health program can be found in *Industrial Safety and Health for Infrastructure Services*.

7 Safety Hazards

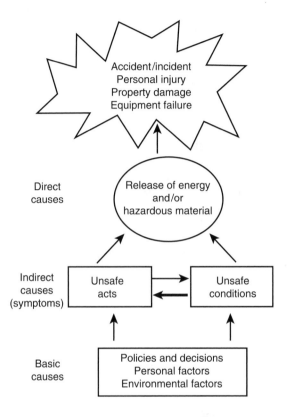

Accident-causes levels. (Courtesy of the Mine Health and Safety Administration.)

Potential safety hazards come from a large number of sources, each posing unique dangers, and also vary greatly in the degree of risk as well as the type of energy that each can release when not prevented or controlled. Table 7.1 provides a list of a wide range of equipment, tools, sources, etc. that can cause safety hazards.

7.1 EMPHASIS ON HAZARDS

The emphasis in this chapter is upon safety hazards. Hazards are defined as sources of danger that could result in a chance event such as an accident. A danger itself is a potential exposure or a liability to injury, pain, or loss. Not all hazards and dangers are the same. Exposure to hazards may be dangerous, but this is dependent on the

TABLE 7.1

Potential Sources of Safety Hazards

Acids	Hot processes	Power sources
Abrasives	Forklifts	Power tools
Biohazards	Fumes	Pressure vessels
Bloodborne pathogens	Generators	Radiation
Blasting	Gases	Rigging
Caustics	Hand tools	Respirators
Chains	Hazardous chemical processes	Scaffolds
Chemicals	Hazardous waste	Slings
Compressed gas cylinders	Heavy equipment	Solvents
Conveyors	Hoists	Stairways
Cranes	Hoses	Storage facilities
Confined spaces	Hot items	Stored materials
Derricks	Hot process	Transportation equipment
Electrical equipment	Housekeeping/waste	Transportation vehicles
Elevators and manlifts	Ladders	Trucks
Emergencies	Lasers	Unsafe conditions
Environmental factors	Lifting	Unsafe act
Excavations	Lighting	Ventilation
Explosives	Loads	Walkways and roadways
Falls	Machines	Walls and floor openings
Fibers	Materials	Warning devices
Fires	Mists	Welding and cutting
Flammables	Noise	Wire ropes
Hazardous waste	Platforms	Working surfaces
High voltage	Personal protective equipment	

amount of risk that accompanies it. The risk of water contained by a dam is different from being caught in a small boat in rapidly flowing water. Risk is the possibility of loss or injury or the degree of the possibility of such loss. Accidents do not occur in a hazardless environment. If the potential exposure is high, there is a greater risk that an undesired event will occur. An accident is an unplanned or undesirable event whose outcome is normally a trauma. Trauma is the injury to living tissue caused by some outside or extrinsic agent. Trauma is caused by an agent, force, or mechanism impinging on the human body (Figure 7.1).

The emphasis here will be to identify the hazard and its danger, and suggest ways to remove, intervene, or mitigate its risk to prevent accidents resulting from the errant uncontrolled release of energy that has a traumatic effect on those who are exposed to that hazard.

7.2 ACCIDENT CAUSES

Experts who study accidents often do a breakdown or analysis of the causes. They analyze them at three different levels:

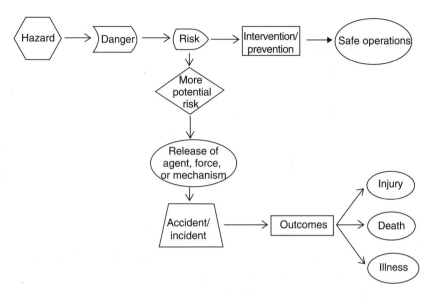

FIGURE 7.1 Potential outcomes from hazards.

1. Direct causes (unplanned release of energy and/or hazardous material)
2. Indirect causes (unsafe acts and unsafe conditions)
3. Basic causes (management safety policies and decisions, and personal factors)

7.2.1 DIRECT CAUSES

Most accidents are caused by the unplanned or unwanted release of large amounts of energy, or of hazardous materials. In a breakdown of accident causes, the direct cause is the energy or hazardous material released at the time of the accident. Accident investigators are interested in finding out what the direct cause of an accident is, because this information can be used to help prevent other accidents, or to reduce the injuries associated with them.

Energy is classified in one of two ways. It is either potential or kinetic energy. Potential energy is defined as stored energy such as a rock on the top of a hill. There are usually two components to potential energy: the weight and height of the object. The rock resting at the bottom of the hill has little potential energy as compared to the one at the top of the hill. Some examples of potential energy are represented in Table 7.2.

The other classification is kinetic energy that is best described as energy motion. Kinetic energy is dependent upon the mass of the object. Mass is the amount of matter making up an object; for example, an elephant has more matter than a mouse, therefore more mass. The weight of an object is a factor of the mass of an object and the pull of gravity on it. Kinetic energy is a function of an object's mass and its speed of movement or velocity. A bullet thrown at you has the same mass as one shot at you, but the difference is in the velocity and there is no doubt as to which has the

TABLE 7.2
Examples of Potential Energy

Compressed gases	Hand or power tool
Object at rest	Liquefied gas
Effort to move an object	Dust
Spring loaded objects	Unfallen tree
Electrically charged component	Radiation source
Idling vehicle	Chemical source
Disengaged equipment	Biological organism
Flowable material	

most kinetic energy or potential to destroy. Some examples of kinetic energy are represented in Table 7.3.

Energy has many forms and each has its own unique potential for danger. The forms of energy are pressure, biological, chemical, electrical, thermal, light, mechanical, and nuclear. Table 7.4 depicts examples of each form of energy.

If the direct cause is known, then equipment, materials, and facilities can be redesigned to make them safer, personal protection can be provided to reduce injuries, and workers can be trained to protect themselves in hazardous situations.

7.2.2 INDIRECT CAUSES

Indirect causes, or symptoms, may be considered as contributing factors. In most cases, the release of excessive amounts of energy or hazardous materials is caused by unsafe acts or unsafe conditions. Unsafe acts and unsafe conditions trigger the release of large amounts of energy or hazardous materials, which directly cause the accident. This chapter refers to indirect causes as symptoms or contributing factors. That is because unsafe acts and unsafe conditions do not themselves cause accidents. These are just symptoms or indicators of poor management policy, inadequate controls, lack of or insufficient knowledge of existing hazards, or other personal factors. Tables 7.5 and 7.6 depict some examples of unsafe acts and unsafe conditions.

TABLE 7.3
Examples of Kinetic Energy

Operating tools or equipment	Moving conveyors
Flow of materials	Running machines
Falling objects	Running equipment
Lifting a heavy object	Moving dust
Moving vehicles or heavy equipment	Tree falling
Release of energy from radiation, chemical or biological sources	Pinch area from moving objects
Energy transfer devices such as pulleys, belts, gears, shears, edgers	Running power tools

TABLE 7.4

Forms of Energy and Examples of Their Sources

Pressure energy	Chemical energy	Nuclear energy
Pressurized vessel	Corrosive materials	Alpha particles
Caisson work	Flammable/combustible	Beta particles
Explosives	materials	High energy nuclear
Noise	Toxic chemicals	particles
Compressed gases	Compressed gases	Neutrons
Steam source	Carcinogens	Gamma rays
Liquefied gases	Confined spaces	X-rays
Air under pressure	Oxidizing materials	
Diving	Reactive materials	**Thermal (heat) energy**
Confined spaces	Poisonous chemicals	Chemical reactions
	and gases	Combustible materials
Light energy	Explosives	Cryogenic materials
Intense light	Acids and bases	Fire
Lasers	Oxygen deficiency	Flames
Infrared sources	atmosphere	Flammable materials
Microwaves	Fuels	Friction
Sun	Dusts or powders	Hot processes
Ultraviolet light		Hot surfaces
Welding	**Electrical energy**	Molten metals
RF fields	Capacitors	Steam
Radio frequency	Transformers	Solar
	Energized circuits	Weather phenomena
Biological energy	Power lines	Welding
Allergens	Batteries	
Biotoxins	Exposed conductors	
Pathogens	Static electricity	
Poisonous plants	Lightning	

7.2.3 Basic Causes

The cause of most accidents is indeed a release of energy, an unsafe condition, or an unsafe act, but the basic or root causes of most accidents are found to be more a result of failure to address some very specific underlying causes. These causes fall into three groups: policies and decisions, personal factors, and environmental factors depicted in Tables 7.7, 7.8, and 7.9, respectively.

While we often think of hazardous acts and conditions as the basic causes of accidents, they are actually symptoms of failure on another level. Unsafe acts and unsafe conditions can usually be traced to the basic causes: poor management policies and decisions, and personal factors.

The first category of basic causes—management safety policies and decisions—includes such things as management's intent (relative to safety); production and safety goals; staffing procedures; use of records; assignment of responsibility, authority, and accountability; employee selection; training, placement, direction,

TABLE 7.5
Unsafe Acts

(95% of all accidents)

1. Operating or using equipment without authorization
2. Failure to prevent unexpected movement
3. Working or operating at unsafe speeds
4. Failure to warn or signal
5. Removing, nullifying, or not using guards
6. Using defective tools or equipment
7. Using tools or equipment unsafely
8. Taking an unsafe position
9. Failure to shut down and lockout
10. Riding equipment
11. Horseplay, startling, or distracting
12. Failure to wear or use personal protective equipment
13. Failure to warn coworkers or to secure equipment
14. Improper lifting
15. Alcohol or drug use
16. Violation of safety and health rules

TABLE 7.6
Unsafe Conditions

(5% of all accidents)

1. Lack of or inadequate guards
2. Lack of or inadequate warnings or signaling systems
3. Improper storage of flammable or explosives
4. Unexpected start-up conditions
5. Poor housekeeping conditions
6. Protruding objects
7. Congestion conditions
8. Atmospheric conditions
9. Improper placement or stacking
10. Defective tools or equipment
11. General working conditions
12. Improper clothing
13. Radiation exposure
14. Poor illumination
15. Excessive noise
16. Unstable work areas or platforms
17. No firefighting equipment
18. Dangerous soil
19. Hazardous conditions
20. Radiation

TABLE 7.7
Policies and Decisions

Safety policy is not
- In writing
- Signed by top management
- Distributed to each employee
- Reviewed periodically

Safety procedures do not provide for
- Written manuals
- Safety meetings
- Job safety analysis
- Housekeeping
- Medical surveillance
- Accident investigations
- Preventive maintenance
- Reports
- Safety audits/inspections

Safety is not considered in the procurement of
- Supplies
- Equipment
- Services

Safety is not considered in the personnel practices of
- Selection
- Authority
- Responsibility
- Accountability
- Communication
- Training
- Job observations

TABLE 7.8
Personal Factors

Physical
- Inadequate size
- Inadequate strength
- Inadequate stamina

Experiential
- Insufficient knowledge
- Insufficient skills
- Accident records
- Unsafe work practices

(continued)

TABLE 7.8 (continued)
Personal Factors

Motivational
- Needs
- Capabilities

Attitudinal
- Toward others
  - People
  - Company
  - Job
- Toward self
  - Alcoholism
  - Drug use
  - Emotional upset

Behavioral
- Risk taking
- Lack of hazard awareness

---

and supervision; communications procedures; inspection procedures; equipment, supplies, and facility design; purchasing; maintenance; standard and emergency job procedures; and housekeeping.

The second category—personal factors—includes motivation, ability, knowledge, training, safety awareness, assignments, performance, physical and mental state, reaction time, and personal care.

The third category is the actual physical facility design, the unsafe procedures being used, and the geological and climatic conditions.

TABLE 7.9
Environmental Factors

Unsafe facility design
- Poor mechanical layout
- Inadequate electrical system
- Inadequate hydraulic system
- Crowded limited access ways
- Insufficient illumination
- Insufficient ventilation
- Lack of noise control

Unsafe operating procedures
- Normal
- Emergency

Weather
Geographical area

7.3 SUMMARY

As can be seen, accidents that result because of safety hazards are actually the result of a complex set of events or elements that have come together from nature, human error, and failure of systems that should have protected workers from injury and death. Thus, the emphasis seen in this book and *Industrial Safety and Health for Infrastructure Services, Industrial Safety and Health for Administrative Services,* and *Industrial Safety and Health for People-Oriented Services*, is regarding the need for an organized approach to occupational safety and health and the protections and benefits from implementing a well thought out approach to job safety and health.

The remainder of this book is directed toward managing, preventing, and controlling hazards that occur within the goods and material service sector of the service industry. This includes the wholesale trade, retail trade, and warehousing sectors.

8 Health Hazards

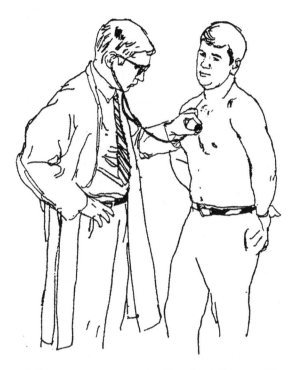

Exposure in the workplace can cause occupationally related illnesses. (Courtesy of the U.S. Environmental Protection Agency.)

8.1 OCCUPATIONAL ILLNESSES

Occupation illnesses are not as easily identified as injuries. According to the Bureau of Labor Statistics, there were 5.7 million injuries and illnesses reported in 1999. Of this number only 372,000 cases of occupational illnesses were reported. The 372,000 occupational illnesses included repeat trauma such as carpal tunnel syndrome, noise-induced hearing loss, and poisonings. It certainly appears that many occupational illnesses go unreported when the employer or worker is not able to link exposure with the symptoms the employees are exhibiting. Also, physicians fail to ask the right questions regarding the patients employment history, which can lead to the commonest of diagnoses of a cold or flu. This has become very apparent with the recent occupational exposure to anthrax where a physician sent a worker home with anthrax without addressing his/her potential occupational exposure hazards. Unless physicians are trained in occupational medicine, they seldom address work as the potential exposure source.

This is not entirely a physician problem by any means since the symptoms that are seen by the physician are often those of flu and other common illnesses suffered by the general public. It is often up to the employee to make the physician aware of their on-the-job exposure. If, I have continuously used the term exposure since, unlike trauma injuries and deaths, which are usually caused by the release of some source of energy, occupational illnesses are often due to both short- and long-term exposures. If the result of an exposure leads to immediate symptoms, it is said to be acute. If the symptoms come at a later time, it is termed a chronic exposure. The time between exposure and the onset of symptoms is called the latency period. It could be days, weeks, months, or even years, as in the case of asbestos where asbestosis or lung cancer appears 20–30 years after exposure.

It is often very difficult to get employers, supervisors, and employees to take seriously the exposures in the workplace as a potential risk to the workforce both short and long term, especially long term. "It cannot be too bad if I feel alright now." This false sense of security is that the workplace seems safe enough. The question is how bad could it be in our workplace? Everyone seems well enough now.

8.2 IDENTIFYING HEALTH HAZARDS

Health-related hazards must be identified (recognized), evaluated, and controlled to prevent occupational illnesses, which come from exposure to them. Health-related hazards come in a variety of forms, such as chemical, physical, ergonomic, or biological:

- Chemical hazards arise from excessive airborne concentrations of mists, vapors, gases, or solids that are in the form of dusts or fumes. In addition to the hazard of inhalation, many of these materials may act as skin irritants or may be toxic by absorption through the skin. Chemicals can also be ingested although this is not usually the principle route of entry into the body.
- Physical hazards include excessive levels of nonionizing and ionizing radiations, noise, vibration, and extremes of temperature and pressure.
- Ergonomic hazards include improperly designed tools or work areas. Improper lifting or reaching, poor visual conditions, or repeated motions in an awkward position can result in accidents or illnesses in the occupational environment. Designing the tools and the job to be done to fit the worker should be of prime importance. Intelligent application of engineering and biomechanical principles is required to eliminate hazards of this kind.
- Biological hazards include insects, molds, fungi, viruses, vermin (birds, rats, mice, etc.), and bacterial contaminants (sanitation and housekeeping items such as potable water, removal of industrial waste and sewage, food handling, and personal cleanliness can contribute to the effects from biological hazards). Biological and chemical hazards can overlap.

TABLE 8.1
Reported Nonfatal Occupational Illnesses

Type of Illness	Total Illnesses Reported (%)
Skin disease or disorders	17
Respiratory conditions because of toxic agents	8
Poisoning	1
Hearing loss	11
All other diseases	62

Source: From Bureau of Labor Statistics. United States Department of Labor. *Workplace Injuries and Illnesses in 2004.* Available at http://bls.gov.

Health-related hazards can often be elusive and difficult to identify. A common example of this is a contaminant in a building that has caused symptoms of illness. Even the evaluation process may not be able to detect the contaminant that has dissipated before a sample can be collected. This leaves nothing to control and possibly no answer to what caused the illnesses. Table 8.1 depicts the most common reported illnesses in the workplace.

8.3 HEALTH HAZARDS

Health hazards are caused by any chemical or biological exposure that interacts adversely with organs within our body causing illnesses or injuries. The majority of chemical exposures result from inhaling chemical contaminants in the form of vapors, gases, dusts, fumes, and mists, or by skin absorption of these materials. The degree of the hazard depends on the length of exposure time and the amount or quantity of the chemical agent. This is considered to be the dose of a substance. A chemical is considered a poison when it causes harmful effects or interferes with biological reactions in the body. Only those chemicals that are associated with a great risk of harmful effects are designated as poisons (Figure 8.1).

Dose is the most important factor determining whether or not you will have an adverse effect from a chemical exposure. The longer you work at a job and the more chemical agent that gets into the air or on your skin, the higher the dose potential. Two components that make up dose are as follows:

1. The length of exposure, or how long you are exposed—1 h, 1 day, 1 year, 10 years, etc
2. The quantity of substance in the air (concentration), how much you get on your skin, and/or the amount eaten or ingested

Another important factor to consider about the dose is the relationship of two or more chemicals acting together that cause an increased risk to the body. This interaction of chemicals that multiply the chance of harmful effects is called a

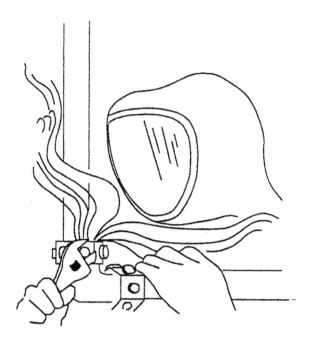

FIGURE 8.1   Chemical exposure poses real health issues for workers. (Courtesy of the U.S. Environmental Protection Agency.)

synergistic effect. Many chemicals can interact and although the dose of any one chemical may be too low to affect you, the combination of doses from different chemicals may be harmful. For example, the combination of chemical exposures and a personal habit such as cigarette smoking may be more harmful than just an exposure to one chemical. Smoking and exposure to asbestos increase the chance of lung cancer by as much as 50 times.

The type and severity of the body's response is related to dose and the nature of specific contaminant present. Air that looks dirty or has an offensive odor may, in fact, pose no threat whatsoever to the tissues of the respiratory system. In contrast, some gases that are odorless or at least not offensive can cause severe tissue damage. Particles that normally cause lung damage cannot even be seen. Many times, however, large visible clouds of dust are a good indicator that smaller particles may also be present.

The body is a complicated collection of cells, tissues, and organs having special ways of protecting itself against harm. We call these the body's defense systems. The body's defense system can be broken down, overcome, or bypassed. This can result in injury or illness. Sometimes, job-related injuries or illnesses are temporary, and you can recover completely. At other times, as in the case of chronic lung diseases like silicosis or cancer, these are permanent changes that may lead to death.

8.3.1 ACUTE HEALTH EFFECTS

Chemicals can cause acute (short-term) or chronic (long-term) effects. Whether or not a chemical causes an acute or chronic reaction depends both on the chemical and the dose you are exposed to. Acute effects are seen quickly, usually after exposures to high concentrations of a hazardous material. For example, the dry cleaning solvent perchloroethylene can immediately cause dizziness, nausea, and at higher levels, coma and death. Most acute effects are temporary and reverse shortly after being removed from the exposure. But at high enough exposures permanent damage may occur. For most substances, neither the presence nor absence of acute effects can be used to predict whether chronic effects will occur. Dose is the determining factor. Exposures to cancer-causing substances (carcinogens) and sensitizers may lead to both acute and chronic effects.

An acute exposure may occur, for example, when we are exposed to ammonia while using another cleaning agent. Acute exposure may have both immediate and delayed effects on the body. Nitrogen dioxide poisoning can be followed by signs of brain impairment (such as confusion, lack of coordination, and behavioral changes), days or weeks after recovery.

Chemicals can cause acute effects on breathing. Some chemicals irritate the lungs and some sensitize the lungs. Fluorides, sulfides, and chlorides are all found in various welding and soldering fluxes. During welding and soldering, these materials combine with the moisture in the air to form hydrofluoric, sulfuric, and hydrochloric acids. All three can severely burn the skin, eyes, and respiratory tract. High levels can overwhelm the lungs, burning and blistering them, and causing pulmonary edema. (Fluid building up in the lungs will cause shortness of breath and if severe enough can cause death.)

In addition, chemicals can have acute effects on the brain. When inhaled, solvent vapors enter the bloodstream and travel to other parts of the body, particularly the nervous system. Most solvents have a narcotic effect. This means they affect the nervous system by causing dizziness, headaches, inebriation, and tiredness. One result of these symptoms may be poor coordination, which can contribute to falls and other accidents on a worksite. Exposure to some solvents may increase the effects of alcoholic beverages.

8.3.2 CHRONIC HEALTH EFFECTS

A chronic exposure occurs during longer and/or repeated periods of contact, sometimes over years and often at relatively low concentrations of exposure. Perchloroethylene or alcohol, for example, may cause liver damage or other cancers 10–40 years after first exposure. This period between first exposure and the development of the disease is called the latency period. An exposure to a substance may cause adverse health effects many years from now with little or no effects at the time of exposure. It is important to avoid or eliminate all exposures to chemicals that are not part of normal ambient breathing air. For many chemical agents, the toxic effects following a single exposure are quite different from those produced by repeated exposures. For example, the primary acute toxic effect of benzene is central nervous system damage, while chronic exposures can result in leukemia.

There are two ways to determine if a chemical causes cancer: studies conducted on people and studies on animals. Studies on humans are expensive, difficult, and near impossible. This type of long-term research is called epidemiology. Studies on animals are less expensive and easier to carry out. This type of research is sometimes referred to as toxicology. Results showing increased occurrences of cancer in animals are generally accepted to indicate that the same chemical causes cancer in humans. The alternative to not accepting animal studies means we would have a lot less knowledge about the health effects of chemicals. We would never be able to determine the health effects of the more than 100,000 chemicals used by the industry.

There is no level of exposure to cancer-causing chemicals that is safe. Lower levels are considered safer. One procedure for setting health standard limits is called risk assessment. Risk assessment on the surface appears very scientific yet the actual results are based on many assumptions. It is differences in these assumptions that allow scientists to come up with very different results when determining an acceptable exposure standard. The following are major questions that assumptions are based on:

- Is there a level of exposure below which a substance would not cause cancer or other chronic diseases? (Is there a threshold level?)
- Can the body's defense mechanisms inactivate or break down chemicals?
- Does the chemical need to be at a high enough level to cause damage to a body organ before it will cause cancer?
- How much cancer should we allow? (One case of cancer among 1 million people, or one case of cancer among 100,000 people, or one case of cancer among 10 people?)

For exposures at the current permissible exposure limit (PEL), the risk of developing cancer from vinyl chloride is about 700 cases of cancer for each million workers exposed. The risk for asbestos is about 6,400 cases of cancer for each million workers exposed. The risk for coal tar pitch is about 13,000 cases for each million workers exposed. PELs set for current federal standards differ because of these different risks.

The dose of a chemical-causing cancer in human or animal studies is then used to set a standard PEL below which only a certain number of people will develop illness or cancer. This standard is not an absolute safe level of exposure to cancer-causing agents, so exposure should always be minimized even when levels of exposure are below the standard. Just as the asbestos standard has been lowered in the past from 5 to 0.2 fibers/cm^3, and now to 0.1 fibers/cm^3 (50 times lower). It is possible that other standards will be lowered in the future as new technology for analysis is discovered and public outrage insists on fewer deaths for a particular type of exposure. If a chemical is suspected of causing cancer, it is best to minimize exposure, even if the exposure is below accepted levels.

8.3.3 CHRONIC DISEASE

Chronic disease is not always cancer. There are many other types of chronic diseases, which can be as serious as cancer. These chronic diseases affect the function of different organs of the body. For example, chronic exposure to asbestos or silica dust (fine sand) causes scarring of the lung. Exposure to gases such as nitrogen oxides or ozone may lead to destruction of parts of the lung. No matter what the cause, chronic disease of the lungs will make the individual feel short of breath and limit their activity. Depending on the extent of disease, chronic lung disease can kill. In fact, it is one of the top 10 causes of death in the United States.

Scarring of the liver (cirrhosis) is another example of chronic disease. It is also one of the 10 causes of death in the United States. The liver is important in making certain essential substances in the body and cleaning certain waste products. Chronic liver disease can cause fatigue, wasting away of muscles, and swelling of stomach from fluid accumulation. Many chemicals such as carbon tetrachloride, chloroform, and alcohol can cause cirrhosis of the liver.

The brain is also affected by chronic exposure. Chemicals such as lead can decrease IQ and memory, and/or increase irritability. Many times these changes are small and can only be found with special medical tests. Workers exposed to solvents, such as toluene or xylene in oil-based paints, may develop neurological changes over a period of time.

Scarring of the kidney is another example of a chronic disease. Individuals with severe scarring must be placed on dialysis to remove the harmful waste products or have a kidney transplant. Chronic kidney disease can cause fatigue, high blood pressure and swollen feet, as well as many other symptoms. Lead, mercury, and solvents are suspect causes of chronic kidney disease.

8.3.4 BIRTH DEFECTS/INFERTILITY

The ability to have a healthy child can be affected by chemicals in many different ways. A woman may be unable to conceive because a man is infertile. The production of sperm may be abnormal, reduced, or stopped by chemicals that enter the body. Men working in an insecticide plant manufacturing 1,3-dibromo-3-chloropropane (DBCP) realized after talking among themselves that none of their wives had been able to become pregnant. When tested, all the men were found to be sterile.

A woman may be unable to conceive or may have frequent early miscarriages because of mutagenic or embryotoxic effects. Changes in genes in the woman's ovaries or man's sperm from exposure to chemicals may cause the developing embryo to die. A woman may give birth to a child with a birth defect because of a chemical with mutagenic or teratogenic effects. When a chemical causes a teratogenic effect, the damage is caused by the woman's direct exposure to the chemical. When a chemical causes a mutagenic effect, changes in genes from either the man or woman have occurred.

Many chemicals used in the workplace can damage the body. Effects range from skin irritation and dermatitis to chronic lung diseases such as silicosis and asbestosis or even cancer. The body may be harmed at the point where a chemical touches or enters it. This is called a local effect. When the solvent benzene touches the skin, it can cause drying and irritation (local effect).

A systemic effect develops at some place other than the point of contact. Benzene can be absorbed through the skin, breathed into the lungs, or ingested.

Once in the body, benzene can affect the bone marrow, leading to anemia and leukemia. (Leukemia is a kind of cancer affecting the bone marrow and blood.) Adverse health effects may take years to develop from a small exposure or may occur very quickly to large concentrations.

8.4 BIOLOGICAL MONITORING

Biological monitoring is the analysis of body systems such as blood, urine, finger-nails, teeth, etc. that provide a baseline level of contaminants in the body. Medical testing can have several different purposes, depending on why the worker is visiting a doctor. If it is a preemployment examination, it is usually considered a baseline to use as a reference for future medical testing. Baselines are a valuable tool to measure the amount of toxic substances in the body and often give an indication of the effectiveness of personal protective equipment (PPE) (Figure 8.2).

Occupational Safety and Health Administration (OSHA) regulations allow the examining physician to determine most of the content reviewed in the examination. Benefits received from an examination will vary with content of the examination. No matter what tests are included in the examination, there are certain important limitations of medical testing:

- Medical testing cannot prevent cancer. Cancer from exposure to chemicals or asbestos can only be prevented by reducing or eliminating an exposure.
- For many conditions, there are no medical tests for early diagnosis. For example, the routine blood tests conducted by doctors for kidney functions do not become abnormal until half the kidney function is lost. Nine of ten

FIGURE 8.2   Biological monitoring is a part of medical assessment. (Courtesy of the U.S. Environmental Protection Agency.)

people with lung cancer die within 5 years because chest x-rays do not diagnose lung cancer in time to save the individual.

- No medical test is perfect. Some tests are falsely abnormal and some falsely normal.

8.4.1 MEDICAL QUESTIONNAIRE

A medical and work history, despite common perceptions, is probably the most important part of an examination. Most diagnoses of disease in medicine are made by the work history. Laboratory tests are used to confirm past illnesses and injuries. Doctors are interested in the history of lung, heart, kidney, liver, and other chronic diseases for the individual and family. The doctor will also be concerned about symptoms indicating heart or lung disease and smoking habits.

A physical examination is very beneficial for routine screening. Good results are important but an individual may be physically fit and still have a serious medical problem. Blood is tested for blood cell production (anemia), liver function, kidney function, and if taken while fasting, for increased sugar, cholesterol, and fat in the blood. Urine is tested for kidney function and diabetes (sugar in the urine). It is possible to measure in the blood and urine chemicals that get into the body from exposures on a jobsite. This type of testing is called biological monitoring.

8.4.2 PULMONARY FUNCTION TESTS

A spirometer measures the volume of air in an individual's lungs and how quickly he/she can breathe in and out. This is called pulmonary function testing. This is useful for diagnosing diseases that cause scarring of the lungs that affects the expandability (asbestosis). Emphysema or asthma may also be diagnosed with pulmonary function testing. It is vital for evaluating the ability of an individual to wear a respirator without additional health risk.

8.4.3 ELECTROCARDIOGRAM

An electrocardiogram is a test used to measure heart injury or irregular heart beats. Work can be extremely strenuous, particularly when wearing protective equipment in hot environments. A stress test utilizing an electrocardiogram while exercising is sometimes a help in determining fitness, especially if there are indications from the questionnaire that an individual has a high risk of heart disease (Figure 8.3).

8.4.4 CHEST X-RAY

X-rays are useful in determining the cause of breathing problems or to use as a baseline to determine future problems. A chest x-ray is used to screen for scarring of the lungs from exposure to asbestos or silica. It should not be performed routinely, unless the history indicates a potential lung or heart problem and the physician thinks a chest x-ray is necessary. Some OSHA regulations require chest x-rays as part of the medical surveillance program. Unnecessary x-ray screening should be eliminated. For work-related biological monitoring, it is sufficient to have chest x-rays every 5 years.

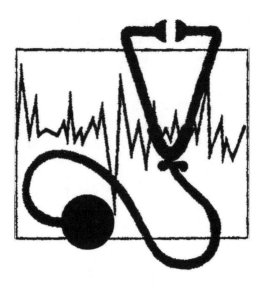

FIGURE 8.3 Work is often a strain on the heart. (Courtesy of the U.S. Environmental Protection Agency.)

8.5 HAZARDOUS CHEMICALS

Hazardous and toxic (poisonous) substances can be defined as harmful chemicals present in the workplace. In this definition, the term "chemicals" includes dusts, mixtures, and common materials such as paints, fuels, and solvents. OSHA currently regulates exposure to approximately 400 substances. The OSHA chemical sampling information file contains a listing for approximately 1500 substances. The Environmental Protection Agency's (EPA) Toxic Substance Chemical Act Chemical Substances Inventory lists information on more than 62,000 chemicals or chemical substances. Some libraries maintain files of material safety data sheets (MSDSs) for more than 100,000 substances. It is not possible to address the hazards associated with each of these chemicals.

Since there is no evaluation instrument that can identify the chemical or the amount of chemical contaminant present, it is not possible to be able to make a real-time assessment of a worker's exposure to potentially hazardous chemicals. Additionally, threshold limit values (TLVs) provided by the American Conference of Governmental Industrial Hygienist (ACGIH) in 1968 are the basis of OSHA's PELs. In the early 2000s, workers are being provided protection with chemical exposure standards that are 40 years old. The ACGIH regularly updates and changes its TLVs based upon new scientific information and research.

The U.S. EPA allows for one death or one cancer case per million people exposed to a hazardous chemical. Certainly, the public needs these kinds of protections. Using the existing OSHA PELs, risk factor is only as protective as one death because of exposure in 1000 workers. This indicates that there exists a fence line mentality which suggests that workers can tolerate higher exposures than what the public would be subjected to. As one illustration of this, the exposure to sulfur dioxide for the public is set by the EPA at 0.14 ppm average over 24 h, while the OSHA PEL is 5 ppm average over 8 h. Certainly, there is a wide margin between what the public can be subjected to and what a worker is supposed to be able to tolerate. The question is, "Is there a difference between humans in the public arena and those in the work arena?" Maybe workers are assumed to be more immune to the effects of chemicals when they are in the workplace than when they are at home, because of workplace regulations and precautions.

A more significant issue is that regarding mixtures. The information does not exist to show the risk of illnesses, long-term illnesses, or the toxicity of combining these hazardous chemicals. At present, it is assumed that the most dangerous chemical of the mixture has the most potential to cause serious health-related problems, then the next most hazardous, and so on. However, little consideration is given regarding the increase in toxicity, long-term health problems, or present hazards. Since most chemicals used in industry are mixtures formulated by manufacturers, it makes it even more critical to have access to the MSDSs and take a conservative approach to the potential for exposure. This means that any signs or symptoms of exposure should be addressed immediately, worker complaints should be addressed with sincerity and true concern, and employers should take precautions beyond those called for by the MSDSs if questions persist.

Actually, the amount of information that exists on dose/response for chemicals and chemical mixtures is limited. This is especially true for long-range effects. If a chemical kills or makes a person sick within minutes or hours, the dose response is easily understood. But, if chemical exposure over a long period results in an individual's death or illness, then the dose needed to do this is, at best, a guess. It most certainly does not take into account other chemicals the worker was exposed to during his/her work life and whether they exacerbated the effects or played no role in the individual's death or illness. This is why it is critical for individual workers to keep their exposure to chemicals as low as possible. Even then, there are no guarantees that they may not come down with an occupational disease related to chemical exposure.

Many employers and workers as well as physicians are not quick or trained to identify the symptoms of occupational exposure to chemicals. In one case, two men painted for 8 h with a paint containing 2-nitropropane in an enclosed environment. At the end of their shift, one of the workers felt unwell and stopped at the emergency center at the hospital. After examination, he was told to take rest and was assured he would be better the next morning. Later that evening, he returned to the hospital and died of liver failure from 2-nitropropane exposure. The other worker suffered irreparable liver damage but survived. No one asked the right questions regarding occupational exposure. The symptoms were probably similar to a common cold or flu which is often the case unless some investigation is done. Often those who suffer from chemical poisoning go home and start excreting the contaminant during the 16 h where they have no exposure. They feel better the next day and return to work and are reexposed. Thus, the worker does not truly recognize this as a poisoning process. Being aware of the chemicals used, reviewing the MSDSs, and following the recommended precautions are important to the safe use of hazardous chemicals.

With this point made, it becomes critical that employers should be aware of the dangers posed to their workforce by the chemicals that they use. Employers need to get and review the MSDSs for all chemicals in use on their worksite and take proper precautions recommended by the MSDSs. Also, it behooves workers to get copies of MSDSs for chemicals they use. Examples of MSDSs can be found in Appendix B.

MSDSs can also provide information for training employees in the safe use of materials. These data sheets, developed by chemical manufacturers and importers, are supplied with manufacturing or construction materials and describe the ingredients of a product, its hazards, protective equipment to be used, safe handling procedures, and emergency first-aid responses. The information contained in these sheets can help employers identify employees in need of training (i.e., workers handling substances described in the sheets) and train employees in safe use of the substances. MSDSs are generally available from suppliers, manufacturers of the substance, large employers who use the substance on a regular basis, or they may be developed by employers or trade associations. MSDSs are particularly useful for those employers who are developing training in safe chemical use as required by OSHA's hazard communication standard.

8.5.1 CARCINOGENS

Carcinogens are any substances or agents that have the potential to cause cancer. Whether these chemicals or agents have been shown to only cause cancer in animals should make little difference to employers and their workers. Employers and their workers should consider these as cancer causing on a precautionary basis since all is not known regarding their effects upon humans on a long-term basis. Since most scientists say that there is no known safe level of a carcinogen, zero exposure should be the goal of workplace health and safety. Do not let the label "suspect" carcinogen or agent fool you. This chemical or agent can cause cancer. The OSHA has identified 13 chemicals as carcinogens. They are as follows:

1. 4-Nitrobiphenyl, Chemical Abstracts Service Register Number (CAS No.) 92933
2. α-Naphthylamine, CAS No. 134327
3. Methyl chloromethyl ether, CAS No. 107302
4. 3,3′-Dichlorobenzidine (and its salts), CAS No. 91941
5. Bis-chloromethyl ether, CAS No. 542881
6. β-Naphthylamine, CAS No. 91598
7. Benzidine, CAS No. 92875
8. 4-Aminodiphenyl, CAS No. 92671
9. Ethyleneimine, CAS No. 151564
10. β-Propiolactone, CAS No. 57578
11. 2-Acetylaminofluorene, CAS No. 53963
12. 4-Dimethylaminoazo-benzene, CAS No. 60117
13. N-Nitrosodimethylamine, CAS No. 62759

There are many other chemicals that probably should be identified as carcinogens, but have escaped the scrutiny of the regulatory process. This is probably, in many cases, due to special interests of manufacturers and other groups.

The OSHA regulation 29 CFR 1910.1003 pertains to solid or liquid mixtures containing less than 0.1% by weight or volume of 4-nitrobiphenyl, methyl chloromethyl ether, bis-chloromethyl ether, β-naphthylamine, benzidine, or 4-aminodiphenyl and solid or liquid mixtures containing less than 1.0% by weight or volume of α-naphthylamine, 3,3'-dichlorobenzidine (and its salts), ethyleneimine, β-propiolactone, 2-acetylaminofluorene, 4-dimethylaminoazo-benzene, or N-nitrosodimethylamine.

The specific nature of the previous requirements is an indicator of the danger presented by exposure to, or work with, carcinogens that are regulated by OSHA. There are other carcinogens that OSHA regulates (not part of the original 13). These carcinogens are as follows:

- Vinyl chloride (1910.1017)
- Inorganic arsenic (1910.1018)
- Cadmium (1910.1027 and 1926.1127)
- Benzene (1910.1028)
- Coke oven emissions (1910.1029)
- 1,2-Dibromo-3-chloropropane (1910.1044)
- Acrylonitrile (1910.1045)
- Ethylene oxide (1910.1047)
- Formaldehyde (1910.1048)
- Methylenedianiline (1910.1050)
- 1,3-Butadiene (1910.1051)
- Methylene chloride (1910.1052)

Recently, OSHA has reduced the PEL for methylene chloride from 400 to 25 ppm. This is a huge reduction in the PEL, equal to a 15-fold decrease in what a worker can be exposed to. This reduction indicates the potential of methylene chloride to cause cancer and should highlight the serious consequences of cancer-causing chemicals. Information and research are continuously evolving and providing new insight into the dangers of these chemicals and agents. Make sure to comply with any warning signs regarding cancer-causing chemical such as in Figure 8.4.

8.6   IONIZING RADIATION

Ionizing radiation has always been a mystery to most people. Actually, much more is known about ionizing radiation than the hazardous chemicals that constantly bombard the workplace. After all, there are only four types of radiation (alpha particles, beta particles, gamma rays, and neutrons) rather than thousands of chemicals. There are instruments that can detect each type of radiation and provide an accurate dose-received value. This is not so for chemicals, where the detection of the presence of a chemical, leave alone its identification, is the best that can be achieved. With radiation detection instruments, the boundaries of contamination can be detected and set, while detecting such boundaries for chemicals is near impossible except for a solid.

It is possible to maintain a lifetime dose for individuals exposed to radiation. Most workers wear personal dosimetry, which provides reduced levels of exposure.

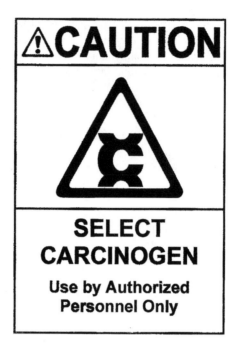

FIGURE 8.4   Cancer-causing chemical warning label.

The same is impossible for chemicals where no standard unit of measurement, such as the roentgen equivalent in man (rem), exists for radioactive chemicals. The health effects of specific doses are well known such as 20–50 rems, when minor changes in blood occur; 60–120 rems, when vomiting occurs but no long-term illness; or 5,000–10,000 rems, certain death within 48 h. Certainly, radiation can be dangerous, but one or a combination of three factors, distance, time, and/or shielding, can usually be used to control exposure. Certainly, distance is the best since the amount of radiation from a source drops off quickly as a factor of the inverse square of the distance; for instance, at 8 ft away, the exposure is 1/64th of the radiation emanating from the source. As for time, many radiation workers are only allowed to stay in a radiation area for a limited period, and then they must vacate. Shielding often conjures up lead plating or lead suits (similar to when x-rays are taken by a physician or dentist). Wearing a lead suit may seem appropriate but the weight alone can be prohibitive. Lead shielding can be used to protect workers from gamma rays (similar to x-rays). Once they are emitted, they could pass through anything in their path and continue on their way, unless a lead shield is thick enough to protect the worker.

For beta particles, aluminum foil will stop its penetration. Thus, a protective suit will prevent beta particles from reaching the skin, where they can burn and cause surface contamination. Alpha particles can enter the lungs and cause the tissue to become electrically charged (ionized). Protection from alpha particles can be obtained with the use of air-purifying respirators with proper cartridges to filter out radioactive particles. Neutrons are found around the core of a nuclear reactor and are absorbed by both water and the material in the control rods of the reactor. If a worker is not in, close to the core of the reactor, then no exposure can occur.

Ionizing radiation is a potential health hazard. The area, where potential exposure can occur, is usually highly regulated, posted, and monitored on a continuous basis. There is a maximum yearly exposure that is permitted. Once it has been reached, a worker can have no more exposure. The general number used is 5 rems/year. This is 50 times higher than what U.S. EPA recommends for the public on a yearly basis. The average public exposure is supposed to be no more than 0.1 rems/year. A standard of 5 rems has been employed for many years and seems to reasonably protect workers. Exposure to radiation should be considered serious since overexposure can lead to serious health problems or even death.

8.7   NOISE-INDUCED HEARING LOSS

Occupational exposure to noise levels in excess of the current OSHA standards places hundreds of thousands of workers at risk of developing material hearing impairment, hypertension, and elevated hormone levels. Workers in some industries (i.e., construction, oil and gas well drilling and servicing) are not fully covered by the current OSHA standards and lack the protection of an adequate hearing conservation program. Occupationally induced hearing loss continues to be one of the leading occupational illnesses in the United States. OSHA is designating this issue as a priority for rule-making action to extend hearing conservation protection, provided in the general industry standard, to the construction industry and other uncovered industries.

According to the U.S. Bureau of the Census, statistical abstract of the United States, there are over 7.2 million workers employed in the construction industry (6% of all employment). The National Institute for Occupational Safety and Health's (NIOSH) National Occupational Exposure Survey (NOES) estimates that 421,000 construction workers are exposed to noise above 85 dBA. NIOSH estimates that 15% of workers exposed to noise levels of 85 dBA or higher will develop material hearing impairment.

Research demonstrates that construction workers are regularly overexposed to noise. The extent of the daily exposure to noise in the construction industry depends on the nature and duration of the work. For example, rock drilling, up to 115 dBA; chain saw, up to 125 dBA; abrasive blasting, 105–112 dBA; heavy equipment operation, 95–110 dBA; demolition, up to 117 dBA; and needle guns, up to 112 dBA. Exposure to 115 dBA is permitted for a maximum of 15 min for an 8 h workday. No exposure above 115 dBA is permitted. Traditional dosimetry measurement may substantially underestimate noise exposure levels for construction workers since short-term peak exposures may be responsible for acute and chronic effects. Hearing can be lost in lower, full-shift time-weighted average (TWA) measurements.

There are a variety of control techniques, documented in the literature, to reduce the overall worker exposure to noise. Such controls reduce the amount of sound energy released by the noise source, divert the flow of sound energy away from the receiver, or protect the receiver from the sound energy reaching him/her. For example, types of noise controls include proper maintenance of equipment, revised

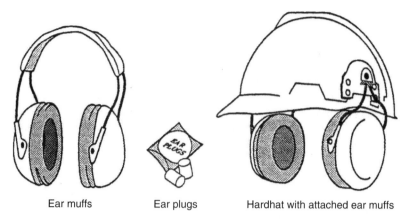

Ear muffs   Ear plugs   Hardhat with attached ear muffs

FIGURE 8.5   Hearing protection devices. (Courtesy of the Department of Energy.)

operating procedures, equipment replacements, acoustical shields and barriers, equipment redesign, enclosures, administrative controls, and PPE. Figure 8.5 provides some examples of hearing protection.

Under OSHA's general industry standard, feasible administrative and engineering controls must be implemented whenever employee noise exposures exceed 90 dBA (8 h TWA). In addition, an effective hearing conservation program (including specific requirements for monitoring noise exposure, audiometric testing, audiogram evaluation, hearing protection for employees with a standard threshold shift, training, education, and recordkeeping) must be made available whenever employee exposures equal or exceed an 8 h TWA sound level of 85 dBA (29 CFR 1910.95). Similarly, under the construction industry standard, the maximum permissible occupational noise exposure is 90 dBA (8 h TWA), and noise levels in excess of 90 dBA must be reduced through feasible administrative and engineering controls. However, the construction industry standard includes only a general minimum requirement for hearing conservation and lacks the specific requirements for an effective hearing conservation program included in the general industry standard (20 CFR 1926.52). NIOSH and the ACGIH have also recommended exposure limits (NIOSH: 85 dBA TWA, 115 dBA ceiling; ACGIH: 85 dBA).

Noise, or unwanted sound, is one of the most pervasive occupational health problems. It is a by-product of many industrial processes. Sound consists of pressure changes in a medium (usually air), caused by vibration or turbulence. These pressure changes produce waves emanating away from the turbulent or vibrating source. Exposure to high levels of noise causes hearing impairment and may have other harmful health effects as well. The extent of damage depends primarily on the intensity of the noise and the duration of the exposure. Noise-induced hearing loss can be temporary or permanent. Temporary hearing loss results from short-term exposures to noise, with normal hearing returning after a period of rest. Generally, prolonged exposure to high noise levels over a period of time gradually causes permanent damage.

Sometimes, the loss of hearing because of industrial noise is called the silent epidemic. Since this type of hearing loss is not correctable by either surgery or the use of hearing aids, it is certainly a monumental loss to the worker. It distorts communication both at work and socially. In cases where hearing needs to be at its optimum, it may result in a loss of job. The loss of hearing is definitely a handicap to the worker.

8.8 NONIONIZING RADIATION

Nonionizing radiation is a form of electromagnetic radiation, and it has varying effects on the body, depending largely on the particular wavelength of the radiation involved. In the following paragraphs, in approximate order of decreasing wavelength and increasing frequency, are some hazards associated with different regions of the nonionizing electromagnetic radiation spectrum. Nonionizing radiation is covered in detail by 29 CFR 1910.97.

Low frequency, with longer wavelengths, includes power line transmission frequencies, broadcast radio, and shortwave radio. Each of these can produce general heating of the body. The health hazard from these radiations is very small, however, since it is unlikely that they would be found in intensities great enough to cause significant effect. An exception can be found very close to powerful radio transmitter aerials.

Microwaves (MWs) have wavelengths of 3 m to 3 mm (100–100,000 MHz). They are found in radar, communications, some types of cooking, and diathermy applications. MW intensities may be sufficient to cause significant heating of tissues. The effect is related to wavelength, power intensity, and time of exposure. Generally, longer wavelengths produce greater penetration and temperature rise in deeper tissues than shorter wavelengths. However, for a given power intensity, there is less subjective awareness to the heat from longer wavelengths than there is to the heat from shorter wavelengths because absorption of longer wavelength radiation takes place beneath the body's surface.

An intolerable rise in body temperature, as well as localized damage to specific organs, can result from an exposure of sufficient intensity and time. In addition, flammable gases and vapors may ignite when they are inside metallic objects located in an MW beam. Power intensities for MWs are given in units of milliwatts per square centimeter (mW/cm^2), and areas having a power intensity of over 10 mW/cm^2 for a period of 0.1 h or longer should be avoided.

Radiofrequency (RF) and MW radiations are electromagnetic radiation in the frequency range of 3 kHz–300 GHz. Usually, MW radiation is considered a subset of RF radiation, although an alternative convention treats RF and MW radiations as two spectral regions. MWs occupy the spectral region between 300 GHz and 300 MHz, while RF or radio waves are in the 300 MHz to 3 kHz region. RF/MW radiation is nonionizing in that there is insufficient energy (<10 eV) to ionize biologically important atoms.

The primary health effects of RF/MW energy are considered to be thermal. The absorption of RF/MW energy varies with frequency. MW frequencies produce a skin effect; you can literally sense your skin starting to feel warm. RF radiation may penetrate the body and be absorbed in deep body organs without the skin effect that can warn an individual of danger. A great deal of research has turned up other nonthermal effects. All the standards of Western countries have, so far, based their exposure limits solely on preventing thermal problems. In the meantime, research continues. Use of RF/MW radiation includes aeronautical radios, citizen's band (CB) radios, cellular phones, processing and cooking of foods, heat sealers, vinyl welders, high-frequency welders, induction heaters, flow solder machines, communications transmitters, radar transmitters, ion implant equipment, MW drying equipment, sputtering equipment, glue curing, power amplifiers, and metrology.

Infrared radiation does not penetrate below the superficial layer of the skin so that its only effect is to heat the skin and the tissues immediately below it. Except for thermal burns, the health hazard upon exposure to low-level conventional infrared radiation sources is negligible.

Visible radiation, which is about midway in the electromagnetic spectrum, is important because it can affect both the quality and accuracy of work. Good lighting conditions generally result in increased product quality with less spoilage and increased production. Lighting should be bright enough for easy visibility and directed so that it does not create glare. The light should be bright enough to permit efficient visibility.

Ultraviolet radiation in industry may be found around electrical arcs, and such arcs should be shielded by materials opaque to the ultraviolet. The fact that a material may be opaque to ultraviolet has no relation to its opacity to other parts of the spectrum. Ordinary window glass, for instance, is almost completely opaque to the ultraviolet in sunlight; at the same time, it is transparent to the visible light waves. A piece of plastic, dyed a deep red-violet, may be almost entirely opaque in the visible part of the spectrum and transparent in the near-ultraviolet. Electric welding arcs and germicidal lamps are the most common, strong producers of ultraviolet rays in industry. The ordinary fluorescent lamp generates a good deal of ultraviolet rays inside the bulb, but it is essentially all absorbed by the bulb and its coating.

The most common exposure to ultraviolet radiation is from direct sunlight, and a familiar result of overexposure—one that is known to all sunbathers—is sunburn. Almost everyone is also familiar with certain compounds and lotions that reduce the effects of the sun's rays, but many are unaware that some industrial materials, such as cresols, make the skin especially sensitive to ultraviolet rays. So much so that after having been exposed to cresols, even a short exposure in the sun usually results in severe sunburn. Nonionizing radiation, although perceived not to be as dangerous as ionizing radiation, does have its fair share of adverse health effects.

8.9   TEMPERATURE EXTREMES

8.9.1   COLD STRESS

Temperature is measured in degrees Fahrenheit (°F) or Celsius (°C). Most people feel comfortable when the air temperature ranges from 66°F to 79°F and the relative humidity is about 45%. Under these circumstances, heat production inside the body equals the heat loss from the body, and the internal body temperature is kept around 98.6°F. For constant body temperature, even under changing environmental conditions, rates of heat gain and heat loss should be balanced. Every living organism produces heat. In cold weather, the only source of heat gain is the body's own internal heat production, which increases with physical activity. Hot drinks and food are also a source of heat.

The body loses heat to its surroundings in several different ways. Heat loss is greatest if the body is in direct contact with cold water. The body can lose 25–30 times more heat when in contact with cold wet objects than under dry conditions or with dry clothing. The higher the temperature differences between the body surface and cold objects, the faster the heat loss. Heat is also lost from the skin by contact with cold air. The rate of loss depends on the air speed and the temperature difference between the skin and the surrounding air. At a given air temperature, heat loss increases with air speed. Sweat production and its evaporation from the skin also cause heat loss. This is important when performing hard work.

Small amounts of heat are lost when cold food and drink are consumed. Heat is also lost during breathing by inhaling cold air, and through evaporation of water from the lungs.

The body maintains heat balance by reducing the amount of blood circulating through the skin and outer body parts. This minimizes cooling of the blood by shrinking the diameter of blood vessels. At extremely low temperatures, loss of blood flow to the extremities may cause an excessive drop in tissue temperature resulting in damage such as frostbite, and by shivering, which increases the body's heat production. This provides a temporary tolerance for cold but cannot be maintained for long periods.

Overexposure to cold causes discomfort and a variety of health problems. Cold stress impairs performance of both manual and complex mental tasks. Sensitivity and dexterity of fingers lessen in cold. At still lower temperatures, cold affects deeper muscles, resulting in reduced muscular strength and stiff joints. Mental alertness is reduced due to cold-related discomfort. For all these reasons accidents are more likely to occur in very cold working conditions.

The main cold injuries are frostnip, frostbite, immersion foot, and trench foot, which occur in localized areas of the body. Frostnip is the mildest form of cold injury. It occurs when ear lobes, noses, cheeks, fingers, or toes are exposed to cold. The skin of the affected area turns white. Frostnip can be prevented by warm clothing and is treated by simple rewarming.

Immersion foot occurs in individuals whose feet have been wet, but not freezing cold, for days or weeks. The primary injury is to nerve and muscle tissue. Symptoms are numbness, swelling, or even superficial gangrene. Trench foot is wet cold disease resulting from exposure to moisture at or near the freezing point for one to several days. Symptoms are similar to immersion foot, swelling, and tissue damage.

Hypothermia can occur in moderately cold environments; the body's core temperature does not usually fall more than 2°F–3°F below the normal 98.6°F because of the body's ability to adapt. However, in intense cold without adequate clothing, the body is unable to compensate for the heat loss, and the body's core temperature starts to fall. The sensation of cold, followed by pain, in exposed parts of the body is the first sign of cold stress. The most dangerous situation occurs when the body is immersed in cold water. As the cold worsens or the exposure time increases, the feeling of cold and pain starts to diminish because of increasing numbness (loss of sensation). If no pain is felt, serious injury can occur without the victim noticing it. Next, muscular weakness and drowsiness are experienced. This condition is called hypothermia and usually occurs when body temperature falls below 92°F. Additional symptoms of hypothermia include interruption of shivering, diminished consciousness, and dilated pupils. When body temperature reaches 80°F, coma (profound unconsciousness) sets in. Heart activity stops at around 68°F and the brain stops functioning at around 63°F. The hypothermia victim should be immediately warmed, either by being moved to a warm room or by the use of blankets. Rewarming in water at 104°F–108°F has been recommended in cases where hypothermia occurs after the body was immersed in cold water.

Although people easily adapt to hot environments, they do not acclimatize well to cold. However, frequently exposed body parts can develop some degree of tolerance to cold. Blood flow in the hands, for example, is maintained in conditions that would cause extreme discomfort and loss of dexterity in unacclimatized persons. This is noticeable among fishermen who are able to work with bare hands in extremely cold weather.

In the United States, there are no OSHA exposure limits for cold working environments. It is often recommended that work warm-up schedules be developed. In most normal cold conditions, a warm-up break every 2 h is recommended, but, as temperatures and wind increase, more warm-up breaks are needed.

Protective clothing is needed for work at or below 40°F. Clothing should be selected to suit the cold, level of activity, and job design. Clothing should be worn in multiple layers which provide better protection than a single thick garment. The layer of air between clothing provides better insulation than the clothing itself. In extremely cold conditions, where face protection is used, eye protection must be separated from respiratory channels (nose and mouth) to prevent exhaled moisture from fogging and frosting eye shields.

8.9.2 Heat Stress

Operations involving high air temperatures, radiant heat sources, high humidity, direct physical contact with hot objects, or strenuous physical activities have a high potential for inducing heat stress in employees engaged in such operations. Such places include iron and steel foundries, nonferrous foundries, brick-firing and ceramic plants, glass products facilities, rubber products factories, electrical utilities (particularly boiler rooms), bakeries, confectioneries, commercial kitchens, laundries, food canneries, chemical plants, mining sites, smelters, and steam tunnels. Outdoor operations, conducted in hot weather, such as construction, refining, asbestos removal, and hazardous waste site activities, especially those that require workers to wear semipermeable or impermeable protective clothing, are also likely to cause heat stress among exposed workers.

Age, weight, degree of physical fitness, degree of acclimatization, metabolism, use of alcohol or drugs, and a variety of medical conditions, such as hypertension, all affect a person's sensitivity to heat. However, even the type of clothing worn must be considered, before heat injury predisposes an individual to additional injury. It is difficult to predict just who will be affected and when, because individual susceptibility varies. In addition, environmental factors include more than the ambient air temperature. Radiant heat, air movement, conduction, and relative humidity all affect an individual's response to heat.

There is no OSHA regulation for heat stress. The ACGIH (1992) states that workers should not be permitted to work when their deep body temperature exceeds 38°C (100.4°F).

Complications arise when workers suffer from heat exposure. The main anomalies are as follows:

- Heat stroke
- Heat exhaustion
- Heat cramps
- Fainting
- Heat rash

The human body can adapt to heat exposure to some extent. This physiological adaptation is called acclimatization. After a period of acclimatization, the same activity will produce fewer cardiovascular demands. The worker will sweat more efficiently (causing better evaporative cooling), and thus will more easily be able to maintain normal body temperatures. A properly designed and applied acclimatization program decreases the risk of heat-related illnesses. Such a program basically involves exposing employees to work in a hot environment for progressively longer periods. NIOSH (1986) says that, for workers who have had previous experience in jobs where heat levels are high enough to produce heat stress, the regimen should be 50% exposure on day 1, 60% on day 2, 80% on day 3, and 100% on day 4. For new workers who will be similarly exposed, the regimen should be 20% on day 1, with a 20% increase in exposure each additional day.

8.10 VIBRATION

Vibrating tools and equipment at frequencies between 40 and 90 Hz can cause damage to the circulatory and nervous systems. Care must be taken with low frequencies, which have the potential to put workers at risk for vibration injuries. One of the most common cumulative trauma disorders (CTDs) resulting from vibration is Raynaud's syndrome. Its most common symptoms are intermittent numbness and tingling in the fingers; skin that turns pale, ashen, and cold; and eventual loss of sensation and control in the fingers and hands. Raynaud's syndrome occurs due to the use of vibrating hand tools such as palm sanders, planners, jackhammers, grinders, and buffers. When such tools are required for a job, an assessment should be made to determine if any other methods can be used to accomplish the desired task. If not, other techniques, such as time/use limitations, alternating workers, or other such administrative actions, should be considered to help reduce the potential for a vibration-induced CTD. The damage caused by vibrating tools can be reduced by

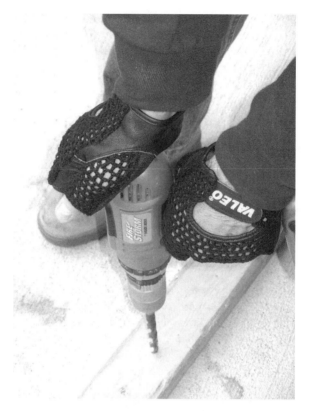

FIGURE 8.6   The use of anti-vibration gloves.

- Using vibration dampening gloves (Figure 8.6)
- Purchasing low vibration tools and equipment
- Putting anti-vibration material on handles of existing tools
- Reducing length of exposure
- Changing the actual work procedure if possible
- Using balanced and dampening tools and equipment
- Rotating workers to decrease exposure time
- Decreasing the pace of the job as well as the speed of tools or equipment

Individuals subject to whole-body vibration have experienced visual problems; vertebrae degeneration; breathing problems; motion sickness; pain in the abdomen, chest, and jaw; backache; joint problems; muscle strain; and speech problems. Although many questions remain regarding vibration, it is certain that physical problems can transpire from exposure to vibration.

REFERENCE

Bureau of Labor Statistics, United States Department of Labor. *Workplace Injuries and Illnesses in 2004.* Available at http://bls.gov.

9 Chemical Hazards

The handling, storage, and sale of chemical is part of the goods and materials service sectors.

9.1 CHEMICALS

Chemical hazards have been addressed in Chapter 8 with regard to their impact upon the workplace and its workforce. Chapter 8 speaks in some detail of the health effects of hazardous or toxic (poisonous) chemicals. Chapter 9 provides information on the means by which chemicals enter the body, the exposure guidelines, and the forms in which chemicals present themselves to the body as contaminants. Also, the chapter lists the categories of chemicals that most often are seen in the workplace and also describes why based upon their composition they may pose a hazard.

9.2 ROUTES OF ENTRY AND MODES OF ACTION

Chemicals enter the human body via many routes. The nature of the chemical often determines how the chemical enters the body. Once into the body the chemical tends to target certain systems and organs of the body. The entry may be through the eyes, skin, lungs, or ingestion and at times by injection (penetration).

9.2.1 EYES

The importance of the human visual system is evident. Good eyesight is a must for performing tasks where man and machine interact. Of all the major body organs prone to worksite injuries, the eye is probably the most vulnerable. Consequently, protection against eyes and face injuries is of major concern and importance for workers. The eye is an organ of sight and is not designed for the demands of prolonged viewing at close distances as is commonplace in today's workplace. Although the eye does have some natural defenses, it has none to compare with the healing ability of the skin, the automatic cleansing abilities of the lungs, or the recuperative powers of the ear. This is why an eye injury is the most traumatic loss to the human body.

The eyeball is housed in a case of cushioning fatty tissue that insulates it from the skull's bony eye socket. The skull, brow, and cheek ridges serve to help protect the eyeball, which is comprised of several highly specialized tissues.

The front of the eyeball is protected by a smooth, transparent layer of tissue called the conjunctiva. A similar membrane covers the inner surface of the eyelids. The eyelids also contain dozens of tiny glands that secrete oil to lubricate the surfaces of the eyelids and the eyeball. Another gland located at the outer edge of the eye socket secretes tears to clean the protective membrane and keep it moist.

The most common injury to the eye is when foreign particles enter into it. Its effects are as follows:

- Pain, because the cornea is heavily covered with nerves and an object sitting on the surface of the cornea will hurt constantly and that may obscure vision and stimulate or damage the nerves
- Infection, because a foreign particle may carry bacteria or fungi, or may be carried by fingers used to rub the eye
- Scarring, from tissue that has healed and may obscure the vision
- Damage, depending on the angle and point of entry and speed of the particle

Heat can destroy eye and eyelid tissues just as it does other body tissues. High-intensity light may have sufficient energy to damage the eye tissue. Exposure to ultraviolet light from welding operations (known as welder's flash) may severely damage the eye. Also, the effects of accidental exposure of the eye to chemicals can vary from mild irritation to complete loss of vision. In some cases, a chemical that does not actually damage the eye may be absorbed through the eye tissue in sufficient quantities so as to cause systemic poisoning. Splash goggles shown in Figure 9.1 help protect the eyes from chemicals.

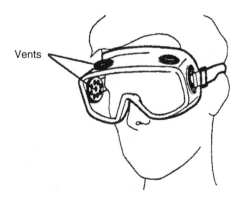

FIGURE 9.1   Example of splash goggles. (Courtesy of the Department of Energy.)

Exposure to caustic chemicals is much more injurious to the eyes than acids. An eye that has been exposed to a caustic may not look too bad on the first day after exposure. It may, however, deteriorate markedly on succeeding days. This is in contrast to acid burns where the initial appearance is a good indication of the ultimate damage.

9.2.2 LUNGS AND INHALATION

The respiratory system consists of all the organs of the body that contribute to normal breathing. This includes the nose, mouth, upper throat, larynx, trachea, and bronchi, all airways that lead to the lungs. It is in these airways that the first defense against contaminants exists. The adult human lung has an enormous area (75 sq yd total surface area) where the body exchanges waste carbon dioxide for needed oxygen. This large surface, together with the blood vessel network (117 sq yd total surface area) and continuous blood flow, makes it possible for an extremely rapid rate of absorption of oxygen from the air in the lungs to the bloodstream. Some highly soluble substances such as gases may pass through the lungs and into the bloodstream so fast that it is not detected by the worker until ill effects set in. On the other hand, there are substances, such as asbestos that are insoluble in body fluids, that remain in our lungs for extended periods of time. Bodily attempts to destroy or remove these substances may result in irritation, inflammation, edema, emphysema, fibrosis, cancer, or allergic reactions and sensitization. Impairment of the lungs will not be noticed in the day-to-day activities of a worker. It does, however, reduce a worker's ability to withstand future exposures.

Air enters through the nostrils and passes through a web of nasal hairs. Air is warmed and moistened as some particles are removed by compacting on the nasal hairs and at the bends in the air path. Interior walls of the nose are covered with membranes that secrete fluid called mucus. The mucus drains slowly into the throat and serves as a trap for bacteria and dust in the air. It also helps dilute toxic substances that enter the airway.

Cilia, another important air cleaner, are hair-like filaments that vibrate 12 times per second. Millions of cilia lining the nose and nasal airway help the mucus clean, moisten, and heat the air before it reaches the lungs. As the air moves into the bronchi it is divided and subdivided into smaller, finer, and more numerous tubes, much like those of the branches of a tree. There are two main branches, each getting smaller until they reach the lungs located on each side of the chest cavity. The respiratory tract branches from the trachea to some 25–100 million branches. These branches terminate in about 300 million air sacs called alveoli, which have access to the blood.

The lungs are suspended within the chest by the trachea, arteries, veins running to and from the heart, and by the pulmonary ligaments. The ability of the lungs to function properly can be adversely affected in many ways. There may be blocked or restricted passageways, reduced elasticity, and/or damaged membranes. The first line of defense is the nose. It filters the air and prevents many contaminants from reaching lower portions. However, we often bypass this filtering defense system by breathing through our mouth. Coughing is another mechanism that expels foreign particles from the trachea and bronchi. Hair cells (called cilia) serve as a continuous cleaning mechanism for the nose, trachea, bronchi, and bronchioles. These hair-like extensions move like an escalator to sweep foreign particles back to the trachea where it is swallowed or spat out. Macrophages also help reduce particle levels by engulfing or digesting bacteria and viruses.

9.2.2.1 Respiration

The process by which the body combines oxygen with food nutrients to produce energy is called metabolism. To produce energy the body must exchange oxygen for carbon dioxide via respiration. Often, gases are not blocked or restricted by the filtering defense system. One of the most common types of inhalation hazard found in the workplace is carbon monoxide, which is present in exhaust from fossil fuel equipment, generators, or compressors. It is also produced as a by-product of welding and soldering operations. Carbon monoxide's main effect is to rob the body of its oxygen supply. After inhalation, carbon monoxide mixes more readily with the blood's oxygen carrier, hemoglobin, than oxygen. So exposures to high levels of carbon monoxide can prevent the body from getting enough oxygen, severely affecting the heart and brain. First symptoms may be headache, dizziness, and nausea. Higher exposures can result in fainting, coma, or even death. Persons with existing heart conditions are more likely to worsen their condition if exposed to carbon monoxide. And smokers already have higher than normal levels in their bloodstream as a burning cigarette produces fairly high carbon monoxide levels.

The fate of substances that reach the lungs depends on their solubility and reactivity. The more soluble the contaminant, the more likely it will be an upper respiratory irritant, such as sulfur dioxide (SO_2). Soluble reactive particles may cause acute inflammatory reactions and build-up of fluid (pulmonary edema). The less soluble gases and materials reach the lower lungs causing lung dysfunction or the particles that stick in the alveoli are engulfed by macrophages that move them back to the mouth, where they are expectorated or swallowed. Some chemicals that reach the digestive tract by this method are then absorbed and may still cause adverse health effects. The size of the particle greatly influences where it will be deposited in the air passage.

An atmosphere containing toxic contaminants, even at very low concentrations, could be a hazard to the lungs and the body. A concentration large enough to decrease the percentage of oxygen in the air can lead to asphyxiation or suffocation, even if the contaminant is an inert gas.

Inhaled contaminants that adversely affect the lungs or body fall into three categories:

1. Aerosols and dusts that, when deposited in the lungs, may produce tissue damage, tissue reaction, disease, or physical plugging.
2. Toxic gases that produce adverse reaction in the tissue of the lungs themselves. For example, hydrogen fluoride is a gas that causes chemical burns.
3. Toxic aerosols or gases that do not affect the lung tissue, but are passed from the lungs into the bloodstream. From there they are carried to other organs, or have adverse affects on the oxygen-carrying capacity of the bloodstream itself.

Four things must be known about inhaled contaminants before the toxic effects can be determined. There are as follows:

- Identification of the contaminant (What chemical or material?)
- Concentration inhaled (How much?)
- Duration of exposure (How long?)
- Frequency of exposure (How often?)

9.2.3 SKIN ABSORPTION

The skin is the largest organ of the body, covering about 19 sq ft of surface area. It is often the first barrier to come in contact with hazardous contaminants. The skin must protect the worker from heat, cold, moisture, radiation, bacteria, fungus, and penetrating objects. The skin is the organ that senses touch or hurt for the central nervous system. One square inch of skin contains about 72 ft of nerves. Contact with a substance may initiate the following actions:

- The skin and its associated layer of fat (lipid) cells can act as an effective barrier against penetration, injury, or other forms of irritation.
- The substance can react with the skin surface and cause a primary irritation (dermatitis).
- The substance can penetrate the skin and accumulate in the tissue, resulting in allergic reactions (skin sensitization).
- The substance can penetrate the skin, enter the bloodstream, and act as a poison to other body organs (systemic action).
- The substance can penetrate the skin, dissolve the fatty tissues, and allow other substances to penetrate skin layers.

Most job-related skin conditions are caused by repeated contact with irritants such as solvents, soap detergents, particulate dusts, oils, grease, and metal working fluids. This is called contact dermatitis, and the symptoms are red, itchy skin, swelling ulcers, and blisters. The length of exposure and the strength of the irritant will affect the severity of the reaction as well as abrasions, sores, and cuts, which open a pathway through the skin and into the body. The skin performs a number of important functions:

- Against invasion by bacteria
- Against injury to other organs that are more sensitive
- Against radiation such as from the sun
- Against loss of moisture
- Providing a media for the nervous system

Serious and even fatal poisoning has occurred from brief skin exposures to highly toxic substances such as parathion or other related organic phosphates (weed and insect killers), phenol, and hydrocyanic acid. Compounds that are good solvents for grease or oil, such as toluene and xylene, may cause problems by being readily absorbed through the skin. Abrasions, lacerations, and cuts may greatly increase the absorption, thus increasing the exposure to toxic chemicals.

9.2.4 INGESTION

Workers on the jobsite may unknowingly eat or drink harmful toxic chemicals. These toxic chemicals, in turn, are then capable of being absorbed from the gastro-intestinal tract into the blood. Lead oxide, found in red paint on steel surfaces, can cause serious problems if workers eat or smoke on the jobsite. Good personal hygiene habits, such as thoroughly washing face and hands before eating or smoking, are essential to prevent exposure.

Inhaled toxic dusts can also be swallowed and ingested in amounts large enough to cause poisoning. Toxic materials that are easily dissolved in digestive fluids may speed absorption into the bloodstream. Ingestion toxicity is normally lower than inhalation toxicity for the same material, because of relatively poor absorption of many chemicals from the intestines into the bloodstream.

After absorption from the intestinal tract into the bloodstream, the toxic material generally targets the liver, which may alter or break down the material. This detoxification process is an important body defense mechanism. It involves a sequence of reactions such as the following:

- Deposition in the liver
- Conversion to a nontoxic substance
- Transportation to the kidney via the bloodstream
- Excretion through the kidney and urinary tract

Sometimes, this process will have a reverse effect by breaking down a chemical into components that are much more toxic than the original compound. These components may stay in the liver causing adverse effects, or they may be transported to other body organs damaging them.

9.2.5 PERSONAL EXPOSURE GUIDES

A variety of hazard guidelines exist to evaluate worker exposure to chemical or other hazardous conditions at worksites. Most of these guidelines can be used to evaluate the dangers present at sites and determine the appropriate level of protection to be worn or other action necessary to protect workers' health. Personal exposure guides are indications that hazardous conditions may exist. Workers should watch for the following personal signs of exposure to toxic chemicals or work stress. If any of these occur, they should leave the site and report the problem immediately. They should not return until the cause of the symptoms has been checked by a qualified person. Warning signs of chemical exposure may be as follows:

- Breathing difficulties—breathing faster or deeper, soreness and a lump in the throat
- Dizziness, drowsiness, disorientation, difficulty in concentration

- Burning sensation in the eyes or on the skin, redness, or soreness
- Weakness, fatigue, lack of energy
- Chills, upset stomach
- Odors and/or a strange taste in the mouth

9.3 CHEMICAL EXPOSURE GUIDELINES

Exposure guidelines are set by reviewing previous experience with hazards from several sources, including actual experience in dealing with hazards, results of studies of human exposure to toxic chemicals, and laboratory studies on animals. Because we do not have absolute knowledge about most hazards and opinions vary about the degree of hazards posed by different chemicals, guidelines will vary, even for the same chemical. Guidelines can and do change as new information is discovered. The goal is to minimize any worker exposure to hazardous conditions.

OSHA regulations require the employee to know about chemicals to which they are being exposed. General guidelines do not require that you know the amount of chemical present or its concentrations in the air. These are often found on labels or placards on chemicals containers. General guidelines often use short phrases, a word, numbers, or symbols to communicate hazards such as "Avoid skin contact" or "Avoid breathing vapors." MSDSs and labels provide information on chemical hazards as seen in Figure 9.2.

Specific OSHA regulations also require the employer to know both the identity and air concentration of the chemicals that may be present at the worksite. The results of air monitoring are compared to specific permissible levels to make decisions about worker exposure. Three different organizations have developed

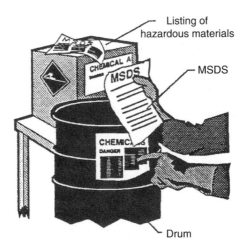

FIGURE 9.2 Chemical labels and MSDSs provide needed hazard information. (Courtesy of the Department of Energy.)

specific chemical exposure levels that are widely used at worksites to reduce worker exposures to levels thought to be safe. They are as follows:

- Permissible exposure limit (PEL) (set by the OSHA)—PELs are legal enforceable standards. PELs are meant to be minimum levels of protection. Employers may use more protective exposure levels for chemicals. In many cases, current PELs are derived from TLVs published in the 1998 ACGIH TLV list. Many PELs are not set to protect workers from chronic effects such as cancer. In addition, most PELs that apply to the construction industry were established in 1969 and are rather outdated.
- Recommended exposure limit (REL) (set by the National Institute for Occupational Safety and Health, NIOSH)—These are advisory levels and are not legally enforceable. RELs are sometimes more protective than PELs. Long-term or chronic health effects are considered when setting the RELs.
- Immediately dangerous to life and health (IDLH) (set by the NIOSH)—These values are established to recognize serious exposure levels that could cause death and serve as a blueprint for selecting specific types of respiratory protection.
- Threshold limit value (TLV) (set by the ACGIH)—TLVs are advisory and are not legally enforceable. A revised list of TLVs is published every year making them more current than PELs. However, chronic effects such as cancer are not always given consideration when setting TLVs. Ways to list chemical hazard guidelines are time-weighted average (TWA), short-term exposure limit (STEL), ceiling values, and skin absorption hazard.

9.3.1 Time-Weighted Average

TWA is the average concentration of a material over a full work shift (set as 8 h/day and 40 h/week). The changes in exposure that occur during the work shift are averaged out. In addition, if the worker is exposed to more than one substance or a mixture of substances, mixture calculations must be conducted.

9.3.2 Short-Term Exposure Limits

STELs are the maximum concentration level that workers can be exposed to for a short period of time (usually 15–30 min) without suffering from irritation; chronic or irreversible tissue damage; and dizziness sufficient to increase the risk of accidents, impair self-rescue, or reduce work efficiency.

9.3.3 Ceiling Limit

Workers often experience acute health effects if the level exceeds the ceiling limit listed in OSHA's PEL. If a ceiling limit is not assigned to a substance or chemical, it is generally recommended that exposures never exceed five times their PEL.

9.3.4 SKIN ABSORPTION NOTATION

The notation "skin" listed in OSHA's PELs indicates that the chemical can be absorbed through the skin as a route of entry into the body. Remember that PELs, RELs, and TLVs refer only to inhalation exposure. No concentration guidelines for skin exposure exist.

9.4 TYPES OF AIRBORNE CONTAMINANTS

Many of the worker exposures are the result of airborne contaminants such as dusts, fumes, gases, mists, or vapors. Each of these contaminants has different actions and physical properties, which will be covered in the following sections. These contaminants are instrumental in creating respiratory hazards such as asbestosis or silicosis.

9.4.1 DUSTS

Dusts are solid particles suspended in air. They may be produced by crushing, grinding, sanding, sawing, or the impact of materials against each other. Some dusts have no effect on the body. They do not seem to harm the body or are not changed by the body's chemistry into other harmful substances. Most harmful dusts cause damage after inhalation. Some dusts, such as cement and arsenic, can also directly affect the skin.

When considering health effects from inhaled dust, we must be concerned about a solid material that is small enough to reach the air sacs in our lungs where oxygen and carbon dioxide exchange takes place. This area is called the alveoli. Only particles smaller than about 5 μm or 5 μ (about 1/100th the size of a speck of pepper) are likely to reach this area of the lung. Particles in the range from 5 to 10 μm will be deposited in the upper respiratory tract airways (nose, throat, trachea, and major bronchial tubes) and cause bronchitis. Particles larger than 10 μm, like wood dusts, can deposit in the nasal airways with the possibility of causing nasal ulcerations and cancer. Particles smaller than about 1 μm are likely to be exhaled during normal breathing.

9.4.2 FUMES

Fumes, like dust, are also solid particles in the air. They are usually formed when metals are heated to their melting points, especially during welding or soldering. Fumes are produced when metal is welded. Solder, electrode, welding rod, or metallic coating on materials may be vaporized generating additional fumes. Chromium and nickel exposures are possible when fumes are generated from stainless steel during arc welding. Sometimes plumbers generate lead fumes when molten lead is used for joining black pipe. Lead fumes are also generated by melting lead to make fishing sinkers or burning lead paint off surfaces.

Although many fumes can irritate the skin and eyes, these fine particles primarily affect the body when they are inhaled. This type of exposure sometimes results in an acute health effect, referred to as metal fume fever, especially if the fumes are from metals such as zinc, cadmium, or magnesium. Workers often generate a lot of lead and metal fumes during demolition projects when using torches to cut and burn I beams. Dangerous fumes may also be produced by heating asphalt during hot-tar roofing or road paving. An ingredient used in this process is called coal tar pitch. These hazardous fumes are regarded as a serious cancer threat.

9.4.3 GASES

Gases are formless at room temperature and always expand to fill their containers. They can be changed into liquids or solids by increasing the pressure and/or decreasing their temperature. It is in these changed forms that gases are normally stored and/or transported. Toxic gases can directly irritate the skin, throat, eyes, or lungs, or they may pass from the lungs into the bloodstream to damage other parts of the body. Some gases such as methane can also cause a worker to suffocate by displacing oxygen in the air. Many fatalities have occurred due to the improper entry of confined spaces such as underground silos containing manure. As the manure decays, it generates methane gas displacing the oxygen.

The body's defenses against some gases include smelling, tearing eyes, and coughing. Ammonia's irritating effects and odor warn workers of exposure. However, workers may be exposed to some gases unknowingly. Carbon monoxide is the most widespread gas risk. It can be found whenever heavy equipment or motors are being used. It is a colorless, odorless gas formed by burning carbon-containing materials such as coal, oil, gasoline, wood, or paper.

9.4.4 MISTS

Mists and fogs are drops of liquid suspended in the air. Fogs may be created by vapors condensing to the liquid state, while mists are droplets being splashed or sprayed. Examples of mists used in industry include paint spray mists and acid mists produced by fluxes used in soldering. Many mists and fogs can damage the body if they are inhaled or if they make direct contact with skin or eyes. Like fumes, mists are small enough to bypass the respiratory system's defenses and go deep inside the lungs from where they pass easily into the bloodstream, and eventually to other parts of the body.

9.4.5 VAPORS

Vapors are gaseous forms of certain materials that are usually solid or liquid at room temperatures. Vapors may be formed when liquids or solids are heated. Some materials, such as solvents, form vapors without being heated. Solvent vapors are one of the most common exposures at a hazardous waste and/or construction site. Mercury is an example of a metal that vaporizes at room temperature and can be a serious health hazard.

Many directly affect the skin causing dermatitis, while some can be absorbed through the skin. As with gases and fumes, most vapors when inhaled pass to the bloodstream and damage other parts of the body. Some of these materials can damage the liver, kidneys, blood, or cause cancer.

9.5 TYPICAL HAZARDOUS CHEMICALS

There are many different types of hazardous chemicals used in all industry that you may be exposed to. Many of these chemicals can be grouped into a set of general categories because they pose the same types of hazards. In this way, it simplifies the general hazards that may be encountered on the worksite. Hazards associated with some common materials found in industry are reviewed in the following sections. They are solvents; acids, bases, and alkalines; cleaners; adhesives and sealants; paints; and fuels.

9.5.1 SOLVENTS

A solvent is a liquid that dissolves another substance without changing the basic characteristic of either material. When the solvent evaporates, the original material is the same. In construction, we most often see them as cleaners, degreasers, thinners, fuels, and glues. Solvents are lumped into three main types or classes: those containing water (aqueous solutions) such as acids, alkalines, and detergents, and those containing carbon (organic solvents) such as acetone, toluene, and gasoline. The third group contains chlorine in their chemical makeup and is called chlorinated solvents like methylenechloride and trichloroethylene.

Solvents can enter into your body in two ways: by inhalation or by absorption through the skin. Any solvent inhaled may cause dizziness or headaches as it affects the central nervous system. If breathing solvent vapors continues over time, the development of nose, throat, eye, and lung irritation and even damage to the liver, blood, kidneys, and digestive system may result. Most solvents in contact with skin can be absorbed into the body. Because solvents dissolve oils and greases, contact with skin can also dry it out producing irritation, cracking, and skin rashes. Once a solvent penetrates through the skin, it enters into the bloodstream and can attack the central nervous system or other body organs.

Like all chemicals, the effect on the body will depend on a number of factors: levels of toxicity, duration of exposure, sensitivity of the body, and levels of concentration of the solvent. Solvent hazards may be minimized by following a few simple rules:

- Know what chemicals you are working with.
- Use protective equipment like gloves, safety glasses, and proper respirators to prevent contact with skin, eyes, and lungs.
- Make sure the work area has plenty of fresh air.
- Avoid skin contact with solvents.
- Wash with plenty of soap and water if contact with skin occurs.
- If a solvent splashes into eyes, flush with running water for a minimum of 15 min and get medical help. Remember, gasoline should never be used as a solvent or cleaning agent.

9.5.2 CLEANERS

Cleaners contain acids, alkalies, aromatics, surfactants, petroleum products, ammonia, and hypochlorite. Because of these ingredients, cleaners are considered to be irritants, and can be harmful if swallowed or inhaled. Many can cause eyes, nose, throat, skin, and lung irritation. Some cleaners are flammable and burn easily. Others may be caustic or corrosive and cause severe skin damage. Because many cleaners used in industrial situations are consumer products commonly found in our homes, you may underestimate the hazard they pose. Close review of precautions listed in the MSDS is needed to protect workers from these chemicals. Often, gloves and eye protection are required. Respirators may be needed to avoid inhaling the vapors and mists. The lack of worker personal hygiene is one of the greatest exposure problems. Hands and face should be washed thoroughly before eating, drinking, or smoking.

Mixing of cleaning chemicals should be avoided unless specifically instructed to do so. For example, a dangerous gas, chlorine, will be created if you mix bleach and ammonia, or bleach and drain cleaner.

9.5.3 ACIDS AND BASES

Acids and bases (caustics) can easily damage the skin and eyes. The seriousness of the damage depends on concentration of chemical, duration of contact, and actions taken after an exposure. Acids and bases can be in the form of liquids, solid granules, powders, vapors, and gases. A few commonly used acids include sulfuric acid, hydrochloric acid, muriatic acid, and nitric acid. Some common bases (caustics) are lye (sodium hydroxide) and potash (potassium hydroxide). Both acids and bases can be corrosive, causing damage to whatever they contact. The more concentrated the chemical, the more dangerous it can be. Vinegar is a mild form of acetic acid and as such it can be swallowed or rubbed on the skin with no damage, but a concentrated solution of acetic acid can cause serious burns.

Various acids react differently when they contact the skin. Sulfuric acid mixes with water to produce heat, so when it contacts the skin, it reacts with moisture and causes burns. Hydrofluoric acid may not even be noticed if it spills on the skin, but hours later as the acid is absorbed into the muscle tissue, it can cause deep burns that are very painful and take a long time to heal. Most acids in a gas or vapor form when inhaled react with the moisture in the nose and throat causing irritation or damage. Acetic and nitric acids do not react as readily with water, but when these vapors are inhaled, they quickly penetrate into the lungs causing serious damage.

Bases, as a class of chemicals, are slippery or soapy. In fact, soap is made from a mixture of a base (lye) and animal fat. Concentrated bases easily dissolve tissue and, therefore, can cause severe skin damage on contact. Concentrated caustic gases like ammonia vapors can damage the skin, eyes, nose, mouth, and lungs. Even dry powder forms of bases can damage tissue when inhaled because they react with the moisture in your skin, eyes, and respiratory tract. Cement and mortar are alkali compounds in their wet or dry form. Workers should remember the following rules when working with acids and bases:

- Know what chemicals you are working with and how strong (concentrated) they are.
- Use personal protective equipment as noted in the MSDS.

- In the case of skin or eye contact, flush with cool water for at least 15 min but do not rub the skin or eyes.
- Always add acid to the water to prevent splatter.
- Keep acids and bases apart, store separately, and clean up spills promptly. Acids and bases react, often violently, when mixed together.

9.5.4 ADHESIVES AND SEALANTS

Most adhesives and sealants have some type of hazard warning on the label. Because of their common usage at home and on the job, these warnings are sometimes taken lightly or ignored altogether. Many adhesives and sealants are toxic because of their chemically reactive ingredients, or because of the solvent base that permits them to be more easily applied.

Adhesives or sealants that contain solvents may be flammable. Other types of adhesives, such as wood glue, may be eye and skin irritants. When working with any glue, care should be taken to avoid eye and skin contact. If the label indicates the adhesive is flammable, use and store away from sources of ignition. Epoxies contain epoxy amine resins and polyamide hardeners, which cause skin sensitization and respiratory tract irritation. Overexposure to epoxies can result in dizziness, drowsiness, nausea, and vomiting. In instances of extreme or prolonged exposure, kidney and liver damage may occur.

Floor adhesives may contain acrylics that can be irritating to the skin, may cause nausea, vomiting, headache, weakness, asphyxia, and death. Other adhesives or sealants may contain coal tar derivatives that are suspected carcinogens. Prolonged inhalation of vapors and skin contact should be avoided.

9.5.5 PAINTS

Paints used today are complex mixtures of various chemicals including solvents, emulsions, polyurethane, epoxies, adhesives, etc. and can cause any number of symptoms of illness and even cancer in the long term. Extreme care should be taken when painting that includes ventilation and personal protective equipment.

9.5.6 FUELS

The primary hazard posed by fuels is, obviously, fire. Fuels are either flammable or combustible. Whether flammable (a material that easily ignites and burns with a vapor pressure below 100°F) or combustible (a material that ignites with a vapor pressure over 100°F), they should be handled with care. Gasoline is a flammable liquid and diesel fuel is an example of a combustible liquid.

Proper storage and transport of fuels in approved, self-closing, safety containers is extremely important, and should be strictly adhered to at all times. When filling portable containers with flammable materials, proper grounding and bonding is a must to prevent ignition caused by static electricity. Store gasoline in containers marked or labeled "Gasoline." Store kerosene in containers marked "Kerosene." Never use kerosene containers for the transport or storage of gasoline.

Excessive skin contact with fuels can result in dermatitis. Fuels entering the body through the skin and over a long period of time can break down the fatty tissues and possibly build up in the body. Excessive inhalation of fuels may cause central nervous system depression and aggravation of any existing respiratory disease. Leukemia is a potential side effect of chronic exposure to some fuels and may lead to death. Ingestion of fuels may cause poisoning and possible lung damage if aspirated into the lungs when ingested. Short exposures to fuel may cause skin, lung, and respiratory tract irritation.

9.6   EXPOSURE MONITORING

The role of monitoring is to tell you what contaminants are present, and at what levels. Yet the limitations of many instruments mean that you cannot be sure of the readings unless all perimeters are taken into consideration or you already know what is in the air. This seems to be a contradiction. After all, how can you know what is present if the instruments cannot tell you? Often, determining contaminant levels are possible only after extensive diagnostic work with a variety of sampling strategies. Air sampling instruments can provide very important information to clarify the hazards at the workplace. Monitoring surveys can help answer questions like the following:

- What types of air contaminants are present?
- What are the levels of these contaminants?
- How far does the contamination range?
- What type of protective gear is needed for the workers?

Effective monitoring can be difficult work. It is much more than pushing buttons on a high-tech gadget. As you will see, it is more like an investigation. The issues fall into the following three major categories:

- What are the limitations of instruments used?
- What strategy should be used to get useful information?
- How do you evaluate results that you get?

There are two types of air-monitoring methods: (1) direct reading and (2) laboratory sampling. Direct reading instruments have built-in detectors to give on-the-spot results. However, there is a trade-off between sophistication and the weight of the unit. The instruments must be truly portable to be useful. Because of this, it is important to be aware that there are limits to any given instrument. Figure 9.3 shows the many types of air-monitoring instruments in use.

Laboratory sampling emphasis is on collecting a sample in the field, then conducting the actual analysis later back at the laboratory. The disadvantage is the delay in obtaining results. An advantage is that the instruments in the laboratory do not have to be portable.

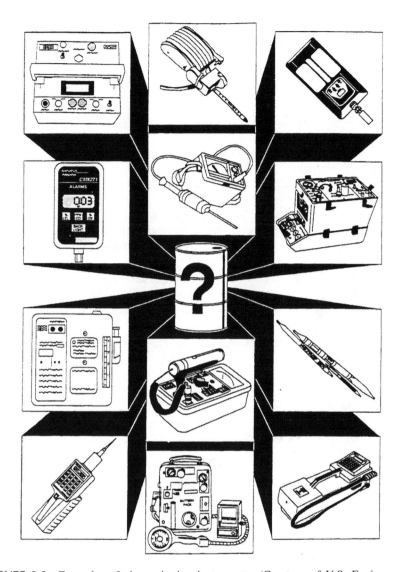

FIGURE 9.3 Examples of air-monitoring instruments. (Courtesy of U.S. Environmental Protection Agency.)

9.7 BIOLOGICAL MONITORING

Biological monitoring is covered in Chapter 8.

9.8 CANCER-CAUSING CHEMICALS

Some chemical are known to be carcinogenic (cancer causing). The safety exposure to carcinogens is zero since there are no known limits that are safe for any cancer-causing chemicals. Some examples are asbestos, benzene, and vinyl chloride. More detail appears in Section 8.5.1.

9.9 HAZARD COMMUNICATIONS (1910.1200)

OSHA has established regulations for the general industry called the Hazard Communication (HAZCOM) (29 CFR 1910.1200) standard. This standard requires that manufacturers of hazardous chemicals inform employers about the hazards of those chemicals. Also, it requires employers to inform employees of the identities, properties, characteristics, and hazards of chemicals they use, and the protective measures they can take to prevent adverse effects. The standard covers both physical hazards (e.g., flammability) and health hazards (e.g., lung damage, cancer). Knowledge acquired under the HAZCOM will help employers provide safer workplaces for workers, establish proper work practices, and help prevent chemical-related illnesses and injuries. Employers are required to do the following:

- The employer must develop a written HAZCOM program.
- The employer must provide specific information and training to workers.
- All employers on a multiple employer site must provide information to each other so that all employees can be protected.
- The owner must provide information to contractors about hazardous materials on the jobsite.

The specific requirements for each of the four main provisions are summarized as follows.

9.9.1 Written HAZCOM Program

The required components of a HAZCOM program are as follows:

- List of hazardous chemicals on the jobsite
- The method the employer will use to inform employees of the hazards associated with nonroutine tasks involving hazardous chemicals
- How the employer plans to provide employees of other companies on the jobsite with the MSDSs, such as making them available at a central location
- The method the employer will use to inform employees of other companies on the jobsite about their labeling system
- How the employer will inform workers about their labeling system
- How the employer plans to provide workers with MSDSs
- How the employer intends to train workers on hazardous chemicals

9.9.2 Information Provided by the Employer

According to the HAZCOM regulation, employers are to supply the following:

- List of hazardous chemicals used on the job
- How to recognize these hazardous chemicals

- How those chemicals might affect worker safety and health
- How workers can protect themselves from those chemicals

9.9.3 TRAINING PROVIDED BY THE EMPLOYER

Training required to comply with the HAZCOM standard is as follows:

- Requirements of the OSHA HAZCOM
- Operations at the worksite where hazardous chemicals are present
- The location and availability of the written HAZCOM program
- List of all hazardous chemicals
- Locations of MSDSs for all hazardous chemicals used on the jobsite
- Methods and observations workers can use to detect the presence or release of hazardous chemicals in your work area (e.g., labels, color, form [solid, liquid, or gas], and order)
- The physical and health hazard workers may be exposed to from the hazardous chemicals on the job
- Methods of protecting oneself, such as work practices, personal protective equipment, and emergency procedure
- Details of the hazardous communication program used by the employer
- Explanation of how workers can obtain and use hazard information

9.9.4 MULTIPLE EMPLOYER SITES

All employers on a multiple employer site must supply information to each other, so that all employees will be protected. The HAZCOM program must specify how an employer will provide other employers with a copy of the MSDSs, or make it available at a central location in the workplace, for each hazardous chemical the other employers' employees may be exposed to while working. The employers must provide the procedures for informing other employers of any precautionary measures that need to be taken to protect employees during the worksite's normal working operating conditions, and of any foreseeable emergencies. An employer must provide the mechanism to inform other employers of his/her labeling system.

9.9.5 CONCLUSIONS

Employers are responsible to develop a HAZCOM program and provide information to employees and other employer's employees and provide training to employees. All workers, as well as other employees on multiple employer worksites, must be provided with information regarding any hazardous chemicals to which workers might be exposed to at the employers' workplace. All employers in the general industry must comply with the hazardous communication regulations.

9.10 SUMMARY

Hazardous chemicals and the dangers that they pose are the primary pieces of information needed to protect workers who have to work with or around potentially dangerous chemicals. Some of these chemical and the hazards that they are likely to cause are as follows:

- Hazardous liquids (caustics or acids)—danger of burns   .
- Hazardous gases—danger of explosion and/or toxic effect
- Inorganic dusts (mineral dusts)—danger of inhalation (asbestos, silica, etc.)
- Metals, metalloids, and their compounds (lead, mercury, arsenic, etc.)—danger of toxic effect
- Organic dusts (dusts produced by grain, wood, cotton, etc.)—danger of explosion
- Organic solvents—hazards dependent on toxicity, vapor pressure, and use (can be absorbed, ingested, or inhaled)
- Pesticides—danger of poisoning through ingestion or inhalation

To try to mitigate the potential chemical hazards, the following should be ensured:

- Proper labeling (signs, color coding, etc.)
- Periodic air sampling
- Close monitoring of employee health
- Safety posters in storage or handling areas

Safe storage of hazardous materials is important to maintain workplace safety since storage facilities and procedures will vary with the type of hazardous material being handled and occupational safety and health standards related to the particular hazards faced by employees. The following should occur or should be taken into consideration:

- Special containers (drums, carboys, cylinders, bins, etc.) and how they should be stacked, piled, or stored
- Material handling equipment (carboy trucks, etc.)
- Ventilation of storage areas
- Proper lighting of storage areas

Safe handling of hazardous materials is a vital part of providing for the safety of workers. The following are steps in handling hazardous chemicals:

- Wear the proper protective equipment (demonstrate).
- Keep floors clean; never allow them to become slippery.
- Know what steps to take in an emergency; know where first-aid equipment is located and how to use it.
- Always read the label before handling a container.
- Follow company rules for showering, changing clothes, etc.
- Be familiar with the symptoms of overexposure to a hazardous material (itching, burns, fever, etc.).

10 Compressed Gases

Compressed gases have a variety of uses and require special handling procedures.

There are two types of hazards associated with the use, storage, and handling of compressed gas cylinders: the chemical hazard associated with the cylinder contents (corrosive, toxic, flammable, etc.) and the physical hazards represented by the presence of a high-pressure vessel in the workplace or laboratory. Figure 10.1 outlines some of the physical attributes of compressed gas cylinders and describes some of the dangers that may result from improper use.

Whether we like it or not, there are always safety rules to follow. But when it comes to safety procedures for compressed gas, these rules are doubly important. They should be practiced daily because the safe way is the only way. When dealing with compressed gases, there are several items that one needs to be aware of at all times so that handling, transporting, storage, and use of compressed gas cylinders can be accomplished efficiently and safely.

Mishandled cylinders may cause a violent rupture, releasing the hazardous contents or the cylinder itself, which can become a dangerous projectile. If the neck of a pressurized cylinder breaks accidentally, the energy released would be sufficient to propel the cylinder to over three-quarters of a mile in height (Figure 10.2).

A standard 250 cu ft cylinder pressurized to 2500 psig can become a rocket attaining a speed of over 30 miles/h in a fraction of a second after venting from the broken cylinder connection.

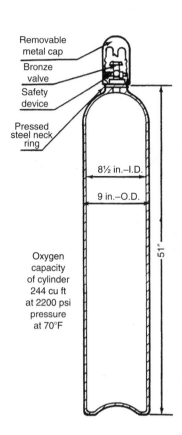

I stand 57 in. tall.

I am 9 in. in diameter.

I weigh in at 155 lb when filled.

I am pressurized at 2200 psi.

I have a wall thickness of about ¼ in.

I wear a label to identify the gas I am holding. My color is not the answer.

I transform miscellaneous stacks of material into glistening plants and many other things, when properly used.

I may transform glistening plants and many other things into miscellaneous stacks of material, when allowed to unleash my fury unchecked.

I can be ruthless and deadly in the hands of the careless or uninformed.

I am frequently left standing alone on my small base without other visible means of support—my cap removed and lost by an unthinking worker.

I am ready to be toppled over—where my naked valve can be damaged or even snapped off—and all my power unleashed through an opening no larger than a lead pencil.

I am proud of my capabilities, here are a few of them:

• I have on rare occasions been known to jetaway, faster than any dragster.

• I might smash my way through brick walls.

• I might even fly through the air.

• I may spin, ricochet, crash, and slash through anything in my path.

You can be my master only under the following terms:

• Full or empty—see it that my cap is on, straight, and snug.

• Never, repeat, never leave me standing alone. Secure me so that I cannot fall.

Treat me with respect—I am a sleeping giant.

FIGURE 10.1 Sleeping giant. (Courtesy of North Carolina Department of Labor, Mine and Quarry Division.)

10.1   CORROSIVE AND TOXIC GASES

Many gases used throughout industry have additional hazards other than those of fire, asphyxiation, or oxygen enrichment. Exposure to some gases may present serious health hazards to unprotected personnel. Before using a corrosive, toxic, or highly toxic gas, read the label and material safety data sheet (MSDS) for the particular gas. Personnel working in the immediate vicinity where exposure to these gases is possible should be informed of their hazards. Exposure to these gases must be kept as low as possible, but in no case should concentrations exceed Occupational Safety and Health Administration (OSHA) permissible exposure limits (PELs) or the current ACGIH threshold limit values (TLVs) set by the American

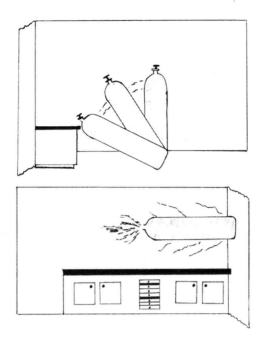

FIGURE 10.2 Compressed gas cylinder can become a missile, which can penetrate a block wall.

Conference of Governmental Industrial Hygienists (ACGIH). Contact an industrial hygienist for information on these exposure limits.

10.1.1 POISONOUS GASES

Poisonous compressed gases represent a significant hazard. Special precautions not otherwise necessary become prudent when using poisonous gases. Common poisonous or highly toxic gases include the following:

- Arsine (AsH_3)
- Ethylene oxide (EtO)
- Hydrogen cyanide (HCN)
- Nitric oxide (NO)
- Phosphine (PH_3)

Certain poisonous gases (e.g., ethylene oxide) can only be used if specific OSHA regulations (1910.1047) and safe practices are followed:

- Emergency procedures should be made clear to all involved, including personnel from adjacent work areas and managers who might be affected.
- Poisonous gas used after normal working hours should require the approval of the chemical hygiene officer for your operation.

- Fume hoods and other ventilation need to be tested before use and checked frequently during the project that involves poisonous gas.
- Notify the environmental health, safety, and risk department before your first use of the poisonous gas.
- Police should also be informed about the locations and types of poisonous gas in use.
- Document the procedures in your work area according to the chemical hygiene plan. As with all chemicals, obtain and review the MSDS for the poisonous gas. Maintain an extra copy of the MSDS in your workplace's chemical hygiene plan.

Disposal of poisonous gas cylinders can often cause problems. If the cylinder cannot be returned to the manufacturer, disposal cost may be as large as $1000 per cylinder, or more. Even cylinders that can be returned must be shipped on a vehicle that does not simultaneously carry any other hazardous materials or foodstuffs.

The energy potential of compressed gas cylinders whether chemical or mechanical can be mitigated by following safe work procedures. These safe procedures include use, handling, storage, transportation, and movement of compressed gas cylinders, and those using them should not fail to follow these procedures.

10.2 PREVENTING COMPRESSED GAS CYLINDER ACCIDENTS

All systems in manned areas have unmodified, DOT-approved, compressed gas cylinders and the appropriate regulators may not require engineering controls if general safety rules are followed. Compressed gas cylinders are the most common source of gas for many operations. As a precaution, these cylinders must be adequately secured when in use or storage. The DOE, ASME, DOT, and OSHA agencies all refer to the Compressed Gas Association (CGA) pamphlet (CGA P-l, 1991) for instructions on how to safely handle compressed gas cylinders. Many factors must be addressed to assure safety in the handling and use of compressed gas cylinders. The great amount of energy stored in the cylinders makes preventing accidents paramount in preventing injury, illnesses, and deaths.

10.2.1 Cylinder Use

Follow these recommendations for safe use of cylinders:

- Make sure all connections are tight. Use soap water to locate leaks.
- Keep cylinder valves, regulators, couplings, hose, and apparatus clean and free of oil and grease.
- Keep cylinders away from open flames and sources of heat.
- Safety devices and valves should not be tampered with, nor repairs attempted.
- Use flashback arrestors and reverse-flow check valves to prevent flashback when using oxy-fuel systems.

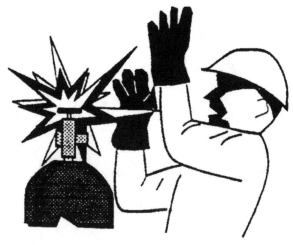

Inproper cracking

FIGURE 10.3 Care must be taken when opening cylinder valves. (Courtesy of Department of Energy.)

- Regulators should be removed when moving cylinders, when work is complete, and when cylinders are empty.
- Cylinders are to be used and stored in the upright position.
- Cylinder valve should always be opened slowly. Always stand away from the face and back of the gauge when opening the cylinder valve (Figure 10.3).
- When a special wrench is required to open a cylinder or manifold valve, the wrench shall be left in place on the valve stem when in use; this precaution is taken so the gas supply can be shut off quickly in case of an emergency, and that nothing should be placed on top of a cylinder that may damage the safety device or interfere with the quick closing of the valve.
- Fire extinguishing equipment should be readily available when combustible materials have a possibility of getting exposed to welding or cutting operations using compressed cylinder gases.

10.2.2 HANDLING

Even though the cylinders are constructed of steel, they must be handled with extreme care to avoid damage. Physical abuse, such as dropping, or violently striking cylinders together, can cause damage to the cylinder, valve, or fuse plug, and in turn present a potential hazard. There are several methods of unloading cylinders from a truck to ground level that help prevent damage. These include the following:

- V-shaped trough—it allows cylinders to be lowered carefully down onto a shock-absorbing mat on the ground.
- Angle-iron cradle—these are used to upend the cylinders and lower them to the ground.

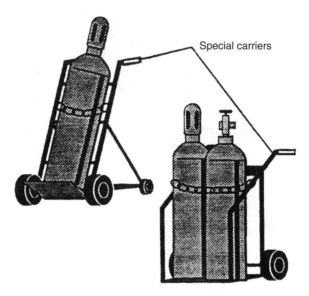

FIGURE 10.4   Carts for safe and secure movement of cylinders. (Courtesy of Department of Energy.)

- Elevator tailgate—this is one of the easiest and safest means of unloading cylinders, and is to be used whenever it is available on the transport truck. But remember, the important thing is to be sure the cylinders are not dropped.
- Use a four-wheel cylinder cart for moving Standard No. 1 and larger gas cylinders. These cylinders are difficult to move manually because of their shape, smooth surface, and weight (Figure 10.4).
- Make sure that the protective valve cover is in place when a cylinder is not connected to a regulator or manifold (Figure 10.5).
- Measure the pressure of contents of half-empty cylinders and mark them.

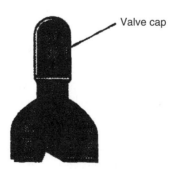

FIGURE 10.5   Cylinders should have valve caps in place when regulators are removed. (Courtesy of Department of Energy.)

- Always assume a cylinder is pressurized; handle it carefully and avoid bumping or dropping.
- Never drop cylinders from trucks or any raised surface to the ground.
- Lifting a standard cylinder, or any cylinder weighing more than 50 lb, requires two people. Never lift a cylinder by the cylinder cap (Figure 10.6).
- Do not handle oxygen cylinders with greasy, oily hands or gloves. The reaction between oxygen and hydrocarbons can be violent, even when small quantities are involved.
- Secure cylinders in suitable cradles or skid boxes before raising them with cranes, fork trucks, or hoists. Do not use ropes or chain slings alone for this purpose.
- Never use a gas cylinder as a roller for moving materials or for supporting other items.

10.2.3 STORAGE

Cylinders are sometimes shipped tied horizontally on wooden pallets, individually contained by saddle blocks, and double-banded to prevent rolling and sliding. These are not recommended methods for cylinder storage. Instead, the work practices prescribed in this section should be followed (from pamphlet CGA P-1-1991):

- Store adequately secured cylinders upright on solid, dry, level footings, preferably outside of occupied buildings and away from traffic lanes.
- Shade cylinders stored in the sun during the summer, whenever possible.
- Store cylinders away from sources of intense heat (furnaces, steam lines, and radiators).

Improper hoisting

FIGURE 10.6 Unsafe hoisting practices for gas cylinders. (Courtesy of Department of Energy.)

- Cylinders should be stored in compatible groups.
  - Flammables from oxidizers
  - Corrosives from flammables
  - Full cylinders from empties
- Empty cylinders should be clearly marked and stored as carefully as full cylinders are because of the presence of residual gas.
- All cylinders should be protected from corrosive vapors.
- Store cylinders in an upright position.
- Keep oxygen cylinders a minimum of 20 ft from flammable gas cylinders or combustible materials. If this cannot be done, separation by a noncombustible barrier at least 5 ft high having a fire-rating of at least 1.5 h is required (Figure 10.7).
- Compressed gas cylinders should be secured firmly at all times. A clamp and belt or chain, securing the cylinder between cylinder waist and shoulder to a wall, are generally suitable for this purpose.
- Cylinders should be individually secured; using a single restraint strap around a number of cylinders is often not effective.

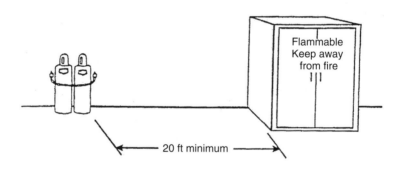

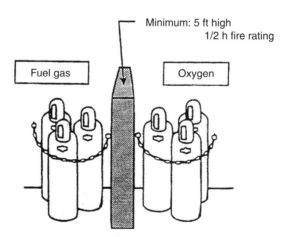

FIGURE 10.7 Maintain required distances for flammable compressed gases. (Courtesy of the Occupational Health and Safety Administration.)

- Keep valve protective caps in place when the cylinder is not in use. Always store cylinders with the protective caps in place.
- Mark empty cylinders EMPTY or MT.
- Keep valves closed on empty cylinders.
- Keep cylinders away from magnetized equipment.
- Cylinders must be kept away from electrical wiring as the cylinder could become part of the circuit.
- Store cylinders in well-ventilated areas designated and marked only for cylinders.
- Do not stockpile gas, especially flammables, poisons, or corrosives, beyond the amount required for immediate use. Consider direct delivery from the distributor when gases are needed.
- Limit the use and storage of poisons and corrosives to less than 1 year to prevent stockpiling. Documentation should be required for these materials. The environmental safety and health (ES&H) department's industrial hygienist should establish and document the maximum quantities of such materials in use and storage to ensure reasonable turnover. The emergency preparedness group should track the materials as an element of its emergency response planning program. Extended use or storage of hazardous materials should occur after discussion by the user, the industrial hygienist, and the emergency preparedness group. The agreed upon storage process should be documented.
- Ensure that containers stored or used in public areas are protected against tampering and damage. Furthermore, containers stored inside or outside shall not obstruct exit routes or other areas that are normally used or intended for the safe exit of people.
- Use a storage basket for smaller cylinders (<5 L). These baskets are available commercially.

10.2.3.1   Outside Storage

Store cylinders outside whenever it is possible. Care must be taken to protect them from bad weather and direct sunlight. Remember, the heat from direct sunlight will cause gas to expand, which creates higher pressure within the cylinder.

10.2.3.2   Inside Storage

It is best not to store cylinders inside, but if you must, here are a few things to remember. Do not place cylinders

- In passageways
- Near elevators
- Near loading platforms
- By entrances or exits where they might be accidentally hit
- Near sources of electricity
- Near sources of excessive heat, such as the sparks resulting from welding or cutting

- Where they may become hotter than 130°F
- Closer than 20 ft from combustibles such as grease, gasoline, paint, oil, and dirty rags

In addition, if a cylinder is frozen to the ground, use warm, not hot, water to free it. If the valve is frozen, again, use warm water, not hot, to thaw it or take the cylinder inside and let it thaw at room temperature.

10.2.4   MOVING CYLINDERS

Here are some pointers that should be remembered when moving compressed gas cylinders:

- Use of a hand truck simplifies moving cylinders from one location to another.
- Cylinders are to be chained or secured in some manner, in an upright position.
- Avoid moving in a horizontal position whenever possible, especially cylinders containing acetylene.
- Protect valves from being damaged or accidentally broken off by the use of properly placed cylinder caps.
- Never drag a cylinder, tilt it sideways, and roll it along on its bottom rim or edge. This gets the job done in an easier and much safer way.
- Use a cylinder cart and secure cylinders with a chain.
- Do not use the protective valve caps for moving or lifting cylinders.
- Do not drop a cylinder or permit them to strike each other violently or be handled roughly.
- Unless cylinders are secured on a cart, regulators are to be removed, valves closed, and protective caps in place before cylinders are moved.

10.2.5   TRANSPORTATION OF CYLINDERS

Cylinders containing compressed gases are primarily shipping containers and should not be subjected to rough handling or abuse. Such misuse can seriously weaken the cylinder and render it unfit for further use or transform it into a rocket having sufficient thrust to drive it through masonry walls:

- To protect the valve during transportation, the cover cap should be screwed on hand tight and remain on until the cylinder is in place and ready to use.
- Cylinders should never be rolled or dragged.
- When moving large cylinders, they should be strapped to a properly design-wheeled cart or cradle to insure stability.
- Only one cylinder should be handled (moved) at a time.

10.2.6   EMPTY CYLINDERS

Leave some positive pressure (a minimum of 20 psig) in empty cylinders to prevent suck-back and contamination. Close the valves on empty cylinders to prevent internal contamination; remove the regulators and replace the protective cap. Use a cylinder status tag to indicate whether the cylinder is full, in service, or if residue is still in the cylinder. This tag is to be installed by the ES&H department and shall remain on the cylinder. Empty cylinders should be stored separately from full cylinders. Properly label and dispose of cylinders. Call the vendor to pick up cylinders that are no longer needed.

10.2.7　Identification and Color Coding

Stencils, DOT shoulder labels, cautionary sidewall labels, or tags are used to identify the contents of all gas cylinders. Do not remove these labels without specific authorization from the ES&H department. Color codes for gas cylinders are not reliable to identify contents since there is no standardization by manufacturers and suppliers:

- Cylinders must be properly labeled, including the gas composition and appropriate hazards (e.g., health, flammability, and reactivity).
- Cylinders have several stamped markings. The top mark is either a DOT or an ICC marking indicating pertinent regulations for that cylinder. The second mark is the serial number. Under the serial number is the symbol of the manufacturer, user, or purchaser. Of the remaining marks the numbers represent the date of manufacture, and retest date (month and year). A (+) sign indicates the cylinder may be 10% overcharged, and a star indicates a 10 year test interval (Figure 10.8).

The hazard classification or the name of the gas being stored shall be prominently marked in container storage areas, and No Smoking signs shall be posted where

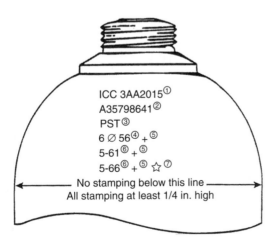

FIGURE 10.8　Markings on compressed gas cylinders. (Courtesy of the Occupational Health and Safety Administration.)

appropriate. Placards, container labels, and markings provide information on the products involved.

The MSDS for the products or other recognized emergency response guides should be consulted for specific hazards, safety precautions, and related emergency response information.

10.2.8 ADEQUATELY SECURING CYLINDERS

All compressed gas cylinders in service or storage at the user's location are to be secured to prevent them from falling. Gas cylinders with a water volume of less than 5 L (305 cu in.) may be stored in a horizontal position, as long as they are prevented from rolling and they would be considered to be adequately secured. Since 1980, cylinder and manifold racks have been fabricated, purchased, or equipped with two chains whenever possible. If available, both chains are to be used to secure these cylinders.

10.2.9 CYLINDER STORAGE SHEDS

Cylinder storage sheds and delivery sheds should be equipped with double chains. Thus, cylinders shall be adequately secured with individual restraining bars or chain restraints (1/4 in. welded chains and safety clips are preferred). The rails on which the restraining bars slide must be pinned and secured to the shed to prevent the bars from sliding off.

10.2.10 COMPATIBILITY

Cylinders are to be segregated by compatibility of contents. For example, oxidizers shall be kept separate from combustibles or flammables by a minimum distance of 20 ft or by a noncombustible barrier that is at least 5 ft high with a fire-resistance rating of at least 1.5 h. Your fire protection engineer or industrial hygienist can provide compatibility evaluations.

10.3 HOSES AND REGULATORS

10.3.1 INSPECTION

Complete the following procedures:

- Inspect hoses and manifolds frequently, and replace worn hoses and connections. Contact the engineering staff for hose or connector replacements.
- Report leaking cylinders that contain hazardous materials to the emergency dispatcher (dial 911). Evacuate the area until the emergency response team arrives.
- Contact your area ES&H department before handling faulty or corroded cylinders; these cylinders should be segregated. Caution: Only the vendor shall alter or repair cylinders or cylinder valves.

10.3.2 GENERAL PRECAUTIONS

General precautions are as follows:

- Secure both ends of the hose with a hose restraint to prevent whipping in the event the hose or fitting fails. For systems in manned areas, support and secure the hose and tubing at least every 7 ft.
- Do not use an open flame to leak-check a gas cylinder; use soapsuds or a leak-detection solution.
- Remove the talc and dust from a new hose before connecting it.
- Do not use white lead, oil, grease, or any other nonapproved joint compound to seal the fittings on an oxygen system; a fire or an explosion could occur if oxygen contacts such materials. Threaded connections in oxygen piping should be sealed with solder, glycerin, or other sealants approved for oxygen service. Gaskets should be made of noncombustible materials.
- Never interchange regulators and hose lines (with one type of gas for another). Explosions can occur if flammable gases or organic materials come in contact with oxidizers (e.g., oxygen) under pressure.
- Never use oxygen to purge lines, operate pneumatic tools, or dust clothing. Remember, oxygen is not a substitute for compressed air. Do not transfer or mix gases in commercial vendor- or laboratory-owned DOT cylinders, or transfer gases from one DOT cylinder to another.
- Do not use vendor-owned cylinders for purposes other than as a source of gas. These cylinders may only be pressurized by the owner.
- Do not strike a welding arc on a cylinder.

10.3.3 OPERATION

The following operations should be completed before using compressed gas cylinders:

- Before installing a regulator on a compressed gas cylinder, vacuum the valve port clean or crack the valve gently to expel any foreign material. Do not perform this task if the gas in the cylinder is toxic, reactive, or flammable.
- After installing the regulator and before opening the cylinder valve, fully release (turning counterclockwise) the regulator pressure-adjusting screw.
- Open the cylinder valves slowly. Never use a wrench on a cylinder valve that will not rotate manually. Stand clear of pressure regulator gauge faces when opening the cylinder valves. If the valves are defective, return the cylinder to the vendor immediately.
- Keep removable keys or handles from valve spindles or stems in place while the cylinders are in service.
- Never leave pressure on a hose or line that is not being used. To shut down a system, close the cylinder valve and vent the pressure from the entire system.

All the rules and practices discussed concerning storage, handling, transportation, and use of compressed gas cylinders apply in all situations. Following these practices completely, along with common sense, will enable the use of these materials in a safe and efficient way. Remember, the safe way is the only way.

10.3.4 SAFE HANDLING AND USAGE GUIDELINES

Plan carefully when setting up an experiment that involves gaseous materials and gas cylinders. The following should be done:

- Ask questions about the suppliers when purchasing gaseous materials, especially with regard to waste disposal and their cylinder return policy. Only purchase cylinders from companies that will accept cylinders back for disposal. The cost of disposal for gas cylinders is dependent upon the material, but even nonhazardous cylinders can be costly to dispose.
- Do not purchase a larger size cylinder than necessary; excess reactant can be a problem for disposal, increases the risk to a larger area if accidentally released, is more difficult to store in a ventilated area if required, and takes up more room in the hood or on the floor.
- Make sure you have adequate ventilation to work with toxic gases. These materials will require constant local ventilation to ensure the safety of personnel. Installing ventilation is not usually a straightforward task; it usually takes considerable money and time, so plan accordingly.
- The National Fire Protection Association (NFPA) sets limitations on the number of cylinders that should not be exceeded in a laboratory. Do not acquire more than the following:
  - Three 10 in. × 50 in. flammable gas or oxygen cylinders.
  - Three 4 in. × 15 in. cylinders containing toxic gases (such as arsine, chlorine, fluorine, hydrogen cyanide, and nitric oxide).
  - NFPA allows the use of liquefied petroleum gas cylinders within the laboratory; however, laws in Texas state that no liquefied petroleum gases (i.e., C3 or C4 such as butanes, propanes, etc.) may be kept within an occupied building (Texas Railroad Commission rules).
- Be familiar with the guidelines on safe transport of high-pressure cylinders:
  - When the cylinder is not in use the valve protection cap must be in place to protect the valve.
  - Never drag, slide, or roll the cylinder, get a cylinder cart or truck and use it.
  - Always have the protective cap covering the valve; never transport with the regulator in place.
  - Make sure the cylinder is secured to the cart during transport.

10.4 COMPRESSED AIR SAFETY GUIDELINES

Compressed air for general shop or laboratory use shall be limited to a maximum of 30 psig (200 kPa) using restricting nozzles (supply pressure from regulator to nozzle can be up to 700 kPa (100 psig); the nozzle reduces pressure). Compressed air at a full-line pressure of up to 700 kPa (without the use of restricting nozzles) may be used only to operate pneumatic tools and certain control instruments. Observe the following safety rules when using compressed air:

- Do not use compressed air to clean clothing; the air jet tends to drive particles into the fabric, where they can cause skin irritation. Keep a cloth brush handy or, preferably, wear a laboratory coat.
- Be sure no one is in the path of the air stream when using compressed air to dry mechanical parts. Always wear goggles or a face shield to protect your eyes.
- Do not use air pressure to transfer liquids from containers with unknown MAWPs. Use a siphon with a bulk aspirator or a pump instead. If a standard 208 L (55 gal) drum is pressurized to 100 kPa (15 psig), the force exerted on the head of the drum is about 25 kN (3 tons). This is not an acceptable practice.
- Limit the transfer pressure of liquid nitrogen Dewars to 100 kPa (15 psig).
- Never apply air pressure to the body.
- Unless an automatic shutoff coupling is used, attach a short chain (or equivalent) between a hose and an air-operated tool to prevent whipping in the event the coupling separates.
- Unless an automatic shutoff coupling is used, vent the pressure in an air line before changing the nozzles or fittings.
- Use Grade D breathing air. This type of air has been specifically approved for use with air respirators, since compressed air contains oil and other contaminants.
- Do not substitute compressed oxygen for air. Clothing saturated with oxygen burns explosively.

10.5 CRYOGENIC SAFETY

Cryogenics may be defined as low-temperature technology, or the science of ultra-low temperatures. To distinguish between cryogenics and refrigeration, a commonly used measure is to consider any temperature lower than $-73.3°C$ ($-100°F$) as cryogenic. Although there is some controversy about this distinction, and some who insist that only those areas within a few degrees of absolute zero may be considered as cryogenic, the broader definition will be used here.

Low temperatures in cryogenics are primarily achieved by the liquefaction of gases, and there are more than 25 gases that are currently in use in the cryogenic area, that is, gases that have a boiling point below $-73.3°C$ ($-100°F$). However, the seven gases that account for the majority of applications in research and industry are helium, hydrogen, nitrogen, fluorine, argon, oxygen, and methane (natural gas). Cryogenics is being applied to a wide variety of research areas, a few of which are food processing and refrigeration, rocket propulsion fuels, spacecraft life support systems, space simulation, microbiology, medicine, surgery, electronics, data processing, and metalworking.

10.5.1 General Precautions

Personnel should be thoroughly instructed and trained in the nature of the hazards and the proper steps to avoid them. This should include emergency procedures, operation of equipment, safety devices, knowledge of the properties of the materials used, and personal protective equipment (PPE).

Equipment and systems should be kept scrupulously clean and contaminating materials should be avoided, which may create a hazard upon contact with the cryogenic fluids or gases used in the system. This is particularly important when working with liquid or gaseous oxygen.

Mixtures of gases or fluids should be strictly controlled to prevent the formation of flammable or explosive mixtures. As the primary defense against fire or explosion, extreme care should be taken to avoid contamination of a fuel with an oxidant, or the contamination of an oxidant with a fuel.

As a further precaution, when flammable gases are being used, potential ignition sources must be carefully controlled. Work areas, rooms, chambers, or laboratories should be suitably monitored to automatically warn personnel when a dangerous situation develops. Wherever practical, it would be advisable to provide facilities for the cryogenic system or equipment to be shut down automatically as well as to sound a warning alarm.

Where there is a possibility of physical contact with a cryogenic fluid, full face protection, an impervious apron or coat, cuffless trousers, and high-topped shoes should be worn. Watches, rings, bracelets, or other jewelry should not be permitted when personnel are working with cryogenic fluids. Personnel should avoid wearing anything capable of trapping or holding a cryogenic fluid in close proximity to skin. Gloves may or may not be worn, but if they are necessary to handle containers or cold metal parts of the system, they should be impervious, and sufficiently large to be easily tossed off the hand in case of a spill. A more desirable arrangement would be hand protection of the potholder type.

When toxic gases are being used, suitable respiratory protective equipment should be readily available to all personnel. They should be aware of the location and use of this equipment.

10.5.2 Storage

Storage of cryogenic fluids is usually in a well-insulated container designed to minimize product loss because of boil-off. The most common container for cryogenic fluids is a double-walled, evacuated container known as a Dewar flask, of either metal or glass. The glass container is similar in construction and appearance to the ordinary thermos bottle. Generally, the lower portion will have a metal base that serves as a stand. Exposed glass portions of the container should be taped to minimize the hazard of flying glass if the container should break or implode.

Metal containers are generally used for larger quantities of cryogenic fluids, and usually have a capacity of 10–100 L (2.6–26 gal). These containers are also of double-walled evacuated construction, and usually contain some adsorbent material in the evacuated space. The inner container is usually spherical in shape because this has been found to be the most efficient in use. Both the metal and glass Dewars should be kept covered with a loose-fitting cap to prevent air or moisture from entering the container, and to allow built-up pressure to escape.

Larger capacity storage vessels are basically the same double-walled containers, but the evacuated space is generally filled with powdered or layered insulating material. For economic reasons, the containers are usually cylindrical with dished ends, which approximate the shape of the sphere but are less expensive to build. Containers must be constructed to withstand the weights and pressures that will be encountered, and adequately vented to permit the escape of evaporated gas. Containers should also be equipped with rupture discs on both inner and outer vessels to release pressure if the safety relief valves should fail.

Cryogenic fluids with boiling point below that of liquid nitrogen (particularly liquid helium and hydrogen) require specially constructed and insulated containers to prevent rapid loss of product from evaporation. These are special Dewar containers that are actually two containers, one inside the other. The liquid helium or hydrogen is contained in the inner vessel, and the outer vessel contains liquid nitrogen, which acts as a heat shield to prevent heat from radiating into the inner vessel. The inner neck as shown in Figure 10.1 should be kept closed with a loose-fitting, nonthreaded brass plug, which prevents air or moisture from entering the container, yet loose enough to vent any pressure that may have developed. The liquid nitrogen fill and vent lines should be connected by a length of gum rubber tubing with a slit approximately 2.54 cm (1 in.) long near the center of the tubing. This prevents the entry of air and moisture, while the slit will permit release of gas pressure. Piping or transfer lines should be double-walled evacuated pipes to prevent product loss during transfer.

Most suppliers are now using a special fitting to be used in the shipment of Dewar vessels. Also, there is an automatic pressure relief valve, and a manual valve to relieve pressure before removing the device. Dewar vessels of this type must be regularly maintained to prevent product loss and to prevent ice plug formation in the neck. The liquid nitrogen outer jacket should be kept filled to maintain its effectiveness as a radiant heat shield. The cap must be kept on at all times to prevent entry of moisture and air, which will form an ice plug. The liquid helium fill (inner neck) should be reamed out before and after transfer, and at least twice daily. Reaming should be performed with a hollow copper rod, with a marker or stop to prevent damage to the bottom of the inner container.

Current designs of Dewar vessels are equipped with a pressure relief valve, a pressure gauge for the inner vessel. Transfer of liquids from metal Dewar vessels should be accomplished with special transfer tubes or pumps designed for the particular application. Since the inner vessel is mainly supported by the neck, tilting the vessel to pour the liquid may damage the container, shorten its life, or create a hazard because of container failure at a later date. Piping or transfer lines should be so constructed that it is not possible for fluids to become trapped between valves or closed sections of the line. Evaporation of the liquid in a section of line may result in pressure buildup and eventual explosion. If it is not possible to empty all lines, then they must be equipped with safety relief valves and rupture discs. When venting storage containers and lines, proper consideration must be given to the properties of the gas being vented. Venting should be to the outdoors to prevent an accumulation of flammable, toxic, or inert gas in the work area.

10.5.3 Hazards

Health hazards involving cryogens include frostbite/burns, skin lesions, asphyxiation, and vision impairment. Immediately call 911 if there is an emergency involving cryogens.

Fighting cryogen fires can be extremely dangerous, as hydrogen burns with a nearly invisible flame. In addition, carbon dioxide fire extinguishers can cause a static discharge energetic enough to reignite a blaze.

10.5.4 Hazards to Personnel

10.5.4.1 Frostbite/Burns and Skin Lesions

Cryogen-induced frostbite/burns and thermal burns have similar characteristics. Burns may be severe where the liquid pools, such as under an eyelid, in a cupped palm, or in a sleeve or cuff. In addition, cryogens can cause blindness if the cornea becomes frozen. Bare skin can instantly bond with unprotected cryogen supply lines or uninsulated equipment and may tear when pulled, causing skin lesions (Figure 10.9).

FIGURE 10.9 Liquid nitrogen used by physicians to freeze skin lesions.

10.5.4.2 Asphyxiation

When a cryogen is spilled in a small area, it will evaporate and expand rapidly, displacing breathing air and eventually causing asphyxiation. Cold gases and gases that are heavier than air concentrate in low places where ventilation is poor, such as sumps or pits.

10.5.4.3 Obscured Vision

Spilled cryogens can condense water vapor from the air, producing a ground-hugging fog that can obscure vision and cause trips and falls.

10.5.5 HAZARDS TO EQUIPMENT

Equipment that comes in contact with cryogens can

- Burst, if it contains a rapidly boiling or evaporating cryogen
- Freeze, causing safety valve dysfunction and subsequent pressure buildup
- Become brittle, causing it to shatter and release its contents

10.5.6 HAZARDS OF CRYOGENS

Cryogenic liquids (or cryogens) are liquefied gases that are cooled below room temperature; most cryogenic liquids are below −150°C. When a small amount of cryogenic liquid is converted into gas, a very large volume of gas is created. Cryogenic liquids are classified as compressed gases.

10.5.6.1 Extreme Cold

Cryogens can freeze skin, causing painful blisters, much like a burn. Prolonged exposure can cause frostbite with pain occurring only when the skin thaws. Cryogen-exposed skin can stick to cold metals.

10.5.6.2 Asphyxiation

Cryogens expand into large volumes of gas that can displace air. For example, 1 L of liquid nitrogen forms nearly a pool of nitrogen gas at room temperature. The gas formed is often cold and pools on the floor or lower areas. In enclosed areas, death or coma from oxygen deficiency may occur. Do not enter an oxygen-deficient atmosphere even to rescue someone. Always store Dewars in well-ventilated areas. Never enter the cryogen facility if the oxygen warning sensor alarm is sounding. The oxygen level alarm and sensor are located on the wall next to the freight elevator in the cryogenic facility.

10.5.6.3 Toxic Hazards

Toxic cryogens will release toxic gases. Read the MSDS that comes with the cryogen.

10.5.6.4 Obscured Vision

The vapor formed from cryogens falling down form a ground level fog that obscures the floor. Beware of trip hazards.

10.5.6.5 High Pressure

Sealed systems containing cryogens may form extremely high pressures, enough to rupture or explode. Always have a relief vent on a cryogen-containing Dewar.

10.5.6.6 Dewars in High Magnetic Fields

Superconducting magnets are routinely filled with cryogens. The Dewars used for this purpose must be nonmagnetic.

10.5.6.7 Liquid Oxygen

Liquid oxygen can make materials burn that are usually noncombustible (Figure 10.10).

FIGURE 10.10 A liquid oxygen container in a secured enclosure.

10.6 PREVENTING CRYOGENIC ACCIDENTS

10.6.1 Dos

The following are some of the proper procedures and practices that should be followed when dealing with cryogenic materials:

- Do wear goggles, cryogen gloves, and loose-fitting clothing with no pockets when handling cryogenic liquids.
- Do read the MSDS that comes with the liquid.
- Do transport cryogenic liquids in containers approved for such use.
- Do avoid activities that will cause splashing of the liquid.
- Do use cryogens in well-ventilated areas.
- Do cover Dewars to prevent liquid oxygen buildup.
- Do wear PPE when handling cryogens; use insulated gloves and face shields or other splash eye/face protection, closed-toed shoes, and lab coats.

10.6.2 Don'ts

The following are procedures and practices that should not be used when dealing with cryogenic materials:

- Do not enclose cryogenic liquids without a vent.
- Do not use large quantities of cryogenic liquids without proper ventilation.
- Do not enter the cryogenic facility if the alarm is sounding.
- Do not tip or spill Dewars.

10.7 COMPRESSED GASES IN THE SERVICE INDUSTRY

The use and distribution of compressed gases for welding, health care, storage, and distribution are standard practices in the service industry. We would expect these gases to be stored and sold by the warehousing, wholesale, and retail sectors and moved extensively by transportation, but also have wide use in other service sectors and utilities. These are possibly less used in the education sector, but extensively used in the health sector in the form of oxygen and as cryogenics. Also, they are used in routine maintenance activities and as a fuel for forklifts. This is also true for the administrative, leisure, and hospitality sectors that employee maintenance personnel who would be using some of these types of gases. Each particular sector would need to address the safe use of flammable and combustible liquids based upon their use in each particular sector.

10.8 OSHA COMPRESSED GAS REGULATIONS: SUMMARY

10.8.1 Compressed Gas Cylinders (29 CFR 1910.101 and .253)

Compressed gas cylinders have exploded and have become airborne. There is a lot of stored energy in a compressed gas cylinder, which is why they should be handled with great care. Cylinders with a water weight capacity over 30 lb (13.5 kg) must be equipped with means for connecting a valve protector device, or with a collar or recess to protect the valve. Cylinders should be legibly marked to clearly identify the gas contained. Compressed gas cylinders stored in areas are to be protected from external heat sources such as flame impingement, intense radiant heat, electric arcs, or high-temperature lines. Inside buildings, cylinders should be stored in a well-protected, well-ventilated, dry location away from combustible materials by 20 ft. Also, the in-plant handling, storage, and utilization of all compressed gases in cylinders, portable tanks, rail tank cars, and motor vehicle cargo tanks should be in accordance with the guidelines laid out in the CGA pamphlet P-1-1965.

Cylinders are to be located or stored in areas where they will not be damaged by passing or falling objects or subject to tampering by unauthorized persons. Cylinders are to be stored or transported in a manner to prevent them from creating a hazard by tipping, falling, or rolling and stored 20 ft away from highly combustible materials. Where a cylinder is designed to accept a valve protection cap, caps are to be in place except when the cylinder is in use or is connected for use.

Cylinders containing liquefied fuel gas are to be stored or transported in a position so that the safety relief device is always in direct contact with the vapor space in the cylinder. All valves must be closed off before a cylinder is moved, when the cylinder is empty, and at the completion of each job. Low-pressure fuel-gas cylinders should be checked periodically for corrosion, general distortion, cracks, or any other defect that might indicate a weakness or render it unfit for service.

There are several hazards associated with compressed gases, including oxygen displacement, fires, explosions, toxic effects from certain gases, as well as the physical hazards associated with pressurized systems. Special storage, use, and handling precautions are necessary to control these hazards. There are specific safety requirements for many of the compressed gases such as acetylene, hydrogen, nitrous oxide, and oxygen.

10.8.2 ACETYLENE (29 CFR 1910.253)

Acetylene cylinders are to be stored and used in a vertical, valve-end-up position only. Under no conditions should acetylene be generated pipeful (except in approved cylinder manifolds) or utilized at a pressure in excess of 15 psi (103 kPa gauge pressure) or 30 psi (206 kPa absolute). The use of liquid acetylene is prohibited. The in-plant transfer, handling, and storage of acetylene in cylinders are to be in accordance with the guidelines laid out in the CGA pamphlet C-1.3-1959.

10.8.3 HYDROGEN (29 CFR 1910.103)

Hydrogen containers must comply with one of the following: (1) designed, constructed, and tested in accordance with appropriate requirements of ASME's *Boiler and Pressure Vessel Code, Section VIII Unfired Pressure Vessels*, 1968 or (2) designed, constructed, tested, and maintained in accordance with the U.S. Department of Transportation's specifications and regulations.

Hydrogen systems are to be located so that they are readily accessible to delivery equipment and to authorized personnel and must be located aboveground, and not be located beneath electric power lines. Systems must not be located close to flammable liquid piping or piping of other flammable gases. Permanently installed containers are to be provided with substantial noncombustible supports on firm noncombustible foundations.

10.8.4   Nitrous Oxide **(29 CFR 1910.105)**

Nitrous oxide piping systems for the in-plant transfer and distribution of nitrous oxide are to be designed, installed, maintained, and operated in accordance with the guidelines laid out in the CGA pamphlet G-8.1-1964.

10.8.5   Oxygen **(29 CFR 1910.253)**

Oxygen cylinders in storage must be separated from fuel-gas cylinders or combustible materials (especially oil or grease) by a minimum distance of 20 ft (6 m) or by a noncombustible barrier at least 5 ft high (1.5 m) having a fire-resistance rating of 1/2 h.

10.8.6   Compressed Air **(29 CFR 1910.242 and 29 CFR 1926.302)**

Pressure of compressed air used for cleaning purposes should be reduced to less than 30 psi (207 kPa) and then used only with effective chip guarding and PPE.

10.9   COMPRESSED GAS AND CYLINDER CHECKLIST

To assure the safe use and handling of compressed gases and their cylinders, a checklist can be used for following safety and compliance procedures for these gases. Figure 10.11 is an example of such a checklist.

10.10   SUMMARY

Remember, the greatest physical hazard represented by the compressed gas cylinder in the workplace or laboratory is the tremendous force that may be released if it is knocked over. Compressed gases present a unique hazard. Depending on the particular gas, there is a potential for simultaneous exposure to both mechanical and chemical hazards. Gases may be as follows:

- Flammable or combustible
- Explosive
- Corrosive
- Poisonous
- Inert
- Combination of these hazards

Safety is a critical part of the use and handling of compressed gas cylinders. Specific rules and guidelines should be followed at all times.

Compressed gas cylinder (CGC) checklist

By answering yes or no to the following checklist items, it will help guide your safety approach to compressed gases and their cylinders as well as OSHA compliance.

Yes ☐   No ☐   Are CGCs kept away from radiators and other sources of heat?

Yes ☐   No ☐   Are CGCs stored in well-ventilated, dry locations at least 20 ft away from materials such as oil, grease, excelsior, reserve stocks of carbide, acetylene, or other fuels as they are likely to cause acceleration of fires?

Yes ☐   No ☐   Are CGCs stored only in assigned areas?

Yes ☐   No ☐   Are CGCs stored away from elevators, stairs, and gangways?

Yes ☐   No ☐   Are CGCs stored in areas where they will not be dropped, knocked over, or tampered with?

Yes ☐   No ☐   Are CGCs stored not in areas with poor ventilation?

Yes ☐   No ☐   Are storage areas marked with signs such as "OXYGEN, NO SMOKING, or NO OPEN FLAMES?"

Yes ☐   No ☐   Are CGCs not stored outside generator houses?

Yes ☐   No ☐   Do storage areas have wood and grass cut back within 15 ft?

Yes ☐   No ☐   Are CGCs secured to prevent falling?

Yes ☐   No ☐   Are stored CGCs in a vertical position?

Yes ☐   No ☐   Are protective caps in place at all times except when in use?

Yes ☐   No ☐   Are threads on cap or cylinder not lubricated?

Yes ☐   No ☐   Are all CGCs legibly marked for the purpose of identifying the gas content with the chemical or trade name of the gas?

Yes ☐   No ☐   Are the markings on CGCs by stenciling, stamping, or labeling?

Yes ☐   No ☐   Are markings located on the slanted area directly below the cap?

Yes ☐   No ☐   Does each employee determine that CGCs are in a safe condition by means of a visual inspection?

Yes ☐   No ☐   Is each portable tank and all piping, valves, and accessories visually inspected at intervals not to exceed $2\frac{1}{2}$ years?

Yes ☐   No ☐   Are inspections conducted by the owner, agent, or approved agency?

Yes ☐   No ☐   On insulated tanks, is the insulation not be removed if, in the opinion of the person performing the visual inspection, external corrosion is likely to be negligible?

Yes ☐   No ☐   If evidence of any unsafe condition is discovered, is the portable tank not be returned to service until it meets all corrective standards?

FIGURE 10.11   Compressed gas cylinder (CGC) safety checklist.

10.10.1   BASIC SAFETY

Some of the basic safety, health rules, and procedures that should be followed when using compressed gas cylinders of any type are as follows:

- Select the least hazardous gas.
- Purchase only the necessary quantities.
- Select gases with returnable containers.
- When receiving gas cylinders:
  - Check for leaks.
  - Visually inspect the cylinder for damage.

- Ensure the valve cover and shipping cap is on.
- Check for proper labeling.
- If a cylinder is damaged, in poor condition, leaking, or the contents are unknown, contact your cylinder vendor. Have the vendor return the damaged cylinder to the manufacturer.
- Wear appropriate foot protection when engaged in moving or transporting cylinders:
  - Sturdy shoes are a minimum.
  - Steel-toed shoes if required by your supervisor or department.
- Proper personal protective clothing and PPE is to be worn.
- Always have an appropriate MSDS available and be familiar with the health, flammability, and reactivity hazards for the particular gas.

10.10.2 Things Not to Do

- Never roll a cylinder to move it.
- Never carry a cylinder by the valve.
- Never leave an open cylinder unattended.
- Never leave a cylinder unsecured.
- Never force improper attachments on to the wrong cylinder.
- Never grease or oil the regulator, valve, or fittings of an oxygen cylinder.
- Never refill a cylinder.
- Never use a flame to locate gas leaks.
- Never attempt to mix gases in a cylinder.
- Never discard pressurized cylinders in the normal trash.

11 Controls and PPE

Safety toed shoes, safety eyewear, head protection, ear protection, and hand protection are all forms of personal protective equipment.

11.1 HAZARD PREVENTION AND CONTROLS

The Occupational Safety and Health Administration (OSHA) require employers to protect their employees from workplace hazards such as machines, work procedures, and hazardous substances that can cause injury or illnesses. It is known from past practices and situations that something must be done to mitigate or remove hazards from the workplace. Actions taken often create other hazards, which had not existed before attempting to address the existing hazard.

Many companies have suggestion programs where workers receive rewards for suggestions that are implemented. It is no surprise that the person who often has the best ideas is the one who suffers most from that particular hazard. It is a sound management process to involve those who are impacted most in decision-making processes.

Several methods have been used over the years to control hazards and these can be segregated into five categories. The preferred methods are engineering controls, awareness devices, predetermined safe work practices, and administrative controls. When these controls are not feasible or do not provide sufficient protection, an alternative or supplementary method of protection is to provide workers with personal protective equipment (PPE) and the know-how to use it properly.

11.2 ENGINEERING CONTROLS

When a hazard is identified in the workplace, every effort should be made to eliminate it so that employees are not harmed. Elimination may be accomplished by designing or redesigning a piece of equipment or process. This could be the installation of a guard on a piece of machinery, which prevents workers from contacting the hazard. The hazard can be engineered out of the operation. Another way to reduce or control the hazard is to isolate the process, such as in the manufacture of vinyl chloride used to make such items as plastic milk bottles, where the entire process becomes a closed circuit. This will result in no one being exposed to vinyl chloride gas, which is known to cause cancer. Thus, any physical controls which are put in place are considered to be the best approach from an engineering perspective. Keep in mind that you are a consumer of products. Thus, at times you can leverage the manufacturer to implement safeguards or safety devices on products that you are looking to purchase. Let your vendor do the engineering for you or do not purchase their product. This may not always be a viable option. To summarize the engineering controls that can be used, the following may be considered:

- Substitution
- Elimination
- Ventilation
- Isolation
- Process or design change

11.3 AWARENESS DEVICES

Awareness devices are linked to the senses. They are warning devices, which can be heard and seen. They act as alerts to workers, but create no type of physical barrier. They are found in most workplaces and carry with them a moderate degree of effectiveness. Such devices are as follows:

- Backup alarms
- Warning signals both audible and visual
- Warning signs

11.4 WORK PRACTICES

Work practices are the means by which a job task or activity is done. This may mean that you create a specific procedure for completing the task or job. It may also mean that you implement special training for a job or task. It also presupposes that you might require inspection of the equipment or machinery before beginning work or when a failure has occurred. An inspection should be done before restarting the process or task. A lockout/tagout procedure may also be required to create a zero potential energy release.

11.5   ADMINISTRATIVE CONTROLS

A second approach is to control the hazard through administrative directives. This may be accomplished by rotating workers, which allows you to limit their exposure, or having workers only work in areas when no hazards exist during that part of their shift. This applies particularly to chemical exposures and repetitive activities that could result in ergonomic related incidents. Examples of administrative controls are as follows:

- Requiring specific training and education
- Scheduling off-shift work
- Worker rotation

11.5.1   MANAGEMENT CONTROLS

Management controls are needed to express the company's view of hazards and their response to hazards that have been detected. The entire program must be directed and supported through the management controls. If management does not have a systematic and set procedure for addressing the control of hazards in place, the reporting/identifying of hazards is a waste of time and money. This goes back to the policies and directives and the holding of those responsible accountable by providing them with the resources (budget) for correcting and controlling hazards. Some aspects of management controls are as follows:

- Policies
- Directives
- Responsibilities (line and staff)
- Vigor and example
- Accountability
- Budget

The attempt to identify the worksite hazards and address them should be an integral part of your management approach. If the hazards are not addressed in a timely fashion, they will not be identified or reported. If money becomes the main criterion for not fixing or controlling hazards, your workforce will lose interest in identifying and reporting them.

11.6   PERSONAL PROTECTIVE EQUIPMENT

Personal protective equipment includes a variety of devices and garments to protect workers from injuries. You can find PPE designed to protect eyes, face, head, ears, feet, hands and arms, and the whole body. PPE includes such items as goggles, face shields, safety glasses, hard hats, safety shoes, gloves, vests, earplugs, earmuffs, and suits for full body protection.

11.6.1 HAZARD ASSESSMENT

Recent regulatory requirements make hazard analysis/assessment part of the PPE selection process. Hazard analysis/assessment procedures shall be used to assess the workplace to determine if hazards are present, or are likely to be present, which may necessitate the use of PPE. As part of this assessment, employees' work environment is to be examined for potential health and physical hazards. If it is not possible to eliminate workers' exposure or potential exposure to the hazard through the efforts of engineering controls, work practices, and administrative controls, then the proper PPE will need to be used. The hazard assessment certification form found in Appendix C may be of assistance in conducting a hazard analysis/assessment.

When employees must be present and engineering or administrative controls are not feasible, it will be essential to use PPE as an interim control and not a final solution. For example, safety glasses may be required in the work area. Far too often, in the scheme of hazard control PPE usage is considered as a last resort. PPE can provide added protection to the employee even when the hazard is being controlled by other means. There are drawbacks to the use of PPE and they are as follows:

- Hazard still looms
- Protection dependent upon worker using PPE
- PPE may interfere with performing task and productivity
- Requires supervision
- Is an ongoing expense

Many forms of PPE need to be addressed and required when a hazard assessment determined that PPE is the only option left for protecting the workforce. PPE includes the following:

- Eye and face protection (29 CFR 1910.133)
- Respiratory protection (29 CFR 1910.134)
- Head protection (29 CFR 1910.135)
- Foot and leg protection (29 CFR 1910.136)
- Electrical protective equipment (29 CFR 1910.137)
- Hand protection (29 CFR 1910.138)
- Respiratory protection of tuberculosis (29 CFR 1910.139)

Any other types of specialized protective equipment needed would be identified as part of the hazard assessment. Such equipment might include body protection for hazardous materials, protective equipment for material handling, protection for welding activities, or protection from exposure to biological agents.

11.6.2 ESTABLISHING A PPE PROGRAM

A PPE program sets out procedures for selecting, providing, and using PPE as part of an organization's routine operation. A written PPE program, although not mandatory, is easier to establish and maintain than a company policy and easier to evaluate than an unwritten one. To develop a written program you should consider including the following elements or information:

- Identify steps taken to assess potential hazards in every employees' workspace and in workplace operating procedures.
- Identify appropriate PPE selection criteria.
- Identify how you will train employees on the use of PPE, including the following:
  - What PPE is necessary.
  - When is PPE necessary.
  - How to properly inspect PPE for wear and damage.
  - How to properly put on and adjust the fit of PPE.
  - How to properly take off PPE.
  - Limitations of PPE.
  - How to properly care for and store PPE.
- Identify how you will assess employee understanding of PPE training.
- Identify how you will enforce proper PPE use.
- Identify how you will provide for any required medical examinations.
- Identify how and when to evaluate the PPE program.

Finally, use PPE for potentially dangerous conditions. Use gloves, aprons, and goggles to avoid acid splashing. Wear earplugs for protection from high noise levels and wear respirators to protect against toxic chemicals. The use of PPE should be the last consideration in eliminating or reducing the hazards the employee is subjected to because PPE can be heavy, awkward, uncomfortable, and expensive to maintain. Therefore, try to engineer the identified hazards out of the job.

11.7 RANKING HAZARD CONTROLS

In determining which hazard control procedures have the best chance of being effective, it is useful to rank them along a continuum. The five hazard controls that were espoused in the earlier part of this chapter are ranked in Figure 11.1. This should assist you in determining, which control, if you have a choice of more than one, would be most effective for your purposes. The ranking goes from most effective to least effective.

11.8 PPE POLICIES

Companies should have policies regarding PPE and clothing that is appropriate for work. Companies must have policies regarding hair length/style and wearing of protective equipment when it impedes upon proper wearing and use. Disciplinary

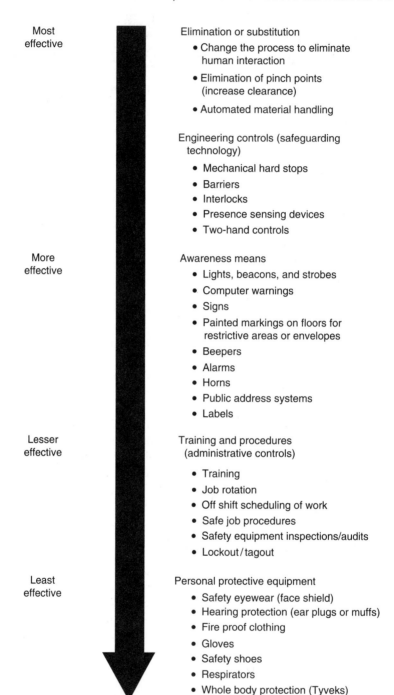

Most effective

Elimination or substitution
- Change the process to eliminate human interaction
- Elimination of pinch points (increase clearance)
- Automated material handling

Engineering controls (safeguarding technology)
- Mechanical hard stops
- Barriers
- Interlocks
- Presence sensing devices
- Two-hand controls

More effective

Awareness means
- Lights, beacons, and strobes
- Computer warnings
- Signs
- Painted markings on floors for restrictive areas or envelopes
- Beepers
- Alarms
- Horns
- Public address systems
- Labels

Lesser effective

Training and procedures (administrative controls)
- Training
- Job rotation
- Off shift scheduling of work
- Safe job procedures
- Safety equipment inspections/audits
- Lockout/tagout

Least effective

Personal protective equipment
- Safety eyewear (face shield)
- Hearing protection (ear plugs or muffs)
- Fire proof clothing
- Gloves
- Safety shoes
- Respirators
- Whole body protection (Tyveks)

FIGURE 11.1   Ranking hazard controls.

measures should be set and be taken if employees fail to adhere to appropriate dress regulations and requirements for wearing PPE.

11.8.1 SAFETY CLOTHING

For some jobs, ordinary clothing—clean, in good repair, and suited to the work involved—may be considered "safe." A few items are important such as the following:

- Good fit is important.
- Trousers should not be too long.
- Cuffs should never be worn while performing operations that produce flying embers, sparks, or other harmful matter that might get caught.
- Loose-fitting garments should be avoided.

Safety precautions regarding regular work clothes should include the following:

- Do not wear anything that could get caught in a machine
- No neckties or scarves that go around the neck
- No wristwatches with buckle or clamp-on bands
- No rings, necklaces, bracelets, or dangling earrings
- No shirts with dangling cuffs or tails
- No gloves around rotating machinery
- No clothing soaked in oil or flammable solvent

Do not buy poorly made or low-quality work clothes; well-made clothes may cost a little more, but they fit better, last longer, and are safer and more comfortable.

Replace or repair torn or worn-out items immediately. Keep your work clothes clean; dusty or dirty clothes can cause skin rash and irritation. Do not cut corners by wearing old dress shoes for work; well-built shoes in good condition with sensible heels are safer, and they cut down on fatigue.

For jobs involving exposure to fire, extreme heat, molten metal, corrosive chemicals, cold temperatures, cuts from handling materials, etc., special protective clothing may be required.

11.8.2 SPECIAL PROTECTIVE CLOTHING

Many types of protective clothing are available and used to protect against certain hazards such as the following:

- Aluminized and reflective clothing: Reflects radiant heat
- Flame-resistant cotton fabric: Often used as hair covering for people who work near sparks and open flames
- Impervious materials (rubber, neoprene, vinyl, etc.): Protect against dust, vapors, mists, moisture, and corrosives
- Leather clothing: Protects against light impact, sparks, molten metal splashes, and infrared and ultraviolet radiation

- Synthetic fibers (Orlon, Dynel, Vinyon, etc.): Resist acids, many solvents, mildew, abrasion and tearing, and repeated launderings
- Water-resistant duck: Protects from water and noncorrosive liquids
- Glass fiber: Used in multilayer construction to insulate clothing

11.8.3 SAFETY AND HAIR PROTECTION

Head guards, caps, nets, etc., are designed to keep hair from catching in machinery. Flame-resistant headgear should be worn for work around sparks or flames. Hair length poses additional problems with regard to safety equipment such as the following:

- Men with extremely long, heavy sideburns may find it difficult to get a proper fit when wearing hearing protection earmuffs.
- Large, bushy mustaches and beards can interfere with the proper fitting of respiratory equipment and breathing apparatus.
- Long or bushy hair may make it difficult to wear a safety hat.

11.8.4 PROTECTING THE HEAD

Head protection is needed by all employees engaged in occupations that pose special hazards to the head. These hazards are as follows:

- Falling objects
- Flying particles
- Electric shock
- Overhead spills of chemicals, acid, or hot liquids

Some of the particularly high-risk industries are tree trimming, construction work, shipbuilding, logging, mining, overhead line construction or maintenance, and metal or chemical production.

Types of protective headgear and what they are designed to protect the head from are as follows:

- Safety helmets or hard hats with full brim from most of the major hazards regarding the head (Figure 11.2).
- Bump caps—for use where a brim might get in the way; in confined spaces where exposure is limited to bumping; should never be worn where there is exposure to more serious hazards.
- Hair protection caps—for use by employees with long hair who work around chains, belts, or other machines.

Some of the key design features are the suspension that maintains the distance between the top of a head and the helmet shell is known as the "crown clearance"; it determines the amount of protection offered against impact and penetration. A suspension that is too rigid can transmit the shock of impact and fracture the neck vertebrae. A suspension that is too flexible permits contact with the head upon

FIGURE 11.2   Hardhat with ear protection and chin strap.

impact, causing skull fracture or concussion. A damaged or worn suspension should be replaced immediately.

A chin strap: made of leather, fabric, or elastic; prevents the hat from falling off or being blown off. During cold weather a liner for warmth can be worn under the hard hat. Also, an eye shield and hearing protection muffs may be attached if required.

Some of the safety precautions that should be followed regarding hard hat protection are as follows:

- Never leave a safety helmet on the rear window shelf of an auto or truck; sunlight may affect its protective quality, and an emergency stop can turn the helmet into a dangerous missile.
- Never keep anything under the safety hat between crown and suspension such as personal items (wallet); it interferes with the suspension.
- Clean the hat and suspension regularly (at least every 30 days).
- Never attempt to repair the shell of a hat once it has been broken or punctured.
- Never drill holes in a safety hat to "improve ventilation" or cut notches in the brim.
- Replace a damaged helmet immediately.
- If the hat is giving you a headache, make sure it is fitted properly.
- Never remove the suspension for any reason.

11.8.5   PROTECTING THE EYES AND FACE

Industrial eye injuries occur at a rate of 2/min and are the costliest in terms of lost production and earning power. Of the more than 1000 industrial eye injuries that occur every working day, over 90% of them are needless and preventable. The primary causes of on-the-job eye injuries are as follows:

- Flying objects (especially those set in motion by hand tools)
- Abrasive wheels (small flying particles)
- Fragments from hammering or sawing
- Corrosive substances
- Injurious light or heat rays
- Splashing metal
- Poisonous gas or fumes

To prevent injury to the eyes workers should wear the proper eye protection such as any of the following:

- Cover goggles
- Protective spectacles/safety glasses
- Protective spectacles with side shields
- Chemical or splash-resistant goggles
- Dust goggles
- Melters' goggles
- Welders' goggles

Make sure the goggles or glasses are comfortable and properly fitted. Fitting, adjusting, and maintaining eye equipment is a part of the wearing process. At times the use of defogging materials helps. Protective eyewear should be cleaned regularly and the use of sweatbands can be helpful.

Today with the variety of styles there are no excuses for failing to wear safety goggles. Get them adjusted or refitted by a professional. Clean protective eyewear regularly; keep them in a case or a place where they would not get scratched. Secure adjustable suspension to make the fit more accommodating on a daily basis; see an eye doctor if headaches or discomfort persist. An employer should not accept any excuse for not wearing protective eyewear.

Do not wear contact lenses where there are considerable amounts of dust, smoke, irritating fumes, or liquid irritants that could splash into the eyes. Never wear contact lenses as a substitute for protective eye equipment. If you must wear contact lenses on the job, get written authorization from your eye doctor and wear safety goggles over your contacts.

Wear your safety goggles at home too (when using power tools, spray painting, etc.). Know the appropriate first-aid measures for eye emergencies; for example, flushing eye with water if a chemical has been splashed.

Face protection is most needed when the following hazards to the face and neck are present:

- Flying particles
- Sprays of hazardous liquids
- Splashes of molten metals
- Hot solutions

Face protective equipment such as face shields should be used when sawing or buffing, sanding or light grinding, and handling chemicals, and helmets must be worn when working with molten metal and radiation. Handheld shield can be used for inspection work, tack welding, etc. At times the use of acid-proof hoods with corrosive chemicals or hoods with air supply for toxic fumes, dusts, mists, and gases may be used as part of face protection.

11.8.6 Ear Protection

Hearing protection is needed when

- Noise standards are exceeded or the company noise levels measurement exceeds acceptable levels.
- Engineering controls currently in use to decrease noise levels (acoustical enclosures, etc.) are not effective.
- Audiometric testing program determines a worker's condition has worsened since the last testing.
- Sources of "noise pollution" in the working environment have remained undetected and effectiveness of existing engineering controls has not been sufficient.

There are two types of hearing protection: aural type—placed inside the ear canal; and super-aural type—sealing the external edge of the ear canal. Rubber, plastic, or wax is used most commonly and cotton offers no protection.

The importance of proper fit is essential when it comes to hearing protection. Some of the procedures that must be considered for fitting are as follows:

- Possible discomfort if points of pressure develop.
- Good seal cannot be obtained without some initial discomfort.
- There should be no lasting problems if earplugs are made of soft material and kept clean.

Complaint that earplugs make it difficult to hear conversation. (Tests show that when noise level is higher than 85 dB, speech is more easily understood with earplugs in place than without them.)

Muff-type protectors' cup or muff covers the external ear to provide an acoustical barrier. Liquid or grease-filled cushions give better noise suppression than plastic or foam rubber types, but may present leakage problems. Head size and shape also affect noise suppression (Figure 11.3).

Helmet protectors completely surround the head. Suppression of sound is achieved through the acoustical properties of the helmet. Cost and bulk normally preclude use of helmet for most jobs.

Commercially available hearing protection is very effective if properly fitted and used; earplugs generally reduce the amount of noise reaching the ear by 25–30 dB

FIGURE 11.3   The use of ear muffs and safety eyewear with side shields.

in the higher frequencies, which are the most harmful. The better type of earmuffs may reduce noise by an additional 10–15 dB. A Combination of earplugs and earmuffs gives an additional 3–5 dB of noise protection.

11.8.7   PROTECTING YOUR HANDS

Certain mechanical actions can trap hands and cause serious injuries or amputations. The following are five such actions that can damage hands and fingers:

- Shearing—Examples include ordinary scissors, guillotine cutters, cleavers, axes, knives, screw or worn conveyors, any two hard-edged objects that pass close together. Keep hands and fingers away from any tight places that can present slicing hazards.
- Rotating (spinning motion may have a horizontal or vertical axis, or it may be at an angle)—Examples include rotary saw, fan blades, lathe, and power drills. Watch out for toothed, spike, or jagged edges that can slash into fingers.
- In-running nip from any parallel wheels, rollers, or shafts turning inward together—Examples include gears, belt and pulley, rack and pinion, chain and sprocket. Learn to recognize and stay clear of the grabbing power of the in-running nip.
- Puncturing from any device or tool that can penetrate flesh if it slips or goes out of control—Examples include screwdriver, awl, and knife points. Remember that a puncture can be twice as dangerous as a superficial cut because it carries the threat of deep infection; get first aid right away.
- Smashing—Examples include hammer, factory presses, "pinch points."

Watch out when putting down heavy objects, moving loads through doorways, and maneuvering drums and cylinders.

Gloves are the most common hand and finger protectors. There are many different types of gloves and each has its own unique benefits. Some examples of types of gloves are as follows:

- Heat resistant gloves that protect against burns and discomfort when hands are exposed to heat.
- Metal mesh gloves, used by those who work with knives, protect against cuts and blows from sharp objects. Kevlar gloves offer some degree of cut resistance as shown in Figure 11.4.
- Rubber gloves are worn by electricians to keep hands insulated from shock.
- Rubber, neoprene, and vinyl gloves are used when handling chemicals and corrosives. These gloves should be selected from a glove chart to protect against the specific chemical being used.
- Leather gloves resist sparks, moderate heat, protect from sharp edges, and cushion against blows.
- Chrome-tanned cowhide gloves with a steel-stapled leather patch on palms and fingers are often used in foundries and steel mills.
- Cotton or fabric gloves provide suitable protection against dirt, chafing, or abrasion.
- Coated fabric gloves protect against moderately concentrated chemicals.
- Hand leathers or hand pads are often better than gloves for protection against heat, abrasion, and splinters.

FIGURE 11.4 Cut resistant Kevlar gloves.

Some basic safety precautions that can help further protect hand, finger, and arms are as follows:

- Never try to "cheat" a guard or any safety device.
- Disconnect power to clean, oil, or adjust a machine; lock it out if work rules require it.
- Remove rings, watchbands, bracelets, etc., when working with machinery.
- Use gloves to protect your hands from chemicals and rough objects; but never wear them around moving machinery.
- Use the right tools and keep your hands out of tight places.
- Use a brush or hook—not bare hands—to clear away filings or shavings from work areas.
- If an accident occurs, get first aid right away.
- Remember that a puncture can be twice as dangerous as a superficial cut because it carries the threat of deep infection; get first aid right away.

11.8.8 PROTECTING THE FEET AND LEGS

About a quarter of a million disabling occupational foot injuries occur each year.

A Bureau of Labor Statistics study conducted in 1981 revealed that over 75% of foot injuries happened to workers who were not wearing safety shoes. The major causes of foot related accidents are as follows:

- 60% from falling objects
- 16% from stepping on a sharp object
- 13% from feet being struck by rolling objects

The types of safety shoes and their uses are as follows:

- Metal-free shoes, boots, etc., are used where there are specific electrical hazards or fire and explosion hazards.
- Gaiter-type shoes protect people from splashes of molten metal or welding sparks.
- Shoes with reinforced soles or innersoles of flexible metal are worn where there are hazards from protruding nails, etc.
- Rubber boots and shoes, leather shoes with wooden soles, or wood-soled sandals are used for wet work conditions.
- Safety shoes with metatarsal guards are worn for operations involving the handling of heavy materials (pig iron, heavy castings, timber, etc.).

Workers do not wear safety shoes, needlessly exposing themselves to injury and disablement, because they complain that they are hot, heavy, and uncomfortable. Many safety shoes nowadays are as comfortable, practical, and attractive as ordinary street shoes. The steel cap weighs about as much as a wristwatch. The toe box is insulated with felt to keep the feet from getting too hot or cold. At times workers object to wearing safety shoes because the steel caps do not cover the smallest toes.

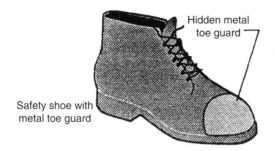

FIGURE 11.5   Example of safety toed shoes. (Courtesy of the Department of Energy.)

Studies show that 75% of all toe fractures happen to the first and second toes. In most accidents, the toe box takes the load of the impact for the entire front part of the foot. Also, workers are afraid that if the toe box were crushed, the steel edge would cut off their toes. In reality, accidents of this type are rare; in the majority of cases, safety shoes give sufficient protection. A blow that would crush the toe cap would certainly smash one's toes since the toe cap is designed to withstand approximately 2500 lb of force (Figure 11.5).

Leg protection, such as leggings that encircle the leg from ankle to knee and have a flap at the bottom to protect the instep, are worn to protect the entire leg. These should be easily removed in case of emergency. Shin guards made of hard fiber or metal are worn to protect the shins against impact. Knee pads protect employees whose work requires a great deal of kneeling, such as cement finishing or tile setting. Ballistic nylon pads are used to protect the thighs and upper legs against injury from chain saws.

11.8.9   RESPIRATORY PROTECTION

It is desirable to wear respiratory protection when air contaminants range from relatively harmless substances to toxic dusts, vapors, mists, fumes, and gases that may be extremely harmful. Standards specify "safe levels" of certain airborne contaminants and they are exceeded. Ideally, these safe levels can be achieved through engineering controls. When engineering controls are not technically feasible, or when the hazardous operation is performed only infrequently (making these controls impractical), respiratory protection is needed. Respiratory equipment should also be regarded as emergency equipment (e.g., in cases of leaks and breakdowns).

While selecting the proper respiratory equipment, the following factors should be taken into consideration:

- Nature of the hazardous operation or process
- Type of air contaminant, including its physical properties, chemical properties, physiological effects on the body, and its concentration
- Period of time for which respiratory protection must be provided
- Location of the hazard with respect to a source of uncontaminated air
- Employee's state of health
- Functional and physical characteristics of the various respiratory devices

Using the wrong kind of respirator for the hazard involved can be dangerous: For example, particulate filter respirators are of no value as protection against solvent vapors, injurious gases, or lack of oxygen.

The types of respirators and their uses commonly dictate the appropriate respirator. Some the most common respirators are as follows:

- Air-purifying respirators remove contaminants from the air being inhaled. Examples of air-purifying respirators are gas masks, chemical cartridge, respirators, particulate filter respirators, and combination respirators (Figure 11.6).
- Air supplied respirators deliver breathing air through a supply hose connected to the wearer's facepiece. Some of the varieties of this type of respirator are hose masks, air line respirators, abrasive blasting respirators, and air supplied suits and hoods.
- Self-contained breathing apparatus (SCBA) devices afford complete respiratory protection in any toxic or oxygen-deficient atmosphere. The types of SCBAs are oxygen re-breathing, self-generating, demand, and pressure-demand.

Care of respiratory devices is important since a worker's life may depend upon the proper function and use of the respirator. It is important that the following practices be followed:

- Schedule for cleaning and repair of respirators
- Inspection procedures and schedule
- Methods of disinfection
- Preventive maintenance steps

It is important that workers keep respirators on at all times when working in a contaminated atmosphere.

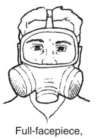 

Full-facepiece,   Half-mask, facepiece-
dual cartridge   mounted cartridge

FIGURE 11.6 Examples of air-purifying respirators. (Courtesy of the National Institute for Occupational Safety and Health.)

11.9 SUMMARY

It is the employers' responsibility to provide the employee with the required PPE. It is the employee's responsibility to take care and keep the PPE clean and assure that it is in good shape. If it is not, it should be returned to the employer for repair or replacement.

12 Emergencies

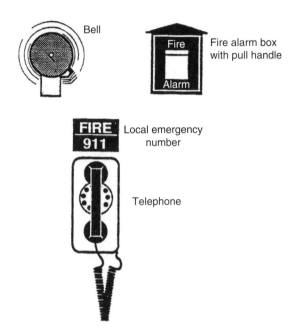

Bell

Fire alarm box with pull handle

Local emergency number

Telephone

Responses to emergencies should be planned in advance and alarms and warning devices in place. (Courtesy of the Department of Energy.)

It is doubtful whether any industry can do away with emergency planning. Wherever there are workers there is always potential for human emergencies such as injury, illness, death, violence, and medical emergencies. The weather and environment can also cause natural emergencies such as storms, floods, earthquakes, and tornados. Man-made emergencies, such as fire, explosion, or chemical spills, cannot be overlooked either. The key is to be the best possibly prepared to react to any emergency. This requires a number of issues to be addressed.

12.1 IDENTIFICATION OF HAZARDS

When looking at what hazards exist in a workplace, it is imperative that a worst-case scenario approach be employed. It is virtually impossible to address each possible hazard, but each industry has areas where it is most vulnerable or most at risk of an unplanned emergency. A risk assessment will allow prioritizing the potential risk. At this point an action plan can be developed to address the hazards that have been identified. Some of the common hazards that might be identified as having an impact on the workplace are as follows:

- Fire
- Explosion
- Natural gas leak
- Chemical spill
- Release of radioactive materials
- Airborne chemical or biological releases
- Power outage
- Loss of communications
- Water leak
- Flooding
- Earthquakes
- Winter storms
- Windstorms
- Hurricanes
- Tornados
- Security issues
- Bomb threat
- Suspicious letter or package
- Civil strife
- War
- Sabotage
- Labor strike
- Accidents (injuries, illnesses, and deaths)
- Mechanical failure
- Transportation incidents (truck, rail, or air)
- Workplace violence

The best way is to prepare to respond to an emergency before it happens. Few people can think clearly and logically in a crisis, so it is important to do this in advance, when you have time to be thorough.

12.2 EMERGENCY ACTION PLANS

An emergency action plan (EAP) covers designated actions employers and employees must take to ensure employee safety from fire and other emergencies. Not all employers are required to establish an EAP.

If an employer has less than 10 employees, the plan can be communicated orally. If, on the other hand, an employer has more than 10 employees, the plan must be written, kept in the workplace, and available for employee review.

It would be unusual for retail, wholesale, or warehousing sectors to not have to comply with portable fire extinguisher regulations (1910.157) or use a fixed fire suppression system. If an employer is regulated as above then the employer would

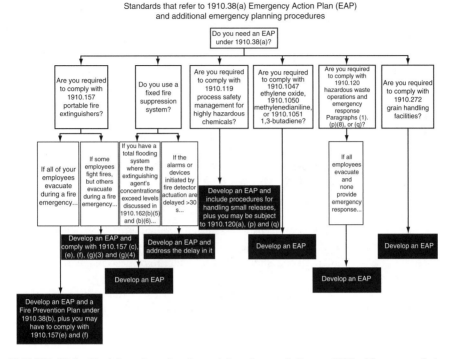

FIGURE 12.1 Decision chart for determining the need for an EAP. (Courtesy of the Occupational Safety and Health Administration.)

need to develop an EAP. Figure 12.1 provides a decision chart for determining if an employer needs an EAP.

When developing your EAP, it is a good idea to look at a wide variety of potential emergencies that could occur in your workplace. It should be tailored to your worksite and include information about all potential sources of emergencies. Developing an EAP means you should do a hazard assessment to determine what, if any, physical or chemical hazards in your workplaces could cause an emergency. If you have more than one worksite, each site should have an EAP.

12.2.1 ELEMENTS OF AN EAP

At a minimum, your EAP must include the following:

- A preferred method for reporting fires and other emergencies
- An evacuation policy and procedure including type of evacuation and exit route assignment
- Procedures to account for all employees after evacuation
- Emergency escape procedures and route assignments, such as floor plans, workplace maps, and safe or refuge areas

- Names, titles, departments, and telephone numbers of individuals both within and outside your company to contact for additional information or explanation of duties and responsibilities under the emergency plan
- Procedures for employees who remain to perform or shut down critical plant operations, operate fire extinguishers, or perform other essential services that cannot be shut down for every emergency alarm before evacuating
- Alarm system to alert workers
- Rescue and medical duties for any workers designated to perform them

In addition, although they are not specifically required by Occupational Safety and Health Administration (OSHA), you may find it helpful to include in your plan the following:

- Designate and train employees to assist in a safe and orderly evacuation of other employees, including those who have handicaps
- The site of an alternative communications center to be used in the event of a fire or explosion
- A secure on- or offsite location to store originals or duplicate copies of accounting records, legal documents, your employees' emergency contact lists, and other essential records

The EAP needs to be reviewed with each employee trained on it when the plan is developed and when an employee is assigned initially to a job, plan is changed, or employees have specific responsibilities under the plan.

12.3 ALARM SYSTEM

Your plan must include a way to alert employees, including disabled workers, to evacuate or take alternative action, and how to report emergencies, as required. Among the steps you must take are the following:

- Make sure alarms are distinctive and recognized by all employees as a signal to evacuate the work area or perform actions identified in your plan.
- Make available an emergency communications system such as a public address system, portable radio unit, or other means to notify employees of the emergency and to contact local law enforcement, the fire department, and others.
- Stipulate that alarms must be audible, seen, or otherwise perceived by everyone in the workplace. It might be good to consider providing an auxiliary power supply in the event of a power breakdown. (29 CFR 1910.165(b)(2) offers more information on alarms.)

The following may also be considered, although it is not specifically required by OSHA:

FIGURE 12.2   Both an audible and visual alarm.

- Using tactile devices to alert employees who would not otherwise be able to recognize an audible or visual alarm (Figure 12.2)
- Providing an updated list of key personnel such as the manager or physician, in order of priority, to notify in the event of an emergency during off-duty hours

12.4   EVACUATION PLAN AND POLICY

A disorganized evacuation can result in confusion, injury, and property damage. That is why when developing your EAP it is important to determine the following:

- Conditions under which an evacuation would be necessary.
- A clear chain of command and designation of the person in your business authorized to order an evacuation or shutdown. You may want to designate an "evacuation warden" to assist others in an evacuation and to account for personnel.
- Specific evacuation procedures, including routes and exits. Post these procedures where they are easily accessible to all employees.
- Procedures for assisting people with disabilities or who do not speak English.
- Designation of what, if any, employees will continue or shut down critical operations during an evacuation. These people must be capable of recognizing when to abandon the operation and evacuate themselves.
- A system for accounting for personnel following an evacuation. Consider employees' transportation needs for community-wide evacuations.

In the event of an emergency, local emergency officials may order you to evacuate your premises. In some cases, they may instruct you to shut off the water, gas, and electricity. If you have access to radio or television, listen to newscasts to keep informed and follow whatever official orders you receive.

In other cases, a designated person within your business should be responsible for making the decision to evacuate or shut down operations. Protecting the health and safety of everyone in the facility should be the first priority. In the event of a fire, an immediate evacuation to a predetermined area away from the facility is the best way to protect employees. On the other hand, evacuating employees may not be the best response to an emergency such as a toxic gas release at a facility across town from your business.

12.5 EMERGENCY RESPONSIBILITY

When drafting your EAP, you may wish to select a responsible individual to lead and coordinate the emergency plan and evacuation. It is critical that employees know who the coordinator is and understand that person has the authority to make decisions during emergencies. The coordinator should be responsible for the following:

- Assessing the situation to determine whether an emergency is present that requires activation of your emergency procedures
- Supervising all efforts in the area, including evacuating personnel
- Coordinating outside emergency services, such as medical aid and local fire departments, and ensuring that they are available and notified when necessary
- Directing the shutdown of work operations when required

It may also be beneficial to coordinate the action plan with other employers when several employers share the worksite, although OSHA standards do not specifically require this.

In addition to a coordinator, it could be a good idea to designate evacuation wardens to help move employees from danger to safe areas during an emergency. Generally, 1 warden for every 20 employees should be adequate, and the appropriate number of wardens should be available at all times during working hours.

Employees designated to assist in emergency evacuation procedures should be trained in the complete workplace layout and various alternative escape routes. All employees and those designated to assist in emergencies should be made aware of employees with special needs who may require extra assistance, how to use the buddy system, and hazardous areas to avoid during an emergency evacuation.

12.6 EXIT ROUTES

Usually, a workplace must have at least two exit routes for prompt evacuation. But more than two exits are required if the number of employees, size of the building, or arrangement of the workplace will not allow a safe evacuation.

Exit routes must be located as far away as practical from each other in case one is blocked by fire or smoke.

12.6.1 REQUIREMENTS FOR EXITS

Exits must be separated form the workplace by fire-resistant materials—that is, 1 h fire-resistance rating if the exit connects three or fewer stories, and a 2 h fire-resistance rating if the exit connects more than three stories.

Exits can have only those openings necessary to allow access to the exit from occupied areas of the workplace or to the exit discharge. Openings must be protected by a self-closing, approved fire door that remains closed or automatically closes in an emergency. Maintain a line-of-sight to exit signs clearly visible always and install "Exit" signs using plainly legible letters (Figure 12.3).

12.6.2 SAFETY FEATURES FOR EXIT ROUTES

The safety features exit routes are as follows:

- Keep exit routes free of explosives or highly flammable furnishings and other decorations.
- Arrange exit routes so employees will not have to travel toward high-hazard areas unless the path of travel is effectively shielded from the high-hazard area.
- Ensure the exit routes are free and unobstructed by materials, equipment, locked doors, or dead-end corridors.
- Provide adequate lighting for exit routes for employees with normal vision.

FIGURE 12.3   A properly labeled exit.

- Keep exit route doors free of decorations or signs that obscure the visibility of exit doors.
- Post signs along the exit access indicating the direction of travel to the nearest exit and exit discharge if that direction is not immediately apparent.
- Mark doors and passages along an exit access that could be mistaken for an exit "Not an exit" or with a sign identifying its use such as "Closet."
- Renew fire-retardant paints or solution when needed.
- Maintain exit routes during construction, repairs, or alterations.

12.6.3 DESIGN AND CONSTRUCTION OF EXITS

The requirements for design and construction of exits are as follows:

- Exit routes must be permanent parts of the workplace.
- Exit discharges must lead directly outside or to a street, walkway, refuge area, public way, or open space with access to the outside.
- Exit discharge areas must be large enough to accommodate people likely to use the exit route.
- Exit route doors must unlock from the inside. They must be free of devices or alarms that could restrict use of the exit route if the device or alarm fails.
- Exit routes can be connected to rooms only by side-hinged doors, which must swing out in the direction of travel if the room may be occupied by more than 50 people.
- Exit routes must support the maximum permitted occupant load for each floor served, and the capacity of an exit route may not decrease in the direction of exit route travel to the exit discharge.
- An exit access must have ceilings at least 7 ft 6 in. high.
- An exit access must be at least 28 in. wide at all points. Objects that project into the exit must not reduce its width (Figure 12.4).

12.7 ACCOUNTING FOR EVACUEES

Accounting for all employees following an evacuation is critical. Confusion in the assembly areas can lead to delays in rescuing anyone trapped in the building, or unnecessary and dangerous search-and-rescue operations. To ensure the fastest, most accurate accountability of your employees, you may want to consider including the following steps in your EAP:

- Designate assembly areas where employees should gather after evacuation.
- Take a head count after the evacuation. Identify the names and last known locations of anyone not accounted for and pass them to the official in charge.
- Establish a method for accounting for non-employees such as suppliers and customers.
- Establish procedures for further evacuation in case the incident expands. This may consist of sending employees home by normal means or providing them with transportation to an offsite location.

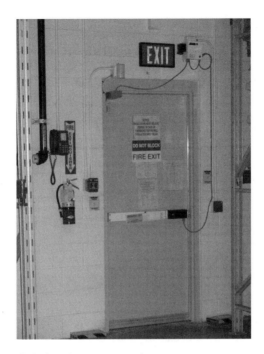

FIGURE 12.4 A well-designed emergency exit.

12.8 RESCUE OPERATIONS

It takes more than just willing hands to save lives. Untrained individuals may endanger themselves and those they are trying to rescue. For this reason, it is generally wise to leave rescue work to those who are trained, equipped, and certified to conduct rescues.

12.9 MEDICAL CARE IN EMERGENCIES

If your company does not have a formal medical program, investigate ways to provide medical and first-aid services. If medical facilities are available near your worksite, make arrangements for them to handle emergency cases. Provide your employees with a written emergency medical procedure to minimize confusion during an emergency.

If an infirmary, clinic, or hospital is not in close proximity to the workplace, ensure that onsite persons have adequate training in first aid. The American Red Cross, some insurance providers, local safety councils, fire departments, or other resources may be able to provide this training. Treatment of a serious injury should begin within 3–4 min of the accident.

Consult with a physician to order appropriate first-aid supplies for emergencies. Medical personnel must be accessible to provide advice and consultation in resolving health problems that occur in the workplace. Establish a relationship with a local ambulance service so transportation is readily available for emergencies.

12.10 TRAINING EMPLOYEES

Educate employees about the types of emergencies that may occur and train them in the proper course of action. The size of your workplace and workforce, processes used, materials handled, and the availability of onsite or outside resources will determine your training requirements. Be sure all employees understand the function and elements of your EAP, including types of potential emergencies, reporting procedures, alarm systems, evacuation plans, and shutdown procedures. Discuss any special hazards you may have onsite such as flammable materials, toxic chemicals, radioactive sources, or water-reactive substances. Clearly communicate to employees who will be in charge during an emergency to minimize confusion. Generally, training for employees should address the following:

- Individual roles and responsibilities
- Threats, hazards, and protective actions
- Notification, warning, and communications procedures
- Means for locating family members in an emergency
- Emergency response procedures
- Evacuation, shelter, and accountability procedures
- Location and use of common emergency equipment
- Emergency shutdown procedures

Also, training employees in first-aid procedures, including protection against bloodborne pathogens, respiratory protection, including use of an escape-only respirator, and methods for preventing unauthorized access to the site might be appropriate.

Once the EAP has been reviewed with employees and everyone has had the proper training, it is a good idea to hold practice drills as often as necessary to keep employees prepared. Include outside resources such as fire and police departments when possible. After each drill, gather management and employees to evaluate the effectiveness of the drill. Identify the strengths and weaknesses of your plan and work to improve it. Review your plan with all employees and consider requiring annual training in the plan. Also offer training when you do the following:

- Develop your initial plan.
- Hire new employees.
- Introduce new equipment, materials, or processes into the workplace that affect evacuation routes.
- Change the layout or design of the facility.
- Revise or update your emergency procedures.

12.11 HAZARDOUS SUBSTANCES

No matter what kind of business exists, the potentiality for an emergency involving hazardous materials such as flammable, explosive, toxic, noxious, corrosive, biological, oxidizable, or radioactive substances is always a possibility.

The source of the hazardous substances could be external, such as a local chemical plant that catches fire or an oil truck that overturns on a nearby freeway. The source may be within the workplace or facility. Regardless of the source, these events could have a direct impact on your employees and your business and should be addressed by your EAP.

If hazardous substances are used or stored at the worksite, there may be an increased risk of an emergency involving hazardous materials, and this possibility should be addressed in your EAP. OSHA's Hazard Communication Standard (29 CFR 1910.1200) requires employers who use hazardous chemicals to keep an inventory, display the relevant manufacturer-supplied material safety data sheets (MSDSs) in a place accessible to workers, label chemical containers with their hazards, and train employees in ways to protect themselves against those hazards. A good way to start is to determine from your hazardous chemical inventory what hazardous chemicals you use and to gather the MSDSs for the chemicals. MSDSs describe the hazards that a chemical may present, list the necessary precautions when handling, storing, or using the substance, and outline emergency and first-aid procedures.

12.12   EMERGENCY EQUIPMENT

Employees may need personal protective equipment (PPE) to evacuate during an emergency. PPE must be based on the potential hazards in the workplace. Assess your workplace to determine potential hazards and the appropriate controls and protective equipment for those hazards. PPE may include items such as the following:

- Safety glasses, goggles, or face shields for eye protection
- Hard hats and safety shoes for head and foot protection
- Proper respirators
- Chemical suits, gloves, hoods, and boots for body protection from chemicals
- Special body protection for abnormal environmental conditions such as extreme temperatures
- Any other special equipment or warning devices necessary for hazards unique to your worksite

Consult with a health and safety professional or an industrial hygienist before making any purchases. Respirators selected should be appropriate to the hazards in your workplace, meet OSHA standards criteria, and be certified by the National Institute for Occupational Safety and Health. Respiratory protection may be necessary if your employees must pass through toxic atmospheres of dust, mists, gases, or vapors, or through oxygen-deficient areas while evacuating.

12.13   SUMMARY

Emergencies are always a possibility. Reacting late to an emergency can be disastrous and be detrimental to any employer's business. The preplanning can actually save a business by decreasing liability and preventing the harm that can occur to the workforce and the workplace itself. Employers should undertake emergency planning seriously and as a good business practice.

13 Ergonomics

Automating the retrieval of shopping carts is a good ergonomic solution.

Ergonomics is by definition fitting the workplace to the worker. It means more than changing a workstation. It means that the whole environment is designed to fit the worker including directions, controls, printed material, warning signals, mental stress, work schedules, the work climate, fatigue and boredom, material handling, noise, vibration, lighting, mental capacity, the worker/machine interface, and the list could go on.

At the present there is no Occupational Safety and Health Administration (OSHA) regulation addressing the hazards caused by poor ergonomic design and problems that result from these issues.

Where there are goods and materials involved in the everyday business as in wholesale, retail, and warehousing sectors of the service industry, the potential for ergonomic issues is very real.

When the potential for ergonomic-related injuries and illnesses exist, action must be taken to address and prevent these occurrences. This would include management commitment and employee involvement (employee involvement is critical in solving ergonomic-related problems); hazard identification and assessment; hazard control and prevention; and education and training.

13.1 IDENTIFYING HAZARDS

Once musculoskeletal disorders' (MSDs) hazards have been identified, the next step is to eliminate or control them. An effective hazard control process involves identifying and implementing control measures to obtain an adequate balance between worker capabilities and work requirements so that MSDs are not reasonably likely to occur.

During the identification and analysis of hazards, you should

- Include in the hazard identification and analysis all of the employees in the problem jobs or those who represent the range of physical capabilities of employees in the job.
- Ask the employees whether performing the job poses physical difficulties, and, if so, which physical work activities or conditions of the job they associate with the difficulties.

An ergonomics hazard identification and analysis is a process for pinpointing the work-related hazards or causes of MSDs. It involves examining the workplace conditions and individual elements or tasks of a job to identify and assess the ergonomic risk factors that are reasonably likely to be causing or contributing to the reported MSDs. This is an important step for those of you whose ergonomics programs include early intervention when employees report MSDs.

Some specific workers need to be evaluated since they may not be indicative of your average worker. This may be especially true of workers performing the same task as others. It is imperative that you look at sizes of workers or handicaps such as the following:

- Shortest employees in the job, because they are likely to have to make the longest reaches or to have a working surface that is too high
- Tallest employees because they may have to maintain the most excessive awkward postures (e.g., leaning over the assembly line, reaching down with the arms) while performing tasks
- Employees with the smallest hands because they may have to exert considerably more force to grip and operate hand and power tools
- Employees who work in the coldest areas of the workplace because they may have to exert more force to perform repetitive motions
- Employees who wear bifocals because they may be exposed to awkward postures (e.g., bending neck back to see)

An assessment tool such as is found in Figure 13.1 can be used to evaluate workers in these categories.

It is also a good idea to conduct a symptom or comfort survey. This allows the worker to tell you where they are experiencing pain or discomfort. They can also tell you what would make it easier to accomplish the work and often suggest very cost-effective solutions. You must remember that there are likely to be situations in which the physical work activities or conditions only pose a risk to the reporting employee. However, other employees who have performed the job for several years do not have (and never have had) difficulties performing the physical work activities of the job. In this case, it might be concluded that the problem is limited to the injured employee.

These efforts may include job–task breakdown, videotaping or photographing the job, job or hazard checklists, employee questionnaires, use of measuring tools, or employee symptom or discomfort surveys, are recognized ergonomic evaluation methods.

Ergonomic hazard identification checklist

Work Area_____ **Employees**_____
Date_____

Conducted by _____ Reviewed by_____
Date_____

Answer the following question based on the primary job activities of the worker at this particular task. Use the following responses to describe how frequently the worker is exposed to the job conditions described below:

   Never—Worker is never exposed to the condition.
   Sometimes—Workers is exposed to the condition less than three times daily.
   Usually—Worker is exposed to the condition three times or more daily.

	Never	Sometimes	Usually	If **usually**, list jobs to which answer applies here
Does worker perform tasks that are externally paced?				
Is the worker required to exert force with their hands (e.g., gripping, pulling, pinching)?				
Does the worker stand continuously for periods of more than 30 min?				
Does the worker sit for periods of more than 30 min without the opportunity to stand or move around freely?				
Does the worker have to stretch to reach the parts, tools, or work area?				
Does the worker use electronic input devices (e.g., keyboards, mice, joysticks, track balls) for continuous periods of more than 30 min?				

FIGURE 13.1 Ergonomic hazard identification checklist. (Courtesy of the Occupational Safety and Health Administration.)

(*continued*)

	Never	Sometimes	Usually	If **usually**, list jobs to which answer applies here
Does the worker kneel (one or both knees)?				
Does the worker perform activities with hands raised above shoulder height?				
Does worker perform activities while bending or twisting at the waist?				
Is the worker exposed to vibration?				
Is the worker required to worker in unnatural body positions?				
Does the worker lift or lower objects between the floor and waist height or above the shoulder?				
Does worker lift, lower, carry large objects that cannot be held close to the body?				
Does the worker lift, lower, or carry objects weighing more than 50 lb?				

TERMS

Primary job activities—Job activities that make up a significant part of the work or are required for safety or contingency. Activities are not considered to be primary job activities if they make up a small percentage of the job (i.e., takes up less than 10% of the worker's time) are not essential for safety or contingency, and can be readily accomplished in other ways (e.g., using equipment already available in the facility).

Externally paced activities—Work activities for which the worker does not have direct control of the rate of work. Externally paced work activities include activities which (1) the worker must keep up with an assembly line or an independently-operating machine, (2) the worker must respond to a continuous queue (e.g., customers standing in line, phone calls at a switch).

FIGURE 13.1 (continued)

While observing the job, employers record a description of each task for use in later risk factor analysis as well as other information that is helpful in completing the analysis:

- Tools or equipment used to perform task
- Materials used in task
- Amount of time spent doing each task
- Workstation dimensions and layout
- Weight of items handled
- Environmental conditions (cold, glare, blowing air)
- Vibration and its source
- Personal protective equipment worn

Hazards cannot be addressed efficiently without an accurate evaluation of the situation. The employee doing the job is one of the best sources of information; they are local process experts. Employees need to be involved in the identification, analysis, and control process because "no one knows the job better than the person who does it." Employees have the best understanding of what it takes to perform each task in a job, and thus, what parts of the job are the hardest to perform or pose the biggest difficulties. Workers can best tell what conditions cause them pain, discomfort, and injuries. They often have easy and practical suggestions on how such problems can be alleviated. Involving workers can make the job process more efficient, and pinpoint the causes of problems more quickly.

13.2   ERGONOMICALLY RISKY ACTIVITIES

13.2.1   WORK ACTIVITIES

Some of the activities that put workers at risk of ergonomically related problems are as follows:

- Exerting considerable physical effort to complete a motion
- Doing same motion over and over again
- Performing motions constantly without short pauses or breaks in between
- Performing tasks that involve long reaches
- Working surfaces are too high or too low
- Maintaining same position or posture while performing tasks
- Sitting for a long time
- Using hand and power tools
- Vibrating working surfaces, machinery, or vehicles
- Workstation edges or objects pressing hard into muscles or tendons
- Using hand as a hammer
- Using hands or body as a clamp to hold objects while performing tasks
- Wearing gloves that are bulky, too large, or too small

Overhead work

FIGURE 13.2   Reaching above the shoulders is an ergonomic hazard. (Courtesy of the Occupational Safety Health Administration.)

13.2.2   MANUAL MATERIAL HANDLING

Since material handling is common in this sector, specific attention should be paid to the following:

- Objects moved are heavy.
- Horizontal reach is long (distance of hands from body to grasp object to be handled).
- Vertical reach is below knees or above the shoulders (distance of hands above the ground when the object is grasped or released) as in Figure 13.2.
- Objects or people are moved significant distance.
- Bending or twisting during manual handling.
- Object is slippery or has no handles.
- Floor surfaces are uneven, slippery, or sloped.

Each of these items presents certain potential risk factors. When evaluating any risky activities, the risk factors in the section need to be considered as contributors to potential ergonomic problems.

13.3   ERGONOMIC RISK FACTORS

Ergonomic risk factors are the aspects of a job or task that impose a biomechanical stress on the worker. Ergonomic risk factors are the synergistic elements of MSD hazards. The following ergonomic risk factors are most likely to cause or contribute to an MSD:

- Force
- Vibration
- Repetition
- Contact stress
- Awkward postures
- Cold temperatures
- Static postures

13.3.1 DESCRIPTION OF RISK FACTORS

13.3.1.1 Force

Force refers to the amount of physical effort that is required to accomplish a task or motion. Tasks or motions that require application of higher force place higher mechanical loads on muscles, tendons, ligaments, and joints. Tasks involving high forces may cause muscles to fatigue more quickly. High forces also may lead to irritation, inflammation, strains and tears of muscles, tendons, and other tissues.

The force required to complete a movement increases when other risk factors are also involved. For example, more physical effort may be needed to perform tasks when the speed or acceleration of motions increases, when vibration is present, or when the task also requires awkward postures. Force can be internal, such as when tension develops within the muscles, ligaments, and tendons during movement. Force can also be external, as when a force is applied to the body, either voluntarily or involuntarily. Forceful exertion is most often associated with the movement of heavy loads, such as lifting heavy objects on and off a conveyor, delivering heavy packages, pushing a heavy cart, or moving a pallet. Hand tools that involve pinch grips require more forceful exertions than those that allow other grips, such as power grips.

13.3.1.2 Repetition

Repetition refers to performing a task or series of motions over and over again with little variation. When motions are repeated frequently (e.g., every few seconds) for prolonged periods (e.g., several hours, a work shift), fatigue and strain of the muscle and tendons can occur because there may be inadequate time for recovery. Repetition often involves the use of only a few muscles and body parts, which can become extremely fatigued while the rest of the body is little used. Table 13.1 shows the frequency of repetition and length of task cycles that are associated with increased risk of injury in repetitive motion jobs.

13.3.1.3 Awkward Postures

Awkward postures refer to positions of the body (e.g., limbs, joints, back) that deviate significantly from the neutral position while job tasks are being performed. For example, when a person's arm is hanging straight down (i.e., perpendicular to the ground) with the elbow close to the body, the shoulder is said to be in a neutral position. However, when employees are performing overhead work (e.g., installing or repairing equipment, grasping objects from a high shelf) their shoulders are far

TABLE 13.1

Repetition and Body Area

Body Area	Frequency Repetition per Minute	Level of Risk	Very High Risk if Modified by Either
Shoulder	More than 2.5	High	High external force, speed, high static load, and extreme posture
Upper arm/elbow	More than 10	High	Jack of training, high output demands, and lack of control
Forearm/wrist	More than 10	High	Long duration of repetitive work
Finger	More than 200	High	

Source: Courtesy of the Occupational Safety and Health Administration.

from the neutral position. Other examples include wrists bent while typing, bending over to grasp or lift an object, twisting the back and torso while moving heavy objects, and squatting. Awkward postures often are significant contributors to MSDs because they increase the work and the muscle force that is required.

13.3.1.4   Static Postures

Static postures (or "static loading") refer to physical exertion in which the same posture or position is held throughout the exertion. These types of exertions put increased loads or forces on the muscles and tendons, which contributes to fatigue. This occurs because not moving impedes the blood flow that is needed to bring nutrients to the muscles and to carry away the waste products of muscle metabolism. Examples of static postures include gripping tools that cannot be put down, holding the arms out or up to perform tasks, or standing in one place for prolonged periods. Antifatigue mats are helpful for cashiers who must stand in one place for long periods of time as seen in Figure 13.3.

13.3.1.5   Vibration

Vibration is the oscillatory motion of a physical body. Localized vibration, such as vibration of the hand and arm, occurs when a specific part of the body comes into contact with vibrating objects such as powered hand tools (e.g., chain saw, electric drill, chipping hammer) or equipment (e.g., wood planer, punch press, packaging machine). Whole-body vibration occurs when standing or sitting in vibrating environments (e.g., driving a truck over bumpy roads) or when using heavy vibrating equipment that requires whole-body involvement (e.g., jackhammers).

13.3.1.6   Contact Stress

Contact stress results from occasional, repeated, or continuous contact between sensitive body tissue and a hard or sharp object. Contact stress commonly affects

FIGURE 13.3 The use of antifatigue mats for those workers who must stand in static positions for periods of time.

the soft tissue on the fingers, palms, forearms, thighs, shins, and feet. This contact may create pressure over a small area of the body (e.g., wrist, forearm) that can inhibit blood flow, tendon and muscle movement, and nerve function. Examples of contact stress include resting wrists on the sharp edge of a desk or workstation while performing tasks, pressing of tool handles into the palms, especially when they cannot be put down, tasks that require hand hammering, and sitting without adequate space for the knees.

13.3.1.7 Cold Temperatures

Cold temperatures refer to exposure to excessive cold while performing work tasks. Cold temperatures can reduce the dexterity and sensitivity of the hand. Cold temperatures, for example, cause the worker to apply more grip force to hold hand tools and objects. Also, prolonged contact with cold surfaces (e.g., handling cold meat) can impair dexterity and induce numbness. Cold is a problem when it is present with other risk factors and is especially problematic when it is present with vibration exposure.

Of these risk factors, force (i.e., forceful exertions), repetition, and awkward postures, especially when occurring at high levels or in combination, are most often associated with the occurrence of MSDs. Exposure to one ergonomic risk factor may be enough to cause or contribute to a covered MSD. However, most often ergonomic risk factors act in combination to create a hazard. Jobs that have multiple risk factors have a greater likelihood of causing an MSD, depending on the duration, frequency, and/or magnitude of exposure to each. Thus, it is important that ergonomic risk factors be considered in light of their combined effect in causing or contributing to an MSD. Table 13.2 depicts tasks and their risk factors.

TABLE 13.2

Tasks and Their Risk factors

Physical work activities and conditions: Ergonomic risk factors that may be present

(1) Exerting considerable physical effort to complete a motion
  (i) Force
  (ii) Awkward postures
  (iii) Contact stress
(2) Doing same motion over and over again
  (i) Repetition
  (ii) Force
  (iii) Awkward postures
  (iv) Cold temperatures
(3) Performing motions constantly without short pauses or breaks in between
  (i) Repetition
  (ii) Force
  (iii) Awkward postures
  (iv) Static postures
  (v) Contact stress
  (vi) Vibration
(4) Performing tasks that involve long reaches
  (i) Awkward postures
  (ii) Static postures
  (iii) Force
(5) Working surfaces are too high or too low
  (i) Awkward postures
  (ii) Static postures
  (iii) Force
  (iv) Contact stress
(6) Maintaining same position or posture while performing tasks
  (i) Awkward posture
  (ii) Static postures
  (iii) Force
  (iv) Cold temperatures

TABLE 13.2 (continued)
Tasks and Their Risk factors

(7) Sitting for a long time
   (i)   Awkward posture
   (ii) Static postures
   (iii) Contact stress
(8) Using hand and power tools
   (i)   Force
   (ii) Awkward postures
   (iii) Static postures
   (iv) Contact stress
   (v)   Vibration
   (vi) Cold temperatures
(9) Vibrating working surfaces, machinery, or vehicles
   (i)   Vibration
   (ii) Force
   (iii) Cold temperatures
(10) Workstation edges or objects press hard into muscles or tendons
   (i)   Contact stress
(11) Using hand as a hammer
   (i)   Contact stress
   (ii) Force
(12) Using hands or body as a clamp to hold object while performing tasks
   (i)   Force
   (ii) Static postures
   (iii) Awkward postures
   (iv) Contact stress
(13) Gloves are bulky, too large or too small
   (i)   Force
   (ii) Contact stress

Manual material handling (lifting/lowering, pushing/pulling, and carrying)

(14) Objects or people moved are heavy
   (i)   Force
   (ii) Repetition
   (iii) Awkward postures
   (iv) Static postures
   (v)   Contact stress
(15) Horizontal reach is long (distance of hands from body to grasp object to be handled)
   (i)   Force
   (ii) Repetition
   (iii) Awkward postures
   (iv) Static postures
   (v)   Contact stress

(*continued*)

TABLE 13.2 (continued)
Tasks and Their Risk factors

(16) Vertical reach is below knees or above the shoulders (distance of hands above the ground when the object is grasped or released)
  (i)   Force
  (ii) Repetition
  (iii) Awkward postures
  (iv) Static postures
  (v)   Contact stress
(17) Objects or people are moved significant distance
  (i)   Force
  (ii) Repetition
  (iii) Awkward postures
  (iv) Static postures
  (v)   Contact stress
(18) Bending or twisting during manual handling
  (i)   Force
  (ii) Repetition
  (iii) Awkward postures
  (iv) Static postures
(19) Object is slippery or has no handles
  (i)   Force
  (ii) Repetition
  (iii) Awkward postures
  (iv) Static postures
(20) Floor surfaces are uneven, slippery, or sloped
  (i)   Force
  (ii) Repetition
  (iii) Awkward postures
  (iv) Static postures

Source: Courtesy of the Occupational Safety and Health Administration.

13.4 PHYSICAL WORK ACTIVITIES AND CONDITIONS

The physical work activities and conditions include the following:

- Physical demands of work
- Workplace and workstation conditions and layout
- Characteristics of objects that are handled or used
- Environmental conditions

Table 13.3 shows the physical work activities and workplace conditions that are associated with the above-mentioned physical aspects.

Employers should examine a job in which an MSD has occurred to identify the physical work activities and workplace conditions and then evaluate the risk factors to make an assessment of the work environment.

TABLE 13.3

Physical Work Activities and Conditions

Physical aspects of jobs and workstations

Physical demands of work

- Exerting considerable physical effort to complete a motion
- Doing the same motion over and over again
- Performing motions constantly without short pauses or breaks in-between
- Maintaining same position or posture while performing tasks
- Sitting for a long time
- Using hand as a hammer
- Using hands or body as a clamp to hold object while performing tasks
- Objects or people are moved significant distances

Layout and condition of the workplace or workstation

- Performing tasks that involve long reaches
- Working surfaces too high or too low
- Vibrating working surfaces, machinery, or vehicles
- Workstation edges or objects press hard into muscles or tendons
- Horizontal reach is long
- Vertical reach is below knees or above the shoulders
- Floor surfaces are uneven, slippery, or sloped

Characteristics of the objects handled

- Using hand and power tools
- Gloves bulky, too large, or too small
- Objects or people moved are heavy
- Object is slippery or has no handles

Environmental conditions

- Cold temperatures
- Temperature extremes and humidity
- Vibration
- Noise
- Illumination
- Colors

Source: Courtesy of the Occupational Safety and Health Administration.

13.5   LIMITS OF EXPOSURE

To determine the real risk, you need to look at the duration, frequency, and magnitude (i.e., modifying factors) of the employee's exposure to the ergonomic risk factors. The risk factors do not always pose a significant risk of injury. This may be because the exposure does not last long enough, is not repeated frequently enough, or is not intensive enough to pose a risk.

13.5.1   DURATION

Duration refers to the length of time an employee is continually exposed to risk factors. The duration of job tasks can have a substantial effect on the likelihood of both localized and general fatigue. In general, the longer the period of continuous work (i.e., the longer the tasks require sustained muscle contraction), the longer the recovery or rest time required. Duration can be mitigated by changing the sequence of activities or recovery time and pattern of exposure. Breaks or short pauses in the work routine help reduce the effects of the duration of exposure.

13.5.2 Frequency

The response of the muscles and tendons to work is dependent on the number of times the tissue is required to respond and the recovery time between activities. The frequency can be viewed at the micro level, such as grasps per minute or lifts per hour. However, often a macro view will be sufficient, such as time in a job per shift, or days per week in a job. Handheld stock inventories devices increase ergonomic stress (Figure 13.4).

13.5.3 Magnitude

Magnitude (or intensity) is a measure of the strength of the risk factors, for example, amount of force, extent of posture deviation, extent of velocity or acceleration of motion, and amount of pressure due to compression. Magnitude can be measured either in absolute terms or relative to an individual's capabilities. There are many qualitative and quantitative ways to determine the magnitude of exposure. Often, all it takes is to ask employees to describe the most difficult part of the job, and the answer will indicate the magnitude of the risk factor. A common practice for

FIGURE 13.4 Handheld inventory devices require repeated repetition.

assessing forceful exertion is to ask the employee to rate the force required to do the task. When magnitude is assessed qualitatively, the employer is making a relative rating, that is, the perceived magnitude of the risk factor relative to the capabilities of the worker. Relative ratings are very useful in understanding whether the job fits the employees currently doing the job.

As mentioned above, ergonomic risk factors are synergistic elements of MSD hazards. Simply put, the total effect of these risk factors is greater than the sum of their parts. As such, employers need to be especially watchful for situations where risk factors occur simultaneously. Levels of risk factors that may pose little risk when found alone are much more likely to cause MSDs when they occur with other risk factors.

13.6 ERGONOMIC CONTROLS

Controls that reduce a risk factor focus on reductions in the risk modifiers (frequency, duration, or magnitude). By limiting exposure to the modifiers, the risk of an injury is reduced. Thus, in any job, the combination of the task, environment, and the worker creates a continuum of opportunity to reduce the risk by reducing the modifying factors. The closer the control approach comes to eliminating the frequency, duration, or magnitude, the more likely it is that the MSD hazard has been controlled. Conversely, if the control does little to change the frequency, duration, or magnitude, it is unlikely that the MSD hazard has been controlled.

In determining control, ask employees in the problem job for recommendations about eliminating or materially reducing the MSD hazards. Second, identify, assess, and implement feasible controls (interim and/or permanent) to eliminate or materially reduce the MSD hazards. This includes prioritizing the control of hazards, where necessary. Thirdly, track your progress in eliminating or materially reducing the MSD hazards. This includes consulting with employees in problem jobs about whether the implemented controls have eliminated or materially reduced the hazard, and last, identify and evaluate MSD hazards when you change, design, or purchase equipment or processes in problem jobs.

13.6.1 IDENTIFY CONTROLS

There are many different methods you can use and places you can go to identify controls. Many employers rely on their internal resources to identify possible controls. These in-house experts may include the following:

- Employees who perform the job and their supervisors
- Engineering personnel
- Workplace safety and health personnel or committee
- Maintenance personnel
- On-site health care professionals
- Procurement staff
- Human resource personnel

Possible controls can also be identified from sources outside the workplace, such as the following:

- Equipment catalogs
- Vendors
- Trade associations or labor unions
- Conferences and trade shows
- Insurance companies
- OSHA consultation services
- Specialists

13.6.2 ASSESS CONTROLS

The assessment of controls is an effort by you, with input from employees, to select controls that are reasonably anticipated to eliminate or materially reduce the MSD hazards. You may find that there are several controls that would be reasonably likely to reduce the hazard. Multiple control alternatives are often available, especially when several risk factors contribute to the MSD hazard. You need to assess which of the possible controls should be tried. Clearly, a control that significantly reduces several risk factors is preferred over a control that only reduces one of the risk factors.

Selection of the risk factors to control and/or control measures to try can be based on numerous criteria. An example of one method involves ranking all of the ergonomic risk factors and/or possible controls according to how well they meet these four criteria:

- Effectiveness—greatest reduction in exposure to the MSD hazards
- Acceptability—employees most likely to accept and use this control
- Timeliness—takes least amount of time to implement, train, and achieve material reduction in exposure to MSD hazards
- Cost—elimination or material reduction of exposure to MSD hazards at the lowest cost

13.6.3 IMPLEMENT CONTROLS

Because of the multifactor nature of MSD hazards, it is not always clear whether the selected controls will achieve the intended reduction in exposure to the hazards. As a result, the control of MSD hazards often requires testing selected controls and modifying them appropriately before implementing them throughout the job. Testing controls verifies whether the proposed solution actually works and whether any additional changes or enhancements need to be made.

13.7 TRACKING PROGRESS

First, evaluating the effectiveness of controls is of utmost importance in an incremental abatement process. Unless they follow up on their control efforts, employers will not know whether the hazards have been adequately controlled or whether the abatement process needs to continue. Simply put, if the job is not controlled, the problem-solving is not complete.

Second, the tracking progress is also essential in those cases where you need to prioritize the control of hazards. It tells you whether they are on schedule with their abatement plans.

Third, tracking the progress of control efforts is a good way of judging the success rate of implementation of the plan. Some of the measures to use include the following:

- Reductions in severity rates, especially at the very start of the program
- Reduction in incidence rates
- Reduction in total lost-workdays and lost-workdays per case
- Reduction in job turnover or absenteeism
- Reduction in workers' compensation costs/medical costs
- Increases in productivity or quality
- Reduction in reject rates
- Number of jobs analyzed and controlled
- Number of problems solved

13.8   EDUCATION AND TRAINING

Education and training can be used in variety of ways. The foremost is to train all employees in ergonomic hazard awareness, your program and procedures, sign and symptom identification, and types of injuries and illnesses. Second, train some of the workforce in ergonomic assessment so you will have teams of both management and labor to evaluate ergonomic hazards and make recommendations for controlling the potential risk factors on the jobs in your workplace. With proper training you will have an educated workforce who can be an asset rather than a liability in solving MSD problems.

13.9   SUMMARY

Ergonomics is a continuous improvement process. If you can show that you have made an organized effort to identify ergonomic stressors, to educate affected employees on ergonomic principles, to implement solutions, and to have a system to identify when a solution is not working and needs to be readdressed, you have taken giant steps toward mitigating ergonomic problems.

14 Fire Hazards Guidelines

Cluttered work areas increase the potential for fires.

14.1 FIRE HAZARDS

Workplace fires and explosions kill 200 and injure more than 5000 workers every year. In 1995, more than 75,000 workplace fires cost businesses more than $2.3 billion. Fires wreak havoc among workers and their families and destroy thousands of businesses each year, putting people out of work and severely impacting their livelihoods. The human and financial toll underscores the gravity of workplace fires.

14.2 CAUSES OF FIRES

The most common causes of workplace fires are as follows:

- Electrical causes—lax maintenance in wiring, motors, switches, lamps, and heating elements
- Smoking—near flammable liquids, stored combustibles, etc.
- Cutting and welding—highly dangerous in areas where sparks can ignite combustibles
- Hot surfaces—exposure of combustibles to furnaces, hot ducts or flues, electric lamps or heating elements, and hot metal
- Overheated materials—abnormal process temperatures, materials in dryers, overheating of flammable liquids
- Open flames—gasoline or other torches, gas or oil burners

- Friction—hot bearings, misaligned or broken machine parts, choking or jamming materials, poor adjustment of moving parts
- Unknown substances—unexpected materials
- Spontaneous heating—deposits in ducts and flues, low-grade storage, scrap waste, oily waste, and rubbish
- Combustion sparks—burning rubbish, foundry cupolas, furnaces, and fire-boxes
- Miscellaneous—including incendiary cases, fires spreading from adjoining buildings, molten metal or glass, static electricity near flammable liquids, chemical action, and lighting

14.3 OSHA STANDARDS REQUIREMENTS

Occupational Safety and Health Administration (OSHA) standards require employers to provide proper exits, firefighting equipment, and employee training to prevent fire deaths and injuries in the workplace. Each workplace building must have at least two exits far from each other to be used in a fire emergency. Fire doors must not be blocked or locked to prevent emergency use when employees are in the buildings. Delayed opening of fire doors is permitted when an approved alarm system is integrated into the fire door design. Exit routes from buildings must be clear and free of obstructions and properly marked with signs designating exits from the building.

Each workplace building must have a full complement of the proper type of fire extinguisher for the fire hazards present, excepting when employers wish to have employees evacuate instead of fighting small fires. Employees expected or anticipated to use fire extinguishers must be instructed on the hazards of fighting fire, how to properly operate the fire extinguishers available, and what procedures to follow in alerting others to the fire emergency. Only approved fire extinguishers are permitted to be used in workplaces, and they must be kept in good operating condition. Proper maintenance and inspection of this equipment is required of each employer. The applicable OSHA standard on fire protection is 29 CFR 1910.157. Figure 14.1 shows a well maintained fire extinguisher.

Where the employer wishes to evacuate employees instead of having them fight small fires there must be written emergency plans and employee training for proper evacuation. Emergency action plans are required to describe the routes to use and procedures to be followed by employees. Also procedures for accounting for all evacuated employees must be part of the plan. The written plan must be available for employee review. Where needed, special procedures for helping physically impaired employees must be addressed in the plan; also, the plan must include procedures for those employees who must remain behind temporarily to shut down critical plant equipment before they evacuate.

The preferred means of alerting employees to a fire emergency must be part of the plan and an employee alarm system must be available throughout the workplace complex and must be used for emergency alerting for evacuation. The alarm system may be voice communication or sound signals such as bells, whistles, or horns. Employees must know the evacuation signal. Fire alarm boxes should be readily accessible as shown in Figure 14.2.

FIGURE 14.1   Adequate fire extinguisher.

Employees must be trained to face emergency situations. Employers must review the plan with newly assigned employees so that they know correct actions in an emergency and with all employees when the plan is changed.

FIGURE 14.2   Example of fire alarm box.

Employers need to implement a written fire prevention plan (FPP) to complement the fire evacuation plan to minimize the frequency of evacuation. Stopping unwanted fires from occurring is the most efficient way to handle them. The written plan shall be available for employee review. Housekeeping procedures for storage and cleanup of flammable materials and flammable waste must be included in the plan. Recycling of flammable waste such as paper is encouraged; however, handling and packaging procedures must be included in the plan. Procedures for controlling workplace ignition sources such as smoking, welding, and burning must be addressed in the plan. Heat-producing equipment such as burners, heat exchangers, boilers, ovens, stoves, fryers, etc., must be properly maintained and kept clean of accumulations of flammable residues; flammables are not to be stored close to these pieces of equipment. All employees are to be apprised of the potential fire hazards of their job and the procedures called for in the employer's fire prevention plan. The plan shall be reviewed with all new employees when they begin their job and with all employees when the plan is changed.

The minimum provisions that make up an FPP are as follows:

- List of all major fire hazards, proper handling and storage procedures for hazardous materials, potential ignition sources and their control, and type of fire protection equipment necessary to control each major hazard
- Procedures to control accumulation of flammable and combustible materials
- Procedure for regular maintenance of safeguards installed on heat-producing equipment to prevent the accidental ignition of combustible materials
- Name or job title of employees responsible for maintaining equipment or control sources of ignition or fires
- Name or job title of employees responsible for the control of fuel source hazards

Any employee assigned to a job must be informed of the fire hazards to which they could be exposed. The employee must have received an explanation of fire prevention plan and how it was designed to protect them.

14.4   AVOIDING FIRES

General safety precautions for avoiding fires caused by smoking are by obeying "No Smoking" signs. By watching for danger spots even if no warning is posted (e.g., temporary storage area that contains combustibles), do not place lighted cigarettes on wooden tables or workbenches, even if smoking is permitted, and do no put ashes in a wastebasket or trash can.

Flammable and combustible liquids can cause fires if they are near open flames and motors that might spark. When you transfer them, bond the containers to each other and ground the one being dispensed from, to prevent sparks from static electricity (Figure 14.3). Clean up spills right away, and put oily rags in a tightly covered metal container. Change clothes immediately if you get oil or solvents on them. Watch out for empty containers that held flammable or combustible liquids; vapors might still be present. Store these liquids in approved containers in well-ventilated areas away

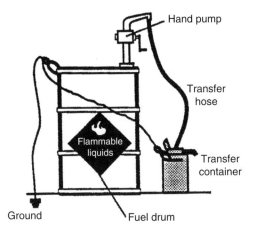

FIGURE 14.3   Safe transfer procedures for flammable liquids. (Courtesy of the Department of Energy.)

from heat and sparks. Be sure all containers for flammable and combustible liquids are clearly and correctly labeled.

Electricity can cause fires if frayed insulation and damaged plugs on power cords or extension cords are not fixed or discarded. Also, electrical conductors should not be damp or wet and there should be no oil and grease on any wires.

A cord that is warm to the touch when current is passing through should warn you of a possible overload or hidden damage. Do not overload motors; watch for broken or oil-soaked insulation, excessive vibration, or sparks; keep motors lubricated to prevent overheating. Defective wiring, switches, and batteries on vehicles should be replaced immediately. Electric lamps need bulb guards to prevent contact with combustibles and to help protect the bulbs from breakage. Do not try to fix electrical equipment yourself if you are not a qualified electrician.

Housekeeping is often a factor in fires in the workplace. Keep your work areas clean. Passageways and fire doors should be kept clear and unobstructed. Material must not obstruct sprinkler heads or be piled around fire extinguisher locations or sprinkler controls. Combustible materials should be present in work areas only in quantities required for the job, and should be removed to a designated storage area at the end of each workday.

Hot work such as welding and cutting should never be permitted without supervision or a hot work permit. Watch out for molten metal; it can ignite combustibles or fall into cracks and start a fire that might not erupt until hours after the work is done. Portable cutting and welding equipment is often used where it is unsafe; keep combustibles at safe distance from a hot work area. Be sure tanks and other containers that have held flammable or combustible liquids are completely neutralized and purged before you do any hot work on them. Have a fire watch (another employee) on hand to put out a fire before it can get out of control.

14.5 FIRE PROTECTION AND PREVENTION

14.5.1 FIRE PROTECTION

To protect workplace from fire the following items should be adhered to:

- Access to all available firefighting equipment will be maintained at all times.
- Firefighting equipment will be inspected periodically and maintained in operating condition. Defective or exhausted equipment must be replaced immediately.
- All firefighting equipment will be conspicuously located at each jobsite.
- Fire extinguishers, rated not less than 2A, will be provided for each 3000 sq ft of the protected work area. Travel distance from any point of the protected area to the nearest fire extinguisher must not exceed 100 ft. One 55 gal open drum of water, with two fire pails, may be substituted for a fire extinguisher having a 2A rating.
- Extinguishers and water drums exposed to freezing conditions must be protected from freezing.
- Do not remove or tamper with fire extinguishers installed on equipment or vehicles, or in other locations, unless authorized to do so or in case of fire. If you use a fire extinguisher, be sure it is recharged or replaced with another fully charged extinguisher. Table 14.1 depicts the types of fire and the classes of fire extinguishers used to extinguish these fires.

14.5.2 FIRE PREVENTION

To prevent fire the following principles should be followed:

- Internal combustion engine powered equipment must be located so that exhausts are away from combustible materials.
- Smoking is prohibited at, or in the vicinity of operations which constitute a fire hazard. Such operations must be conspicuously posted: "No Smoking or Open Flame."
- Portable battery powered lighting equipment must be approved for the type of hazardous locations encountered.
- Combustible materials must be piled no higher than 20 ft. Depending on the stability of the material being piled, this height may be reduced.

TABLE 14.1

Types of Fires and Classes of Extinguishers

Class A (wood, paper, trash)—use water or foam extinguisher
Class B (flammable liquids, gas, oil, paints, grease)—use foam, CO_2, or dry chemical extinguisher
Class C (electrical)—use CO_2 or dry chemical extinguisher
Class D (combustible metals)—use dry powder extinguisher only

---

- Keep driveways between and around combustible storage piles at least 15 ft wide and free from accumulation of rubbish, equipment, or other materials.
- Portable fire extinguishing equipment, suitable for anticipated fire hazards on the jobsite, must be provided at convenient, conspicuously accessible locations.
- Firefighting equipment must be kept free from obstacles, equipment, materials, and debris that could delay emergency use of such equipment. Familiarize yourself with the location and use of the project's firefighting equipment.
- Discard and/or store all oily rags, waste, and similar combustible materials in metal containers on a daily basis.
- Storage of flammable substances on equipment or vehicles is prohibited unless such a unit has adequate storage area designed for such use.

14.6 FLAMMABLE AND COMBUSTIBLE LIQUIDS (29 CFR 1910.106)

Flammable liquids are to be kept in covered containers or tanks when not actually in use. The quantity of flammable or combustible liquid that may be located outside of an inside storage room or storage cabinet in any one fire area of a building cannot exceed the following:

- 25 gal of Class IA liquids in containers
- 120 gal of Class IB, IC, II, or III liquids in containers
- 660 gal of Class IB, IC, II, or III liquids in a single portable tank

Flammable and combustible liquids are to be drawn from or transferred into containers within buildings only through a closed piping system, from safety cans, by means of a device drawing through the top, or by gravity through an approved self-closing valve. Transfer by means of air pressure is prohibited. Not more than 60 gal of Class I or Class II liquids, nor more than 120 gal of Class III liquids may be stored in a storage cabinet. Inside storage rooms for flammable and combustible liquids are to be constructed to meet required fire-resistive rating or wiring for their uses.

Outside storage areas must be grated so as to divert spills away from buildings or other exposures, or be surrounded with curbs at least 6 in. high with appropriate drainage to a safe location for accumulated liquids. The areas shall be protected against tampering or trespassing, where necessary, and shall be kept free of weeds, debris, and other combustible material not necessary to the storage.

Adequate precautions are to be taken to prevent the ignition of flammable vapors. Sources of ignition include, but are not limited to, open flames; lightning; smoking; cutting and welding; hot surfaces; frictional heat; static, electrical, and mechanical sparks; spontaneous ignition, including heat-producing chemical reactions; and radiant heat.

Class I liquids are not to be dispensed into containers unless the nozzle and container are electrically interconnected. All bulk drums of flammable liquids are to be grounded and bonded to containers during dispensing.

14.6.1 Flammable and Combustible Liquids

Some of the more specific rules for flammable and combustible liquids are as follows:

- Explosive liquids, such as gasoline, shall not be used as cleaning agents. Use only approved cleaning agents.
- Store gasoline and similar combustible liquids in approved and labeled containers in well-ventilated areas free from heat sources.
- Handling of all flammable liquids by hand containers must be in approved type safety containers with spring closing covers and flame arrestors (Figure 14.4).
- Approved wooden or metal storage cabinets must be labeled in conspicuous lettering: "Flammable—Keep Fire Away."
- Never store more than 60 gal of flammable, or 120 gal of combustible liquids in any one approved storage cabinet.
- Storage of containers shall not exceed 1100 gal in any one pile or area. Separate piles or groups of containers by a 5 ft clearance. Never place a pile or group within 20 ft of a building. A 12 ft wide access way must be provided within 200 ft of each container pile to permit approach of fire control apparatus.

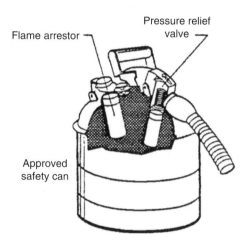

FIGURE 14.4   Example of an approved safety container. (Courtesy of the Department of Energy.)

14.7   FLAMMABLE AND COMBUSTIBLE MATERIALS

Combustible scrap, debris, and waste materials (oily rags, etc.) stored in covered metal receptacles are to be removed from the worksite promptly. Proper storage must be practiced to minimize the risk of fire including spontaneous combustion. Fire extinguishers are to be selected and provided for the types of materials in areas where they are to be used. "No Smoking" rules should be enforced in areas involving storage and use of hazardous materials.

14.8   FIRE SUPPRESSION SYSTEMS

Properly designed and installed fixed fire suppression systems enhance fire safety in the workplace. Automatic sprinkler systems throughout the workplace are among the most reliable firefighting means. The fire sprinkler system detects the fire, sounds an alarm, and sprays water at the source of the fire and heat. Automatic fire suppression systems require proper maintenance to keep them in serviceable condition. When it is necessary to take a fire suppression system out of service while business continues, the employer must temporarily substitute a fire watch of trained employees standing by to respond quickly to any fire emergency in the normally protected area. The fire watch must interface with the employers' fire prevention plan and emergency action plan. Signs must be posted about areas protected by total flooding fire suppression systems which use agents that are a serious health hazard such as carbon dioxide, Halon 1211, etc. Such automatic systems must be equipped with area predischarge alarm systems to warn employees of the impending discharge of the system and thereby provide time to evacuate the area. There must be an emergency action plan to provide for the safe evacuation of employees from within the protected area. Such plans are to be part of the overall evacuation plan for the workplace facility.

The local fire department needs to be well acquainted with your facilities, its location, and specific hazards. The fire alarm system must be certified as required and tested at least once a year. Interior standpipes must be inspected regularly. Outside private fire hydrants must be flushed at least once a year and on a routine preventive maintenance schedule. All fire doors and shutters must be in good operating condition and unobstructed and protected against obstructions, including their counterweights.

14.9   FIREFIGHTING

If an employer expects workers to assist in fighting fires, then he must have them trained to do so. If they are not trained to use fire extinguishing equipment then they should report the fire and sound the alarm followed by evacuation of the premises.

14.10   FIRE HAZARD CHECKLIST

To reduce the chances of fire, all the checklist questions should have an affirmative answer. Figure 14.5 is an example of a fire prevention checklist.

Fire checklist

Yes ☐ No ☐ Are fire extinguishers locations unobstructed?
Yes ☐ No ☐ Are operating instructions on the front of each extinguisher?
Yes ☐ No ☐ Is fire extinguisher locations visibly identified?
Yes ☐ No ☐ Are there fire extinguisher types sufficient to respond to the local area hazards?
Yes ☐ No ☐ Do the fire extinguishers meet the hydrostatic test requirements (every 12 years)?
Yes ☐ No ☐ Are monthly fire extinguisher checks being conducted?
Yes ☐ No ☐ Are fire hose cabinets accessible and unobstructed?
Yes ☐ No ☐ Are fire hose cabinets in good physical condition?
Yes ☐ No ☐ Is flammable liquid stored in approved cabinet?
Yes ☐ No ☐ Are flammable liquid storage cabinets used for only flammable liquids?
Yes ☐ No ☐ Is the volume of flammable liquids stored in cabinets less than the limits stated on the cabinet door?
Yes ☐ No ☐ Are the flammable liquid storage cabinet vent bungs in place and cabinet doors kept closed?
Yes ☐ No ☐ Are the flammable liquids storage cabinets structurally undamaged?
Yes ☐ No ☐ Are caution labels affixed to flammable liquids cabinets, such as "KEEP FIRE AWAY"?
Yes ☐ No ☐ Are electrical panels free and clear for access with a minimum of 3 ft open space in front of the cabinet?
Yes ☐ No ☐ Are electrical disconnects labeled with a description of the equipment they control?
Yes ☐ No ☐ Is the area free of visible exposed wiring?
Yes ☐ No ☐ Are equipment power cords in good condition?
Yes ☐ No ☐ Has your area refrained from using extension cords as long-term power sources?
Yes ☐ No ☐ Are all employees wearing ANSI approved safety glasses with side shields?
Yes ☐ No ☐ Are work areas kept clean and orderly?
Yes ☐ No ☐ Are passageways clearly marked and exit routes visible?
Yes ☐ No ☐ Are exits marked with signs and illuminated?
Yes ☐ No ☐ Are exits easily accessible and unobstructed?
Yes ☐ No ☐ Do self-closing doors operate properly?
Yes ☐ No ☐ Are all work areas properly lighted?

FIGURE 14.5 Fire prevention checklist.

14.11 SUMMARY

Spotting fire hazards in the workplace is the first step in prevention. Become familiar with the most common causes of fires. Inspect on a daily, weekly, monthly basis. (Review briefly the employee's responsibility for fire inspection and prevention.)

When a fire hazard is spotted, eliminate it immediately if you have the ability and the authority to do so. File a fire hazard report form or bring it to your supervisor's attention.

If a fire has started notify the appropriate personnel (company fire brigade, your supervisor, safety director, etc.) or turn in a general alarm following company policies.

If the fire is not out of control, attempt to extinguish it with the appropriate fire extinguishing equipment if you have been trained in the use of fire extinguisher. If the fire is out of control or is not in your area, follow evacuation procedures.

15 Hand Tools

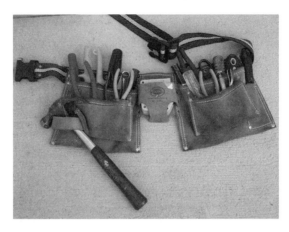

A variety of hand tools are necessary for day-to-day general maintenance.

When we think about hand tools, we normally think of hammers, screwdrivers, or pliers, but a toothbrush, a spoon, a pen, scissors are also hand tools. They are examples of tools that are used to concentrate force and help to carry out a variety of tasks.

For most of us, hand tools are nonpowered tools and include chisels, handsaws, wrenches, shovels, and knives. Many injuries can result when using hand tools including cuts, lacerations, eye injuries, overuse (ergonomic related injuries), and at times slips, trips, and falls. In most instances these injuries occur due to the incorrect use, use of the wrong tool, or improper maintenance of tools. Approximately 8% of all industrial accidents are caused by hand tools. The Mine Health and Safety Administration found that one out of every four accidents was due to hand tools. Some examples of hand tool accidents are as follows:

- Using a screwdriver as a chisel may cause the tip of the screwdriver to break off and fly, hitting the user or other employees.
- If a wooden handle on a tool such as a hammer or ax is loose, splintered, or cracked, the head of the tool may fly off and strike the user or another worker.
- A wrench must not be used if its jaws are sprung, because it might slip.
- Impact tools such as chisels, wedges, or drift pins are unsafe if they have mushroomed heads that might shatter on impact, sending sharp fragments flying (Figure 15.1).

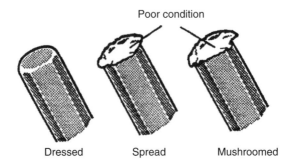

FIGURE 15.1   From safe to unsafe chisels. (Courtesy of the Department of Energy.)

The employer is responsible for the safe condition of tools and equipment used by employees, even personal tools if being used in the workplace. The employer should not issue or permit the use of unsafe hand tools. Damages or broken hand tools should be removed from service and a tag placed on them saying, "Do Not Use" or "Removed from Service" if the tools are not thrown away. Employers should assure that employees are trained in the proper use and handling of hand tools and other equipment.

Employees, when using saw blades, knives, or other tools, should direct tools away from aisle areas and away from other employees working in close proximity. Knives and scissors must be sharp; dull tools can cause more hazards than sharp ones. Cracked saw blades must be removed from service.

Wrenches must not be used when jaws are sprung to the point that slippage occurs (Figure 15.2). Impact tools such as drift pins, wedges, and chisels must be kept free from mushroomed heads. The wooden handle of tools must not be splintered.

Iron or steel hand tools may produce sparks that can be an ignition source around flammable substances. Where this hazard exists, spark-resistant tools made of nonferrous materials should be used where flammable gases, highly volatile liquids, and other explosive substances are stored or used.

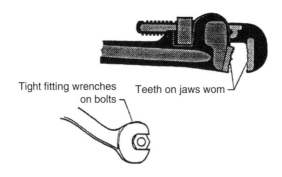

FIGURE 15.2   Wrenches from good to worn. (Courtesy of the Department of Energy.)

Appropriate personal protective equipment (PPE) such as safety eyewear and gloves must be worn to protect against hazards that may be encountered while using hand tools. Workplace floors are to be kept clean and as dry as possible to prevent slips with or around dangerous hand tools.

15.1   PREVENTING HAND TOOL ACCIDENTS

To prevent hand tool accidents certain safe work practices should be followed. Before using hand tools select the correct tool for the job. Do not use tools for jobs they are not intended for. Provide workers with training and information about safer work practices and the correct methods, posture, and use of tools. Other work practices that should be addressed are as follows:

- Always use tools in such a way that a slip or miss does not result in an injury, e.g., when using sharp tools for cutting always cut away from the body or hand.
- Keep the work area free of clutter and waste.
- Ensure the work area has adequate lighting.
- Tools, equipment, and materials should not be thrown or dropped from one employee to another or from one level to another. Hand them, handle first, directly to other workers or use a hand line.
- Ensure workers are wearing appropriate protective clothing and PPE such as goggles, safety shoes, and gloves.

15.1.1   HAND TOOL KEY POINTS

Wear approved PPE such as safety shields, respirators, safety toed shoes, high-top shoes, hard hats, bump caps, leather gloves, leather aprons, coverall, and safety eyewear with side shields of industrial quality that conform to the ANSI Z87.1 standard.

A variety of gloves exist that have different functional use regarding hand tools. Today there are cut resistant gloves made of Kevlar as seen in Figure 15.3, mechanics gloves with rubber gripping surfaces, mesh gloves that protect from cuts, antivibration or shock absorbing gloves, and the common leather glove that is always a good option if it fits properly since too tight a glove tires the hand and fingers and too loose a glove decreases dexterity.

Wearing proper clothing varies depending on the type of hand tools that are being used. Work clothing should not be loose, baggy, or highly flammable. To protect against burns, wear clothing such as coveralls, high-top shoes, leather aprons, and leather gloves. Remove all paper from pockets and wear cuff less pant. When working with heavy metals or items wear hard toed shoes with nonskid soles. Avoid synthetic clothing because they have low flash points which can result in severe burns. Do not wear jewelry especially rings when using hand tools since it may result in rings getting caught or contacting electricity. Jewelry can get caught on moving parts.

FIGURE 15.3   Cut resistant Kevlar gloves.

Protect the hair, scalp, and head by pulling back long hair in a band or a cap to keep it from getting caught in tools. Be extremely careful with long hair when using a rotating tool. When handling carpentry materials wear a hard hat or bump cap to protect the head.

Protect the fingers, hands, and arms by wearing leather gloves or cut resistant gloves and forearm shields. When workers are hammering, care must be taken to strike the object not the hand or fingers.

Avoid horseplay and loud talking so the mind is not distracted from the task at hand. Pushing, running, and scuffling while working with hand tools can result in serious accidents. Be alert and work defensively.

As has been said earlier, the greatest accident potential results from misuse and improper maintenance of hand tools. By adhering to the following procedures we can mitigate this problem:

- Hold supervisors responsible for the safe condition of tools and equipment used by workers, but workers must also use and maintain tools properly.
- Saw blades, knives, or other tools should be directed away from aisle areas and other workers working in close proximity.
- Knives and scissors must be kept sharp. Dull tools can result in the use of more force and slippage and are more dangerous than sharp tools.
- When working with hand knives, boning knives, drawknives, and scissors, workers should use appropriate PPE such as mesh gloves, wrist guards, arm guards, and aprons or belly guards.

- Avoid flammable substances since sparks produced by iron or steel hand tools can be a dangerous ignition source. Where these types of hazards exists use spark-resistant tools made of brass, plastic, aluminum, or wood.
- Do not overwork a tool's capabilities. Probably the most common error is to use a "cheater" to increase leverage of a wrench.
- Avoid striking one tool with another. Certain tools are made to strike other specific tools or materials. Use only the proper striking tools for these jobs. Do not use a wrench to drive a nail or use one hammer to strike another.
- Use the right tools and use them correctly. Even the best made tool will botch a job when used incorrectly. Striking a nail with a hammer cheek instead of its face can cause a nailing problem or accident. Handling a chisel incorrectly such as pushing a chisel with one hand while the other holds the work in front of the cutting edge can cause severe injury. The solution is to clamp in a vise so both hands are free to handle the tool.
- Improper maintenance. A worn tip on a screwdriver can result in a gashed hand. Similarly, a loose or damaged handle can turn a hammer into a deadly flying object.

15.1.2 STORING HAND TOOLS

All tools not in use should be stored where they are not a hazard. Sharp edges or pointed tools should have the edge or point guarded at all times when not in use. Shovels and rakes should have the sharp or pointed edges toward the ground.

15.1.3 OLD HAND TOOLS

Old tools may be unsafe if they lack up-to-date safety features. Instead of buying new good quality hand tools, many workers fill their toolboxes with hand-me-down or cheap (poorly made) tools from discount stores. Although second-hand tools may be cheaper, they can be quite unsafe. Any tool with makeshift repairs should be removed from service and discarded so no one is tempted to use it and be injured.

Any damaged or outdated tool should be removed from service and affixed with a tag that says, "Remove from Service. Do Not Use" or discarded as waste so it cannot be used again.

15.1.4 ERGONOMICS AND HAND TOOLS

The use of hand tools places a great deal of stress on bones, tendon, ligaments, nerves, and soft tissue. Often the use of hand tools can lead to what are called overuse or repetitive injuries. Some of the ways to prevent these injuries are by

- Alternating repetitive and nonrepetitive activities
- Varying or rotating job task
- Taking frequent, short breaks

- Doing gentle exercises during rest breaks
- Arranging work materials or equipment to avoid overreaching or twisting
- Ensuring that hand tools are well-balanced with a comfortable grip and need no more than reasonable force to operate
- Reviewing workloads to ensure they are realistic and within physical and psychological capabilities
- Performing jobs that need precise movements slightly above elbow level
- Performing jobs that need a lot of muscle strength slightly below elbow level

15.1.5 ERGONOMICALLY DESIGNED HAND TOOLS

Hand tools become a problem when workers have to use forceful muscular exertions due to having to hold or guide a tool using a very firm grip, having the wrist bent while using the tool, having a tool that is too heavy, or wearing gloves that are too large. Tools that cause heavy loading of the shoulder while holding the tool, especially when the arm is out from the body, are a problem. Wherever there is a possibility of repetitive movement, there is the potential for stress. At times contact stress occurs due to tools pressing into the palm at the base of the thumb where blood vessels and nerves pass through the hand. If the grip is too wide this can cause a tendon injury known as "trigger finger." Tools that transfer shock to the hands and wrist can also be culprits. Efforts should be made to mitigate these types of problems with hand tools.

There are some specific areas that need to be addressed in designing hand tools. The first is weight and size:

- Tool weight should be kept to less than 4 lb when used in one hand.
- For heavier tools sufficient space should be provided to grasp the tool with two hands.
- Grasping surfaces should be slip-resistant.
- Whenever possible, the edges and corners of tools should be rounded.

The handles on tools should fit the human hand as best as possible. Some of the general guidelines are as follows:

- For a power grip, larger handles are better.
- The thickness of a handle should be between 1 and 2.5 in.
- Hand strength is reduced by up to 30% when wearing gloves.
- If the diameter of a handle is too large, the fingers do not overlap, there is no "locking," and strain is sharply increased. If the diameter is too small, there is an insufficient friction area and the hand cuts into the hand.
- T-handles should be about 1 in. thick.

The length of handles is important since poor handle design can be detrimental to the hands. Some recommendations are as follows:

- The handle should be long enough so that they do not end in the palm of the hand especially pliers (Figure 15.4).
- Without gloves, handles should be 4 in. long at a minimum.
- With gloves, handles should be a minimum of 5 in. in length.

The surfaces and material that compose handles need careful consideration. All handles should be made from nonconductive materials. Thus, they should not conduct heat or electricity. Wood handle is often best for two reasons. Wood releases heat to the hand more slowly than plastic or metal and so it can be of help for a longer period of time before causing an injury. Wood gains heat more slowly than plastic, so it is less likely to reach high temperatures.

Handle should be compressible. Just as a compressible floor is easier on the feet and legs than noncompressible concrete, a compressible handle is easier on the hand. Wood is the best material. Compressible rubber or plastic is acceptable. Rubber-dipped coatings help make the handle more compressible, less conductive, and cover sharp edges that could damage the hand. Textured surfaces add grip as seen in Figure 15.5. Hand serration (finger grips) cut into the fingers since they were only designed to fit the hand that they were modeled from.

Hand tool posture is an important consideration when ergonomically designing tools. Bending the tool is superior to bending the wrist. Tendon movement while the wrist is not bent is less injurious. The most comfortable position is the "handshake" position.

Another alternative to changing the tool angle is to change the orientation of the work itself. Using the appropriate muscle group is less strain and stress.

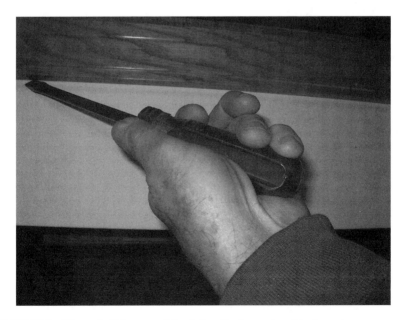

FIGURE 15.4 The end of the screwdriver's handle does not end in the palm.

FIGURE 15.5   The pliers' handles are textured and nonconductive.

Hand-closing muscles are stronger than hand-opening muscles. Use a spring to open hand tools.

15.2   HAND TOOL SPECIFIC SAFETY

15.2.1   ADJUSTABLE WRENCHES

Adjustable wrenches are torsion tools and are used for many purposes. They are not intended, however, to take the place of the standard open-ended, box, or socket wrenches. They are used mainly for nuts and bolts that do not fit a standard wrench. Pressure is applied to the fixed jaw.

15.2.2   AXES

When using an ax, make sure there is a clear circle in which to swing the ax before starting to chop. Remove all vines, brush, and shrubbery, especially overhead vines that may catch or deflect the ax.

Ax blades must be protected with a sheath or metal guard whenever possible. When the blade cannot be guarded, it is safer to carry the ax at one's side. The blade of a single-edged ax must be pointed down. The cutting edges are designed for cutting wood and equally soft metal. Never strike against metals, stone, or concrete. Some other important precautions to take with axes are as follows:

- Never use an ax as a wedge or maul, never strike with the sides, and never use it if the handle is loose or damaged.
- Proper ax grip for a right-handed person is to have the left hand about 3 in. from the end of the handle and the right hand about three-fourth of the way up. Reverse hands for left-handed individuals.

- Sharp, well-honed axes and hatchets are much safer to use because glancing is minimized.
- Safety glasses with side shields and safety shoes must be worn.

15.2.3   BOX AND SOCKET WRENCHES

Box and socket wrenches are used where a heavy pull is necessary and safety is a consideration. Box and socket wrenches completely encircle the nut, bolt, or fitting and grip all corners as opposed to two corners gripped by an open-ended wrench. They will not slip off laterally, and they eliminate the dangers of sprung jaws.

These types of torsion tools are very versatile. This is especially true of socket wrenches having great flexibility in hard-to-reach places. The use of special types must be encouraged where there is danger of injury.

Avoid overloading the capacity of a wrench by using a pipe extension (cheater) on the handle or striking the handle of a wrench with a hammer. Hammering on wrenches weakens the metal of a wrench and causes the tool to break. Special heavy-duty wrenches are available with handles as long as needed. Where possible, use penetrating oil to first loosen tight nuts.

15.2.4   CARPENTER'S OR CLAW HAMMER

This is a shock tool commonly in use and subject to a great deal of wear. The face of the hammer must be kept well dressed at all times to reduce the hazard of flying nails while they are being started into a piece of wood. A checkered face head is sometimes used to reduce this hazard. Eye protection must be worn when nailing and using a hammer to do work that could cause flying debris or material. Never use a common claw/nail hammer to strike other metal objects.

15.2.5   CHISELS

Choose a chisel only large enough for the job so that the blade is used, rather than the point or corner. Never use chisels with dull blades. Also, a hammer heavy enough to do the job should be used. The sharper the tool the better it will perform. Chisels that are bent, cracked, or chipped shall be discarded, Re-dress cutting edges or structure to original contour as needed. When chipping or shearing with a cold chisel, the tool is to be at an angle that permits one level of the cutting edge to be flat against the shearing plane.

Cold chisels should be selected based upon the materials to be cut, the size and shape of the tool, and the depth of the cut to be made. The chisel should be made heavy enough so that it will not buckle or spring when struck. Always wear safety goggles or a face shield when using a chisel. Do not use chisels for prying.

15.2.6   CROWBARS

Crowbars are types of prying tools and come in different sizes. Use the proper size for the job. Never use a makeshift device such as a piece of pipe, since they may slip and cause injury. Crowbars must have a point or toe capable of gripping the object

FIGURE 15.6 A typical crowbar.

to be moved and a heel to act as a pivot or fulcrum point. A block of wood under the heel may prevent slippage and help reduce injuries (Figure 15.6).

15.2.7 CUTTERS

Cutters used on wire, reinforcing rods, or bolts should be sharp enough to cut the material. If this is not the case, the jaws may be sprung or spread. Cutters require frequent lubrication. To keep cutting edges from becoming nicked or chipped, a cutter should not be used as a nail puller or pry bars. Cutter jaws have the hardness specified by the manufacturer for the particular kind of material to be cut. Cutting edges are spaced 0.003 in. apart when closed.

15.2.8 FILES

Selection of the right kind of file for the job will prevent injuries and lengthen the life of the file. Files are to be cleaned only with file-cleaning card or brush; never by striking. Never use a file as a pry bar or hammer, as chipping and breaking could result in user injury. For safe use, grip the file firmly in one hand and use the thumb and forefinger of the other to guide the point. A file should not be made into a center punch, chisel, or any other types of tools because the hardened steel may break.

A file should never be used without a smooth, crack-free handle: were the file to bind, the tang may puncture the palm of the hand, the wrist, or other body parts. Under some conditions, a clamp-on raised offset handle may be useful to give extra clearance for the hands.

Files are not to be used on lathe stock turning at high speeds (faster than three turns per file stroke) because the end of the file may strike the chuck, dog, or faceplate and throw the file (or metal chip) back at the operator hard enough to inflict serious injury.

15.2.9   HACKSAWS

Hacksaws should be adjusted in the frame to prevent buckling and breaking, but should not be tight enough to break off the pins that support the blades. Install blades with teeth pointing forward. Pressure should be applied on the forward stroke, not the back stroke. Lift the saw slightly, pulling back lightly to protect the teeth. If the blades twist or too much pressure is applied, the blades may break and cause injury to the hands or arm of the user. Never continue an old cut with a dull blade.

15.2.10   HAMMERS

A hammer is a shock tool. The head is to be securely affixed wedged handle for the particular type of head. The handle should be smooth, without cracks or splinters, free from oil, shaped to fit the hand, and of the specific size and length. The handles should be straight. Once split, the handles must be replaced. Some other common rules are as follows:

- Do not use a steel hammer on hardened steel surfaces. Instead use a soft-head hammer or one with a plastic, wood, or rawhide head.
- Safety goggles or safety glasses with side shields must be worn to protect against flying chips, nails, or other materials.
- Never strike a hammer with another hammer.
- Discard any hammer that shows chips, dents, etc. Redressing is not recommended.

15.2.11   HATCHETS

Hatchets must not be used for striking hard metal surfaces, since the tempered head may injure the user or others by rebounding or by creating flying chips. When using a hatchet in a crowed area, workers must take special care to prevent injury to themselves and others. Using a hatchet to drive nails is prohibited. Refer to Section 15.2.2 since the hatchet rules are quite similar.

15.2.12   KNIVES

Knives cause more disabling injuries than any other hand tool. The hazards are that the hand may slip from the handle on the blade or that the knife may strike the body or the free hand. Use knives with handle guards if possible. Knives are to be kept sharp and in their holders, cabinets, or sheaths when not in use. Knife strokes should always be away from the body. Use cut resistant gloves when using knives.

Never carry a sheath knife on the front part of a belt. Always carry it over the right or left hip, toward the back. This will prevent severing a leg artery or vein in case of a fall.

Knives must never be left lying on benches or in other places, where they may cause hand injuries. Safe placing and storing of knives are important in knife safety. Supervisors must provide ample room to those who work with knives so they are not in danger of being bumped by other workers. Supervisors should be particularly careful about the hazard of workers leaving knives hidden under a product, under scrap paper, or wiping rags, or among other tools in toolboxes or drawers. Knives are to be kept separate from other tools to protect the cutting edge of the knife as well as to protect the worker. Supervisors must assure that nothing that requires excessive pressure on the knife is undertaken by workers. Knives must not be used as a substitute for can openers, screwdrivers, or ice picks.

Do not wipe dirty or oily knives on clothing. Clean the blade by wiping it with a towel or cloth with the sharp edge away from the wiping hand. Horseplay of any kind (throwing, fencing, etc.) should be prohibited.

15.2.13   PIPE TONGS

Workers should neither stand nor jump on the tongs nor place extensions on the handles to obtain more leverage. They should use larger tongs.

15.2.14   PIPE WRENCHES

The pipe wrench is another example of a torsion tool. Pipe wrenches, both straight and chain tong, must have sharp jaws and be kept clean to prevent slipping. The adjusting nut of the wrench should be inspected frequently. If it is cracked, the wrench must be taken out of service. A cracked nut may break under strain, causing complete failure of the wrench and possible injury to the user (Figure 15.7).

A piece of pipe "cheater" slipped over the handle must not be used to give added leverage because this can strain a pipe wrench to the breaking point. The handle of every wrench is designed to be long enough for the maximum allowable safe pressure. Get a larger pipe wrench to do the job since they come in all sizes.

FIGURE 15.7   Select the proper size pipe wrench for the job.

15.2.15 PLIERS

There are many types and sizes of pliers. Pliers should not be used as a substitute for wrenches. Pliers that are cracked, broken, or sprung should be removed from service.

Pliers should not be used as a hammer nor should they be hammered upon. Pliers' grips should be kept free of grease or oil, which could cause them to slip.

Side-cut pliers sometimes cause injuries when short ends of wire are cut. A guard over the cutting edge and the use of safety glasses with side shields will help prevent eye injuries.

The handles of electricians' pliers must be insulated. In addition, employees must wear the proper electrical rated gloves if they are working on energized lines or circuits.

15.2.16 PUNCHES

Punches are never to be used if the face is mushroomed or with a dull, chipped, or deformed point. Punches that are bent, cracked, or chipped shall be discarded. Safety glasses with side shields should be used when using a punch.

15.2.17 RIVETING HAMMERS

A riveting hammer is another example of a shock tool, often used by sheet metal workers, and must have the same kind of use and care as a ball peen hammer and should be watched closely for cracked or chipped faces.

15.2.18 SHOVELS

The shovel is a useful tool. The edges should be kept trimmed and handles checked for splinters and cracks. Use safety shoes with sturdy soles and gloves when shoveling. Proper shoveling posture requires that the feet be well separated to get good balance and spring in the knees. The leg muscles will take much of the load. To reduce the chance of injury, use the ball of the foot (not the arch) to press the shovel into the ground or other material. Never twist the torso when shoveling, move the feet instead.

Dipping a shovel in water, greasing it, or waxing it will prevent some materials from sticking to it. When not in use, keep them hanging against the wall, or keep them in racks or boxes.

15.2.19 SCREWDRIVERS

A screwdriver is the most commonly used and abused tool. The practice of using screwdrivers as punches, wedges, pinch bars, or pry bars should be discouraged as this practice dulls the blade and causes worker injuries. Screwdrivers should be selected to fit the screw. Sharp-edged bits will not slip as easily as ones that are dull. Re-dress tips to original shape and keep them clean. Always hold work in a vise or lay it on a flat surface to lessen the chance of injury were the screwdriver to slip. Other guidelines to keep in mind are as follows:

- Do not hold work piece against your body while using the screwdriver.
- Do not put your finger near the blade of screwdriver when tightening a screw.
- Do not force a screwdriver by using a hammer or pliers on it.
- Do not use a screwdriver as a hammer or as a chisel.
- Do not use a screwdriver if your hands are wet or oily.
- Discard and replace any screwdriver if it has a broken handle, bent blade, etc.
- Use an insulated screwdriver both handle and blade when performing any electrical work.
- Cross-slot (Phillips-head) screwdrivers are safer than the square bit types, because they slip less. The tip must be kept clean and sharp to permit a good grip on the head of the screw.

15.2.20 SPECIAL CUTTERS

Special cutters include those for cutting banding wire and strap. Claw hammers and pry bars must not be used to snap metal banding material.

15.2.21 TAP AND DIE WORK

Tap and die work requires certain precautions. The work should be firmly mounted in a vise. Only a T-handle wrench or adjustable tap wrench should be used. When threads are being cut with a hard die, hands and arms should be kept clear of the sharp threads coming through the die, and metal cutting should be removed with a brush.

15.2.22 TIN SNIPS

Tin snips should be heavy duty enough to cut the materials such that the worker needs only one hand on the snips and can use the other to hold the material. The material should be well supported before the last cut is made so that cut edges do not press against the hands. The proper snip is to be used for right and left hand cuts and straight cuts. Jaws of snips are to be kept tight and well lubricated.

Workers must wear protective safety eyewear with side shields or goggles when trimming corners or slivers or metal because small particles often fly with considerable force. They must also wear cut resistant gloves or leather gloves.

15.2.23 WOOD CHISELS

Wood chisels are wood cutting tools. Inexperienced workers must be instructed in the proper method of holding and using wood chisels. Handles are to be free of splinters. The wood handle of a chisel struck by a mallet is to be protected by a metal or leather cap to prevent splitting. The object must be free of nails to avoid damage to the blade or cause a chip to fly into the user's face or eye. Drive a wood chisel outward and away from your body. Users should wear safety eyewear with side shields.

15.2.24 WRENCHES

Open-end or box wrenches must be inspected to make sure that they fit properly and are never to be used if the jaws are sprung or cracked. When defective they must be taken out of service and repaired or replaced. Further information regarding wrenches is as follows:

- Select the correct size wrench for the job. Wrenches come in metric and SAE (Society of Automotive Engineers)—the American standard size.
- Never use a pipe as a wrench handle extension (cheater).
- Stand in a balanced position to avoid sudden slips when using a wrench.
- Do not use a wrench if your hands are oily or greasy.

15.3 USE OF HAND TOOLS BY THE SERVICE INDUSTRY

Many industry sectors may require hand tools to accomplish work tasks. This is especially true for those who do repairs, servicing, assembling, and maintenance activities. The sectors that most require hand tools are the utilities, warehousing, wholesale, retail, telecommunications (information), other services, and maintenance personnel in the leisure, hospitality, education, health care, and administration sectors. Each industry sector may use a variety of hand tools or ones specific to their particular industry.

15.4 SUMMARY OF OSHA REGULATION FOR HAND TOOLS (29 CFR 1910.242)

Hand and power tools are a common part of our everyday lives and are present in nearly every industry. These tools help us to easily perform tasks that otherwise would be difficult or impossible. However, these simple tools can be hazardous, and have the potential for causing severe injuries when used or maintained improperly. Special attention toward hand and power tool safety is necessary to reduce or eliminate these hazards.

Hand tools are nonpowered. They include anything from axes to wrenches. The greatest hazards posed by hand tools result from misuse and improper maintenance. The following are some examples. Using a screwdriver as a chisel may cause the tip of the screwdriver to break and fly, hitting the user or other employees; if a wooden handle on a tool such as a hammer or an ax is loose, splintered, or cracked, the head of the tool may fly off and strike the user or another worker; a wrench must not be used if its jaws are sprung, because it might slip; or impact tools such as chisels, wedges, or drift pins are unsafe if they have mushroomed heads. The heads might shatter on impact, sending sharp fragments flying.

The employer is responsible for the safe condition of tools and equipment used by employees but the employees are responsible for properly using and maintaining tools. Employers should caution employees that saw blades, knives, or other tools be directed away from aisle areas and other employees working in close proximity. Knives and scissors must be sharp. Dull tools can be more hazardous than sharp ones. Appropriate PPE, e.g., safety goggles, gloves, etc., should be worn due to hazards that may be encountered while using portable power tools and hand tools.

Safety requires that floors be kept as clean and as dry as possible to prevent accidental slips with or around dangerous hand tools. Around flammable substances, sparks produced by iron and steel hand tools can be a dangerous ignition source. Where this hazard exists, spark-resistant tools made of brass, plastic, aluminum, or wood will provide safety.

Employees who use hand and power tools and who are exposed to the hazards of falling, flying, abrasive, and splashing objects, or exposed to harmful dusts, fumes, mists, vapors, or gases must be provided with the particular personal equipment necessary to protect them from the hazard.

15.5 HAND TOOL CHECKLIST

A checklist ensures that hand tools are in proper working order and being used as intended. It also guides compliance with Occupational Safety and Health Administration (OSHA) regulations and company rules and policies. Figure 15.8 provides a hand tool safety checklist.

15.6 SUMMARY

Employers should provide workers with a variety of hand tools to help them work quickly, reliably, and safely. Some general guidelines need to be followed by those using hand tools:

Hand tools and equipment checklist

Yes☐ No☐ Are all tools and equipment (both company and employee owned) used by employees at their workplace in good condition?

Yes☐ No☐ Are hand tools such as chisels and punches, which develop mushroomed heads during use, reconditioned or replaced as necessary?

Yes☐ No☐ Are broken or fractured handles on hammers, axes, and similar equipment replaced promptly?

Yes☐ No☐ Are worn or bent wrenches replaced regularly?

Yes☐ No☐ Are appropriate handles used on files and similar tools?

Yes☐ No☐ Are employees made aware of the hazards caused by faulty or improperly used hand tools?

Yes☐ No☐ Are appropriate safety glasses, face shields, etc. used while using hand tools or equipment which might produce flying materials or be subject to breakage?

Yes☐ No☐ Are jacks checked periodically to ensure they are in good operating condition?

Yes☐ No☐ Are tool handles wedged tightly in the head of all tools?

Yes☐ No☐ Are tools' cutting edges kept sharp so the tool will move smoothly without binding or skipping?

Yes☐ No☐ Are tools stored in dry, secure locations where they would not be tampered with?

Yes☐ No☐ Is eye and face protection used when driving hardened or tempered spuds or nails?

FIGURE 15.8 Hand tools and equipment checklist.

- Use the right tool for the right job.
- Keep all tools in good condition with regular maintenance.
- Know the application, limitations, and potential hazards of the tool in use.
- Use all tools according to the manufacturer's instructions.
- Use eye protection and appropriate PPE.
- Keep guards in place, in working order, and properly adjusted.
- Maintain clutter free work areas.
- Remain alert to the potential hazards in the working environment such as slippery floors or the presence of highly combustible materials.

Hand tools are so common that it is difficult to always be on the alert for the potential hazards and of the preventive measures required to avert them. Therefore, in an effort to minimize accidents resulting from the use of hand tools, certain precautions need to be taken, such as the following:

- Do not use broken, defective, burned, or mushroomed tools. Report defective tools to your supervisor and turn tools in for replacement.
- Always use the proper tool and equipment for any task you may be assigned to do. For example, do not use a wrench as a hammer, or a screwdriver as a chisel.
- Do not leave tools on scaffolds, ladders, or any overhead working surfaces. Racks, bins, hooks, or other suitable storage space must be provided to permit convenient arrangement of tools.
- Do not strike two hardened steel surfaces together (i.e., two hammers, or a hammer and hardened steel shafts, bearings, etc.).
- Do not throw tools from one location to another, from one worker to another, or drop them to lower levels; this is prohibited. When this type of passing is required, suitable containers and/or ropes must be used.
- Wooden tool handles must be sound, smooth, in good condition, and securely fastened to the tool.
- Sharp-edged or pointed tools should never be carried in an employee's pockets.
- Only nonsparking tools shall be used in locations where sources of ignition may cause a fire or explosion.
- Tools requiring heat treating should be tempered, formed, dressed, and sharpened by workmen experienced in these operations.
- Wrenches, including adjustable, pipe, end, and socket wrenches are not to be used when jaws are sprung to the point that slippage occurs.
- Any defective tool should be removed from service and tagged indicating it is not to be used.

16 Ladder Safety

Ladders are one of the most useful tools in the workplace.

When properly used ladders are one of the most useful tools available to workers. Improper uses, including the following, can result in injuries, falls, and deaths:

- Ladders placed on unstable surfaces
- Personnel reaching too far out to the sides (overreaching)
- Personnel standing too high on the ladder to maintain balance
- Defective or broken ladders (e.g., broken rails or rungs and missing hardware)
- Ladders that were not secured or braced
- Personnel carrying loads while ascending and descending
- Selecting the wrong ladder for the job (Figure 16.1)
- Improper position on the ladder
- Adverse weather such as strong winds, rain, ice, or snow
- Using a ladder to work on electrical conductors or power lines

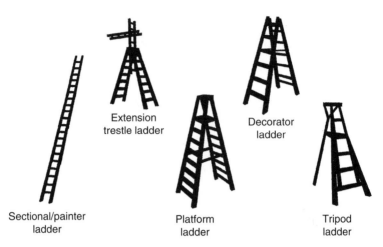

Extension trestle ladder

Decorator ladder

Sectional/painter ladder

Platform ladder

Tripod ladder

FIGURE 16.1   Select the proper ladder for the task. (Courtesy of Department of Energy.)

This type of improper use of ladders results in an estimated 19,000 injuries and 24 fatalities per year according to Occupational Safety and Health Administration (OSHA). Falls or slips account for 80% of accidents and almost half fell at least 8 ft. While falling, 50% held on to other objects while 66% were not trained on how to inspect ladders and 73% were not instructed on the safe use of ladders. In 73% of accidents, the ladders involved were extension or straight ladders and in 20% they were step ladders.

16.1 LADDER SAFETY PROGRAM

16.1.1 MANAGEMENT'S COMMITMENT

Employers need to ensure that they have policies for the use of ladders and workers need to understand that they are expected to comply with them such as the example in Figure 16.2.

16.1.2 TRAINING

Employers have a tendency to think that all workers know how to use ladders in a safe manner. Many times it is the experienced worker who has a ladder accident. All

---

Company ladder safety policy

The _____ company is committed to preventing ladder related accidents. It is expected that the company rules and policies for ladder safety are to be followed by all employees. The company is committed to providing you with safe and appropriate ladders for your work activities. It is your responsibility to use ladders safely in accordance with the company's rules and policies and the ladder training which you have received.

_____Company President's Signature

FIGURE 16.2   Ladder safety and use policy.

Ladder inspection form						
Location:				**Date:**		
Name of Inspector	Ladder Serial #	Type	Broken Parts	Damage to Wood, Metal, Fiberglass	Soundness Structurally	Any Other Problems

FIGURE 16.3 Ladder inspection form.

employees including management and supervisors are to receive ladder safety training. The content of the training shall include the following:

- Company rules and policies
- Accountability and responsibility
- Ladder hazard recognition
- Reporting and responding to ladder hazards
- Safe use of ladders
- Climbing safely

16.1.3 HAZARD IDENTIFICATION

Before using a ladder all employees should inspect the ladder in accordance with the company's ladder checklist (such as the one found at the end of this chapter) and use a ladder inspection form to document the inspection (Figure 16.3). If a faulty ladder having structural defects such as, but not limited to, broken or missing rungs, cleats, steps, broken or split rails, corroded components, or other defective components is found, the ladder should be removed from service and marked prominently as defective or tagged with "Remove from Service. Do Not Use" until repaired or discarded. The supervisor should be notified of the unsafe ladder. The supervisor must assure the ladder is not used and it is either disposed of or repaired to a safe condition. Any ladder accidents should be investigated thoroughly by the supervisor and recommendations to prevent further occurrences should be made and implemented.

16.1.4 LADDER SELECTION

A ladder should be selected based on its use and capacity. Ladders are classified in the following manner relevant to their load capacity (Table 16.1).

TABLE 16.1
Ladder Use and Load Capacities

Type	Grade	Duty (Load) Rating (lb)
III	Household	200
II	Commercial	225
I	Industrial	250
IA	Extra heavy duty industrial	300

TABLE 16.2

Stepladder Size Selection Chart

Maximum Height You Want to Reach (ft)	Purchase This Size Stepladder Level (ft)	Highest Standing Level
7	3	11 in.
8	4	1 ft 11 in.
9	5	2 ft 10 in.
10	6	3 ft 10 in.
11	7	4 ft 9 in.
12	8	5 ft 8 in.
14	10	7 ft 7 in.
16	12	9 ft 6 in.
18	14	11 ft 5 in.
20	16	13 ft 4 in.

Having the correct size ladder is imperative since using too short or too long a ladder is a common mistake that leads to potential hazards. You can use the distance between the rungs (1 ft) to estimate the height requirements. As for stepladders, the highest permitted standing level is two steps from the top. A worker standing high might lose his/her balance and fall. The maximum safe reaching height is about 4 ft higher than the length of the ladder. For example, a typical worker can reach 10 ft with a 6 ft ladder (Table 16.2).

Extension ladders should be 7–10 ft longer than the highest support or contact point, which may be the wall or roof line. This will allow enough length for proper setup, overlap ladder sections (3 ft), height restrictions for the highest standing level, and, where appropriate, the extension of the ladder above the roof line. The highest standing level is four rungs from the top. Never stand on the ladder above the support points. Table 16.3 depicts the selection of length for extension ladders.

The following points must be kept in mind when selecting a ladder: it should have unbroken rungs or steps and safety feet, functional spreaders that lock, it should comply with ANSI standards, must be of the right size, should not have

TABLE 16.3

Extension Ladder Size Selection Chart

Height to Top Support Point (ft)	Buy This Size Extension Ladder (ft)	Maximum Working Ladder Length (ft)	Highest Standing Level
9	16	13	9 ft 2 in.
9–13	20	17	13 ft 1 in.
13–17	24	21	16 ft 11 in.
17–21	28	25	20 ft 10 in.
21–25	32	29	24 ft 8 in.
25–28	36	32	27 ft 7 in.
28–31	40	35	30 ft 6 in.

been painted or varnished, and should be made of fiberglass when used near electrical conductors since metal and wet wooden ladders will conduct electricity.

16.1.5 CONTROL AND PREVENTION

As part of control and prevention of ladder accidents the following guidelines are to be used as constant reminders for the safe use of ladders. The setup of the ladder is very important in preventing accidents such as the following:

- Place it on a level surface.
- Use wide boards under it if you are on soft ground.
- Place the feet parallel with the top support.
- Anchor the top.
- A straight ladder should extend 3 ft past the support point.
- Tie or brace it at the bottom or have someone hold it.
- Keep the ladder the right distance from the wall or support (use the 4 to 1 rule or 75.5) as seen in Figure 16.4. Raise an extension ladder before extending it.
- If you place a ladder in front of a door, make sure the door is locked or blocked.
- Make sure areas of high traffic are barricaded around the ladder.
- Maintain good housekeeping around the bottom of the ladder.

As further guidance, the following dos and don'ts should be adhered to for preventing accidents and injuries.

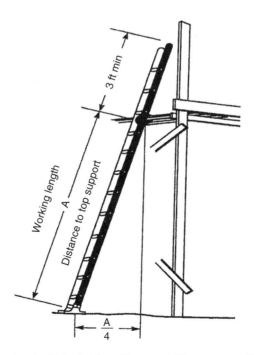

FIGURE 16.4 The 4 to 1 rule for ladders. (Courtesy of Department of Energy.)

16.2   DOS

- Clean your shoes first to remove mud, oil, and debris.
- Allow only one person at a time on a ladder.
- Face the ladder while climbing up or down.
- Have three points of contact when climbing (two hand and one foot or two feet and one hand).
- Use the side rails for grip while climbing instead of the rungs.
- Hold the ladder with both hands while climbing.
- Hold the ladder with one hand while working.
- Use a hanger or tool pouch for tools or a bucket.
- Keep your weight centered between the rails.
- Use nonslip gloves when climbing.

16.3   DON'TS

- Try to use a ladder if it is scaffolding that you really need.
- Carry objects while climbing. Use a special belt, tool pouch, or hoist materials up with a rope.
- Step on the top two stepladder steps or the top three ladder rungs of an extension ladder.
- Lean too far in either direction while working on a ladder.
- Let your belt buckle go outside the rails.
- Go near power lines or electricity with metal or wooden ladders.
- Join or tie ladders together to have it extended.

16.4   USE OF LADDERS BY THE SERVICE INDUSTRY

Many industry sectors may require ladders to extend a worker's reach. The sectors that most require ladders are the utilities, warehousing, wholesale trade, retail trade, telecommunication (information), other service, and maintenance personnel in the leisure, hospitality, education, health care, and administration sectors.

Depending on the frequency of use, the degree of training and attention to ladder safety will be determined by the amount of emphasis that is needed to continue a program for the safe use of ladders.

16.5   OSHA LADDER STANDARDS

The following is a summary of the OSHA standards for ladders and not the complete regulation.

16.5.1   Fixed Ladders (29 CFR 1910.27)

A fixed ladder must be able to support at least two loads of 250 lb each, concentrated between any two consecutive rungs. Fixed ladders must also support added anticipated loads caused by ice buildup, winds, rigging, and impact loads resulting from the use of ladder safety devices. Fixed ladders must be used at a pitch no greater than 90° from the horizontal, measured from the rear of the ladder.

Individual rung/step ladders must extend at least 42 in. above an access level or landing platform either by the continuation of the rung spacings as horizontal grab bars or by providing vertical grab bars that must have the same lateral spacing as the vertical legs of the ladder rails. Each step or rung of a fixed ladder must be able to support a load of at least 250 lb applied in the middle of the step or rung.

The minimum clear distance between the sides of individual rung/step ladders and between the side rails of other fixed ladders must be 16 in. The rungs of individual rung/step ladders must be shaped to prevent slipping off the end of the rungs. The rungs and steps of fixed metal ladders manufactured after March 15, 1991, must be corrugated, knurled, dimpled, coated with skid-resistant material, or treated to minimize slipping. The minimum perpendicular clearance between fixed ladder rungs, cleats, and steps and any obstruction behind the ladder must be 7 in., except that the clearance for an elevator pit ladder must be 4.5 in. The minimum perpendicular clearance between the centerline of fixed ladder rungs, cleats, and steps, and any obstruction on the climbing side of the ladder must be 30 in. If obstructions are unavoidable, clearance may be reduced to 24 in., provided a deflection device is installed to guide workers around the obstruction. The step-across distance between the center of the steps or rungs of fixed ladders and the nearest edge of a landing area must be no less than 7 in. and no more than 12 in. A landing platform must be provided if the step-across distance exceeds 12 in. (30 cm). Fixed ladders without cages or wells must have at least a 15 in. clear width to the nearest permanent object on each side of the centerline of the ladder.

Fixed ladders must be provided with cages, wells, ladder safety devices, or self-retracting lifelines where the length of climb is less than 24 ft but the top of the ladder is at a distance greater than 24 ft above lower levels. If the total length of the climb on a fixed ladder equals or exceeds 24 ft, the following requirements must be met: fixed ladders must be equipped with either (1) ladder safety devices; (2) self-retracting lifelines and rest platforms at intervals not to exceed 150 ft; or (3) a cage or well, and multiple ladder sections, each ladder section not to exceed 50 ft in length. These ladder sections must be offset from adjacent sections, and landing platforms must be provided at maximum intervals of 50 ft (Figure 16.5).

The side rails of through- or side-step-fixed ladders must extend 42 in. above the top level or landing platform served by the ladder. Parapet ladders must have an access level at the roof if the parapet is cut to permit passage through it; if the parapet is continuous, the access level is the top of the parapet. Steps or rungs for through-fixed-ladder extensions must be omitted from the extension; and the extension of side rails must be flared to provide between 24 and 30 in. clearance between side rails. When safety devices are provided, the maximum clearance distance between side rail extensions must not exceed 36 in.

Cages must not extend less than 27 in., or more than 30 in. from the centerline of the step or rung, and must not be less than 27 in. wide. The inside of the cage must be clear of projections.

Horizontal bands must be fastened to the side rails of rail ladders or directly to the structure, building, or equipment for individual rung ladders. Horizontal bands

FIGURE 16.5   Example of a fixed ladder.

must be spaced at intervals not more than 4 ft apart measured from centerline to centerline.

Vertical bars must be on the inside of the horizontal bands and must be fastened to them. Vertical bars must be spaced at intervals not more than 9.5 in., measured centerline to centerline.

The bottom of the cage must be between 7 and 8 ft above the point of access to the bottom of the ladder, and the bottom of the cage must be flared no fewer than 4 in. between the bottom horizontal band and the next higher band. The top of the cage must be a minimum of 42 in. above the top of the platform or the point of access at the top of the ladder. Provisions must be made for access to the platform or any other point of access.

Wells must completely encircle the ladder. Wells must be free of projections. The inside face of the well on the climbing side of the ladder must extend between 27 and 30 in. from the centerline of the step or rung. The inside width of the well must be at least 30 in. The bottom of the well above the point of access to the bottom of the ladder must be between 7 and 8 ft.

All safety devices must be able to withstand, without failure, a drop test consisting of a 500 lb weight dropping 18 in. They must permit the worker to ascend or descend without continually having to hold, push, or pull any part of the device, leaving both hands free for climbing. All safety devices must be activated within 2 ft after a fall occurs, and limit the descending velocity of an employee to 7 ft/s or less. The connection between the carrier or lifeline and the point of attachment to the body harness must not exceed 9 in. in length.

Mountings for rigid carriers must be attached at each end of the carrier, with intermediate mountings, spaced along the entire length of the carrier, to provide the necessary strength to stop workers' falls. Mountings for flexible carriers must be attached at each end of the carrier. Cable guides for flexible carriers must be installed with a spacing between 25 and 40 ft along the entire length of the carrier, to prevent wind damage to the system. The design and installation of mountings and cable guides must not reduce the strength of the ladder. Side rails and steps or rungs for side-step fixed ladders must be continuous in extension.

Fixed ladders with structural defects—such as broken or missing rungs, cleats, or steps, broken or split rails, or corroded components—must be withdrawn from service until repaired. Defective fixed ladders are considered withdrawn from use when they are (1) immediately tagged with "Do Not Use" or something to that effect, (2) marked in a manner that identifies them as defective, or (3) blocked such as with a plywood attachment that spans several rungs.

16.5.2 Portable Ladders (29 CFR 1910.25 and .26)

Non-self-supporting and self-supporting portable ladders must support at least four times the maximum intended load; extra heavy-duty type 1A metal or plastic ladders must sustain 3.3 times the maximum intended load. The ability of a self-supporting ladders to sustain loads must be determined by applying the load to the ladder in a downward vertical direction. The ability of a non-self-supporting ladder to sustain loads must be determined by applying the load in a downward vertical direction when the ladder is placed at a horizontal angle of 75.5°.

When portable ladders are used for access to an upper landing surface, the side rails must extend at least 3 ft above the upper landing surface. When such an extension is not possible, the ladder must be secured, and a grasping device such as a grab rail must be provided to assist workers in climbing up and down. A ladder extension must not deflect under a load that would cause the ladder to slip off its supports.

Ladders must be maintained free of oil, grease, and other slipping hazards. Ladders must not be loaded beyond the maximum intended load for which they were built nor beyond their manufacturer's rated capacity. Ladders must be used only for the purpose for which they were designed. Non-self-supporting ladders must be used at an angle where the horizontal distance from the top support to the foot of the ladder is approximately one-quarter of the working length of the ladder. Wood job-made ladders with spliced side rails must be used at an angle where the horizontal distance is one-eighth the working length of the ladder.

Ladders must be used only on stable and level surfaces unless secured to prevent accidental movement. Ladders must not be used on slippery surfaces unless secured or provided with slip-resistant feet to prevent accidental movement. Slip-resistant feet must not be used as a substitute for the care in placing, lashing, or holding a ladder upon slippery surfaces. Ladders placed in areas such as passageways, doorways, or driveways, or where they can be displaced by workplace activities or traffic, must be secured to prevent accidental movement or a barricade must be used to keep traffic or activities away from the ladder. The area around the top and bottom of the ladders must be kept clear.

The top of a non-self-supporting ladder must be placed with two rails supported equally unless it is equipped with a single support attachment. Ladders must not be displaced or extended while in use. Ladders must have nonconductive side rails if used in proximity to exposed energized electrical equipment.

The top step of a stepladder must not be used as a step. Crossbracing on the rear section of stepladders must not be used for climbing unless the ladders are designed and provided with steps for climbing on both front and rear sections.

Ladders must be periodically inspected by a competent person for visible defects and after any incident that could affect their safe use. Single-rail ladders must not be used.

When ascending or descending a ladder, the worker must face the ladder. Each worker must use at least one hand to grasp the ladder when climbing. A worker on a ladder must not carry any object or load that could cause him/her to lose balance and fall.

A double-cleated ladder or two or more ladders must be provided when ladders are the only way to enter or exit a work area having 25 or more employees, or when a ladder serves simultaneous two-way traffic. Ladder rungs, cleats, and steps must be parallel, level, and uniformly spaced when the ladder is in position for use. Rungs, cleats, and steps of portable and fixed ladders (except as provided below) must not be spaced less than 10 in. apart, nor more than 14 in. apart, along the ladder's side rails. Rungs, cleats, and steps of step stools must not be less than 8 in. apart, nor more than 12 in. apart, between centerlines of the rungs, cleats, and steps.

Ladders must not be tied or fastened together to create longer sections unless they are specifically designed for such use. A metal spreader or locking device must be provided on each stepladder to hold the front and back sections in an open position when the ladder is being used. Two or more separate ladders used to reach an elevated work area must be offset with a platform or landing between the ladders, except when portable ladders are used to gain access to fixed ladders.

Ladder components must be surfaced to prevent injury from punctures or lacerations, and prevent snagging of clothing. Wooden ladders must not be coated with any opaque covering, except for identification or warning labels which may be placed only on one face of a side rail.

Portable ladders with structural defects—such as broken or missing rungs, cleats, or steps, broken or split rails, corroded components or other faulty or defective components—must immediately be marked defective, or tagged with "Do Not Use" or something to that effect and withdrawn from service until repaired. Ladder repairs must restore the ladder to its original design before the ladder can be reused.

Under the provisions of the OSHA standards, employers must provide a training program for each employee using ladders and stairways. The program must enable each employee to recognize hazards related to ladders and stairways and to use proper procedures to minimize these hazards. For example, employers must ensure that each employee is trained by a competent person in the following areas, as applicable:

- The nature of fall hazards in the work area
- The correct procedures for erecting, maintaining, and disassembling the fall protection systems to be used
- The proper construction, use, placement, and care in handling of all stairways and ladders
- The maximum intended load-carrying capacities of ladders used

16.6 LADDER CHECKLIST

A checklist is an excellent method to determine ladder safety and OSHA compliance. Figure 16.6 depicts an example of a ladder checklist.

Ladder checklist

Answer the following questions yes or no to determine compliance or presence of ladder hazards:

Yes ☐ No ☐ Are only Type 1 or Type 1A industrial ladders are used?
Yes ☐ No ☐ Do steps on ladders a minimum load capacity of 250 lb?
Yes ☐ No ☐ Are all ladders inspected for damage before use?
Yes ☐ No ☐ Are ladders shall not placed against movable objects?
Yes ☐ No ☐ Are ladders placed to prevent movement by lashing or other means?
Yes ☐ No ☐ Are employees shoes are free of mud, grease, or other substances that could cause a slip or fall?
Yes ☐ No ☐ Are ladders not placed on unstable bases such as boxes or barrels?
Yes ☐ No ☐ Do employees not on the top two steps of a stepladder?
Yes ☐ No ☐ Is a ladder used to gain access to a roof extends at least 3 ft above the point of support, at eave, gutter, or roof line?
Yes ☐ No ☐ Are stepladders fully opened to permit the spreaders to lock?
Yes ☐ No ☐ Are all labels in place and legible on ladders?
Yes ☐ No ☐ Are ladder always moved to prevent and avoid overreaching?
Yes ☐ No ☐ Are single ladders not more than 30 ft in length?
Yes ☐ No ☐ Do extension ladders up to 36 ft have a 3 ft overlap between sections?
Yes ☐ No ☐ Do extension ladders over 36 ft and up to 48 ft have a 4 ft overlap between sections?
Yes ☐ No ☐ Do extension ladders over 48 ft and up to 60 ft have a 5 ft overlap between sections?
Yes ☐ No ☐ Do two-section extension ladders not exceed 48 ft in total length?
Yes ☐ No ☐ Do ladders ever two-section not exceed 60 ft in total length?
Yes ☐ No ☐ Are ladders not used horizontally as scaffolds, runways, or platforms?
Yes ☐ No ☐ Is the area around the top and base of ladders kept free of tripping hazards such as loose materials, trash, cords, hoses, and leaves?
Yes ☐ No ☐ Is the base of a straight or extension ladders set back a safe distance from the vertical or approximately 1/4 of the working length of the ladder?

FIGURE 16.6 Ladder safety checklist.

(*continued*)

Yes ☐   No ☐   Are ladders that project into passageways or doorways where they could be struck by personnel, moving equipment, or materials being handled, protected by barricades or guards?

Yes ☐   No ☐   Do employees face the ladder when ascending or descending?

Yes ☐   No ☐   Do employees must use both hands when going up or down a ladder?

Yes ☐   No ☐   Are materials or equipment raised or lowered by way of lines?

Yes ☐   No ☐   Are employees trained and educated on the proper use of ladders?

Yes ☐   No ☐   Are repairs done professionally?

Yes ☐   No ☐   Are inspections conducted before each use and defective, broken, or damaged ladders shall be pulled from service tagged and marked "Dangerous. Do Not Use?"

Yes ☐   No ☐   Are the rungs tight in the joint of the side rails?

Yes ☐   No ☐   Do all moving parts operate freely without binding?

Yes ☐   No ☐   Are all pulleys, wheels, and bearings lubricated frequently?

Yes ☐   No ☐   Are rungs kept free of grease and oil?

Yes ☐   No ☐   Is rope that is badly worn or frayed replaced immediately?

Yes ☐   No ☐   Are all ladders equipped with slip-resistant feet, free of grease, and in good condition?

Portable wood ladders

Yes ☐   No ☐   Are all wood ladders free of splinters, sharp edges, shake, wane, compression failures, decay, and other irregularities?

Yes ☐   No ☐   Are portable stepladders no longer than 20 ft?

Yes ☐   No ☐   Is the step spacing no more than 12 in. apart?

Yes ☐   No ☐   Are stepladders which have a metal spreader or locking device of sufficient strength and size to hold the front and back when open?

Portable metal ladders

Yes ☐   No ☐   Are ladders inspected immediately when dropped or tipped over?

Yes ☐   No ☐   Are the step spacing no more than 12 in. apart?

Yes ☐   No ☐   Are metal ladders not for electrical work or in areas where they could contact energized conductors?

Fixed ladders

Yes ☐   No ☐   Are the steps shall no more than 12 in. apart?

Yes ☐   No ☐   Are job made ladders constructed to conform with the established OSHA standards.

Yes ☐   No ☐   Are all fixed ladders painted or treated to prevent rusting?

Yes ☐   No ☐   Do fixed ladders 20 ft or higher have a landing every 20 ft if there is no surrounding cage?

Yes ☐   No ☐   If it has a cage or safety device, a landing is required every 30 ft?

FIGURE 16.6 (continued)

17 Lifting

Lifting is an integral part of goods and material handling.

It is probably safe to assume that workers in retail, wholesale, and warehousing perform a large number of lifting tasks as an integral part of their job duties some of which could be both heavy and at times awkward. Employers in these industry sectors should lift carefully since overexertion and back injuries are the leading causes of injuries. It should not be a foregone conclusion that back injuries are an acceptable part of doing business. Efforts should be directed toward prevention although many employers as well as workers view lifting as a natural activity that everyone knows how to do correctly. This has proven to be a false belief.

17.1 BACK INJURIES

Back disorders can develop gradually as a result of microtrauma brought about by repetitive activity over time or can be caused by a single traumatic event. Because of the slow and progressive onset of this internal injury, the condition is often ignored until the symptoms become acute, often resulting in disabling injury. Acute back injuries can be the immediate result of improper lifting techniques and/or lifting loads that are too heavy for the back to support. While the acute injury may seem to be caused by a single well-defined incident, the real cause is often a combined interaction of the observed stressor coupled with years of weakening of the musculo-skeletal support mechanism by repetitive microtrauma. Injuries can arise in muscle, ligament, vertebrae, and disks, either singly or in combination.

Although back injuries account for no work-related deaths, they do account for a significant amount of human suffering, loss of productivity, and economic burden on compensation systems. Back disorders are one of the leading causes of disability for people in their working years and afflict over 600,000 employees each year with a cost of about $50 billion annually in 1991 according to National Institute for Occupational Safety and Health (NIOSH). The frequency and economic impact of back injuries and disorders on the workforce are expected to increase over the next several decades as the average age of the workforce increases and medical costs go up.

17.2 BACK DISORDERS

17.2.1 FACTORS ASSOCIATED WITH BACK DISORDERS

Back disorders result from exceeding the capability of the muscles, tendons, disks, or the cumulative effect of several contributors:

- Reaching while lifting
- Poor posture—how one sits or stands (Figure 17.1)
- Stressful living and working activities—staying in one position for too long
- Bad body mechanics—how one lifts, pushes, pulls, or carries objects
- Poor physical condition—losing the strength and endurance to perform physical tasks without strain
- Poor design of job or workstation
- Repetitive lifting of awkward items, equipment, or (in health care facilities) patients

Awkward postures

FIGURE 17.1 Stretching and poor posture can cause back injuries. (Courtesy of the Occupational Safety and Health Administration.)

- Twisting while lifting
- Bending while lifting
- Maintaining bent postures
- Heavy lifting
- Fatigue
- Poor footing such as slippery floors or constrained posture
- Lifting with forceful movement
- Vibration, such as with lift truck drivers, delivery drivers, etc.

17.2.2 BEFORE A LIFT

Before a lift is performed some actions need to be taken to assure the lift is a safe one. These are as follows:

- Checking the object before a lift is attempted by testing every load before lifting by pushing the object with your hands or feet to see how easily it moves. This gives an idea of how heavy it is. Remember, a small size does not always mean a light load.
- Check to make sure the load is packed correctly. Make sure the weight is balanced and packed so it would not move around. Loose pieces inside a box can cause accidents if the box becomes unbalanced.
- Check to see if the load can be gripped easily. Be sure that you have a tight grip before lifting. Objects with handles can be lifted in a safer manner.
- Check to see if the load is within easy reach. An injury can occur if the back is arched when lifting a load. To prevent a back injury use a ladder when lifting something over head level.
- Determine that the best way to pick up an object is being used. Use slow and smooth movements. Hurried, jerky movements can strain muscles in the back. Keep the body facing the object while lifting. Twisting while lifting can injure the back. Keep the load close to the body. Having to reach and carry an object may hurt the back. Lifting with the legs should be done only when the load can be straddled. To lift with the legs, bend the knees while keeping the back straight. Try to carry the load in the space between the shoulder and the waist. This puts less strain on the back muscles.
- Follow the steps suggested in Figure 17.2 to make a safe lift.

17.3 SYMPTOMS AND CONTRIBUTORS TO INJURIES

Signs and symptoms include pain when attempting to assume normal posture, decreased mobility, and pain when standing or rising from a seated position. At times the following are contributors to work-related back injuries:

- Congenital defects of the spine
- Increase in static standing or sitting tasks
- An aging workforce
- Decreases in physical conditioning and exercise

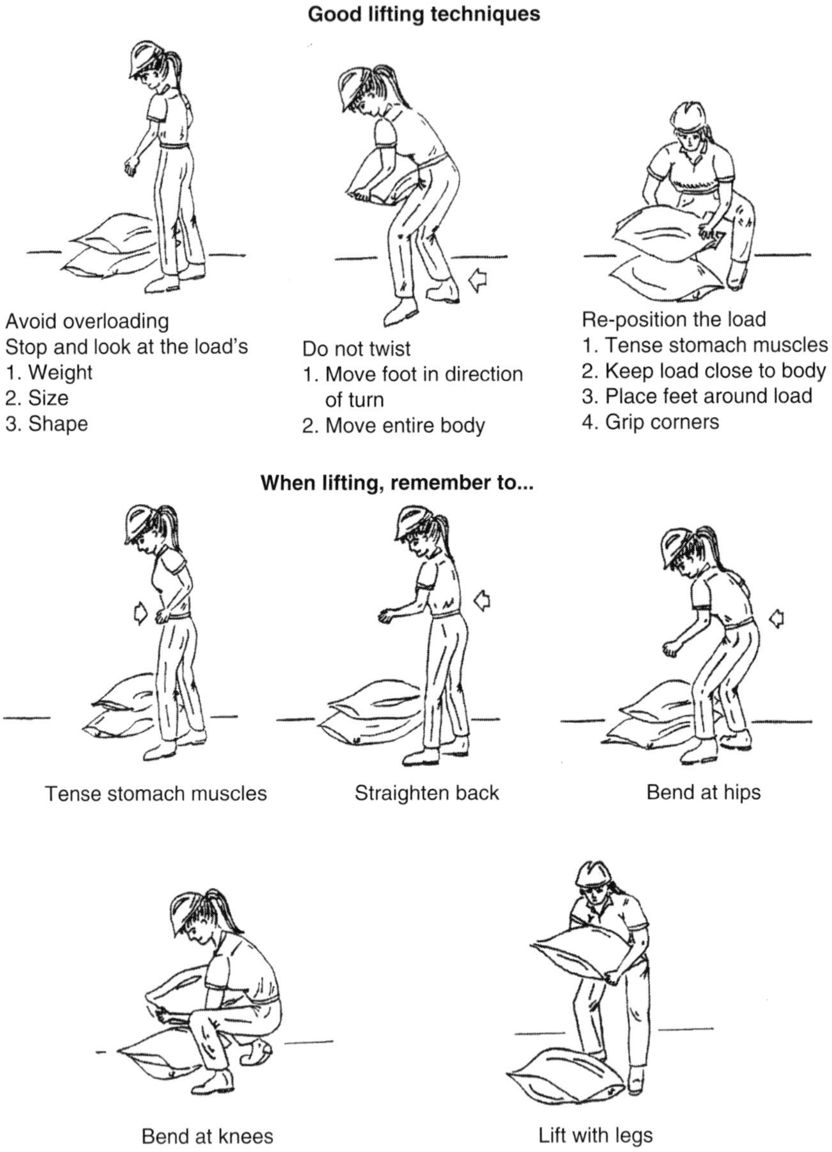

Good lifting techniques

Avoid overloading
Stop and look at the load's
1. Weight
2. Size
3. Shape

Do not twist
1. Move foot in direction
   of turn
2. Move entire body

Re-position the load
1. Tense stomach muscles
2. Keep load close to body
3. Place feet around load
4. Grip corners

When lifting, remember to...

Tense stomach muscles   Straighten back   Bend at hips

Bend at knees   Lift with legs

FIGURE 17.2   Steps to follow in making a safe lift.

- Lack of awareness of workplace hazards
- Job dissatisfaction

Manual materials handling is the principal source of compensable injuries in the American workforce, and four out of five of these injuries will affect the lower back.

17.4   RECORDS REVIEW: OSHA 300 LOG

Note when back or other musculoskeletal disorders appear excessive from lost work day injury and illness (LWDII) rate calculations. Understand that excessiveness is relative, since there is no set limit that delineates safe from unsafe. A better measure is to look for trends of escalating number of injuries or of increasing severity of injuries. Comparing your target population with Bureau of Labor Statistics (BLS) data, other company rates, other lines, departments, wings, or occupational titles can yield a meaningful measuring point to gauge excessiveness.

Back injuries should be treated as an injury on the OSHA (Occupational Safety and Health Administration) 300 log regardless of whether the injury was the result of an acute or chronic exposure. To determine if trends exist, at least several years of the OSHA 300 log will have to be reviewed. Record or copy information, including occupational titles, departments, dates of injury, or illness, from the OSHA 300 log and pertinent OSHA 301 (or equivalent).

17.5   EVALUATING BACK INJURIES

The following techniques can be used to assess why back injuries occurred:

- Use a walkaround.
- Interview employees about their opinion on the difficulty of the task as well as personal experiences of back pain.
- Observe worker postures and lifting.
- Determine weight of objects lifted.
- Determine the frequency and duration of lifting tasks.
- Measure the dimensions of the workplace and lift.
- Videotapes should be taken of the work task for later review and for evidence of recognized musculoskeletal hazards.

17.6   MANUAL LIFTING

Repetitive material handling increases the likelihood of a back disorder. Principal variables in evaluating manual lifting tasks to determine how heavy a load can be lifted are the horizontal distance from the load to the employee's spine, the vertical distance through which the load is handled, the amount of trunk twisting the employee utilized during the lifting, the ability of the hand to grasp the load, and the frequency with which the load is handled. Additional variables include floor and shoe traction, space constraints, two-handed lifts, size, and stability of the load.

17.7   PREVENTION AND CONTROL

17.7.1   ENGINEERING CONTROLS

Generally, the task can be altered to eliminate the hazardous motion and/or change the position of the object in relation to the employee's body—such as adjusting the height of

FIGURE 17.3   Use a handtruck to avoid lifting and carrying task. (Courtesy of the Occupational Safety and Health Administration.)

a pallet or shelf. Manual handling tasks should be designed to minimize the weight, range of motion, and frequency of the activity. Work methods and stations should be designed to minimize the distance between the person and the object being handled. Platforms and conveyors should be built at about waist height to minimize awkward postures. Conveyors or carts should be used for horizontal motion whenever possible. Reduce the size or weight of the objects lifted. High-strength push–pull requirements are undesirable, but pushing is better than pulling. Material handling equipment should be easy to move, with handles that can be easily grasped in an upright posture (Figure 17.3).

Workbench or workstation configurations can force people to bend over. Corrections should emphasize adjustments necessary for the employee to remain in a relaxed upright stance or fully supported seated posture. Bending the upper body and spine to reach into a bin or container is highly undesirable. The bins should be elevated, tilted, or equipped with collapsible sides to improve access. Repetitive or sustained twisting, stretching, or leaning to one side are undesirable. Corrections could include repositioning bins and moving employees closer to parts and conveyors. Store heavy objects at waist level. Provide lift-assist devices and lift tables.

17.8   CONTROLS AND WORK PRACTICES

The following are controls and other methods that address the prevention of back injuries at the workplace:

- Engineering controls are the preferred mechanism to address interventions.
- Worker training and education should include general principles of ergonomics, recognition of hazards and injuries, procedures for reporting hazardous conditions, and methods and procedures for early reporting of

injuries. Additionally, job-specific training should be given on safe work practices, hazards, and controls. The training should include practical sessions where workers are taught and practice safe lifting techniques under expert supervision.

- Strength and fitness training can reduce compensation costs.
- Rotating of employees, providing a short break every hour, or using a two-person lift may be helpful. Rotation is not simply a different job, but must be a job that utilizes a completely different muscle group from the ones that have been over-exerted.
- Standing for extended periods places excessive strain on the back and legs. Solutions include a footrest or rail, resilient floor mats, height-adjustable chairs or stools, and opportunities for the employee to change position.
- Where employees are seated the chairs or stools must be chosen properly. Proper adjustable lumbar support must be provided.
- Static seated postures with bending or reaching should be avoided.

17.9 SUMMARY

To review safe lifting, the legs should be used and not the back. To pick up a load stand close to the load, bend the knees while maintaining the backs natural curve. Grip the load firmly and push the body and load up slowly and smoothly with the legs.

To put down a load do not twist the body. Bend the knees to lower the load and place the load on the edge of a surface, then slide it back.

At times two persons will be needed to lift. Put one person in charge to say when to lift. Both individuals should lift at the same time while keeping the load level and unload at the same time (Figure 17.4).

Lifting bulky loads

FIGURE 17.4 Two person lifts reduce the strain of a single person lift. (Courtesy of the Occupational Safety and Health Administration.)

18 Machine Safety

Material handling equipment is usually guarded during manufacturing, and must be operated safely.

According to National Institute for Occupational Safety and Health's (NIOSH) National Traumatic Occupational Fatality (NTOF) data from 1980 to 1998 occupational injury from machinery was ranked third after motor vehicle and homicide as causes of death. Fatalities from machine-related incidents accounted for approximately 13% of the total. The service industry did not rank among the highest sectors having machine-related incidents. Some of the leading injuries experienced in these industries were as follows: struck by or against an object, caught in or compressed by equipment, and caught in or crushed in collapsing materials.

According to the Bureau of Labor Statistics (BLS), 92,560 private-sector lost-time injuries during the year 2002 were caused by machinery. The median number of lost workdays resulting from these injuries was 7 with 24% of the total incidents resulting in 31 or more lost workdays. The type of machine (source) most often identified included metal, woodworking, and special materials machineries (19,269 injuries); material handling machinery (16,183 injuries); special process machinery (15,576 injuries); heating, cooling, and cleaning machineries (13,330 injuries); unspecified machinery (6148 injuries); and construction, logging, and mining machineries (6069 injuries). Machinery was identified as the primary source of fatal occupational injuries in 483 of 5915 total fatalities during 2002.

The safe operation of all types of equipment takes a variety of approaches since the manufacturer cannot always be depended upon to provide inherently safe machines. It is often up to the owner (employer) to ensure that the machine/equipment is as safe as possible for their employees to operate. This may require that specific redesign of safeguards, unique training, and safe operating procedures (SOPs) be developed.

18.1 GUARDING

Although a more detailed approach to safeguard is found in *Industrial Safety and Health for Administrative Services* and *Industrial Safety and Health for People-Oriented Services*, a short review has been placed in this chapter. Any mobile machine part presents a hazard. Guarding eliminates or controls this danger. The most dangerous machine motions are rotating; reciprocating/ transverse motions; in-running nip points; cutting actions; and punching, shearing, and bending. The types of machine guards and their uses are summarized as follows:

- Enclosure guards are preferable to all other types because they prevent access to dangerous moving parts by enclosing them completely. They are used on power presses, sheet leveling or flattening machines, milling machines, gear trains, drilling machines, etc.
- Fixed guards may be adjustable to accommodate different sets of tools or various kinds of work. However, once they have been adjusted, they should remain "fixed."
- Interlocking guards are the first alternative when fixed guards or enclosures are not practicable. They prevent operation of the control that sets the machine in motion until the guard or barrier is moved into position.
- Barrier that shuts off or disengages power, preventing the machine from starting when the guard is open.

Electric contact or mechanical stop that activates a brake when any part of the operator's body enters the danger zone:

- Two-handed tripping devices are commonly used on bakery machinery, guillotine cutters, power presses, dough mixers, centrifugal extractors, tumblers, and some kinds of pressure vessels.
- Automatic guards must prevent the operator from coming in contact with the dangerous part of the machine while it is in motion, or must be able to stop the machine in case of danger. Examples of these are pull-away or hand-restraint devices; and photoelectric relay switches that stop the power supply to the machine.
- It is important that machines/equipment controls be properly adjusted and maintained.
- Remote control, placement, feeding, ejecting may be used to protect the operator from dangerous points of operation. Examples are two-handed operating devices; chutes, hoppers, conveyors, etc., to feed stock automatically;

and special jugs or feeding devices made of metal or wood; mechanical or air-operated ejecting devices may be used to complement another type of guard or as a substitute.

18.2   SAFE PRACTICES REGARDING MACHINE GUARDS

No guard, barrier, or enclosure should be adjusted or removed for any reason by anyone other than an authorized person. Before removal of safeguards for repairs, adjustments, or servicing, the power must be turned off and the main switch locked out and tagged. No machine should be started unless the guards are in place and in good condition. Defective or missing guards should be reported to the supervisor immediately. Employees should not work on or around mechanical operating equipment while wearing neckties, loose clothing, watches, rings, or other jewelry.

18.3   TRAINING

All workers should receive machine/equipment specific training. They should be trained on the specific piece of equipment that they are going to be operating. If they have not operated the equipment for a long time, they should receive renewed training. It is a supervisors' responsibility to make safe job observation to determine if the employee is still proficient on the safe operation of the machine or equipment. If not, the supervisor must have the employee retrained or give hands-on training to ensure the safe operation of the machine or equipment. One of the most important pieces of information that can be used to both check and retrain workers is an SOP for a piece of equipment or machine. An SOP should exist for all pieces of equipment or machine.

18.4   SAFE OPERATING PROCEDURES

Safe operating procedures or standard operating procedures should include safety as a part of the standard operating practices, which are delineated within it. Workers may not automatically understand a task just because they have experience or training. Thus, many jobs, tasks, and operations are best supported by an SOP. The SOP walks the worker through the steps of how to do a task or procedure in a safe manner and calls attention to the potential hazards at each step.

You might ask why an SOP is needed if the worker has already been trained to do the job or task. As you may remember from Chapter 17, a job safety analysis usually keys in on those particular jobs which pose the greatest risk of injury or death. These are the high-risk types of work activities and definitely merit the development and use of an SOP. There are times when an SOP, or step by step checklist, is useful. This is the case when

- New worker is performing a job or task for the first time.
- Experienced worker is performing a job or task for the first time.
- Experienced worker is performing a job, which he/she has not done recently.

- Mistakes could cause damage to equipment or property.
- Job is done on an intermittent or infrequent basis.
- New piece of equipment or different model of equipment is obtained.
- Supervisors need to understand the safe operation to be able to evaluate performance.
- Procedure or action within an organization is repetitive.
- Procedure is critically important, no matter how seldom performed, be carried out exactly according to detailed, step-wise instructions.
- Need to standardize the way a procedure is carried out for ensuring quality control or system compatibility.

When airline pilots fly, the most critical parts of the job are takeoffs and landings. Since these are two very critical aspects of flying, a checklist for proceeding in a safe manner is used to mitigate the potential for mistakes. It is critical to provide help when a chance for error can result in grave consequences.

Plasticized SOPs should be placed on equipment, machines, and vehicles for those individuals who need a refresher before operation because they have not used the equipment or performed the task on an infrequent basis.

Few people or workers want to admit that they do not know how to perform a job or task. They will not ask questions, let alone ask for help in doing an assigned task. This is the time when a plasticized SOP or checklist could be placed at the worksite or attached to a piece of equipment. This can prove to be a very effective accident-prevention technique. It can safely walk a worker through the correct sequence of necessary steps and thus avoid the exposure to hazards which can put the worker at risk of injury, illness, or death.

These SOPs could be used when, for example, helicopters are used for lifting, industrial forklifts are used, materials are moved manually, etc. These types of SOPs should list the sequential steps required to perform the job or task safely, the potential hazards involved, and the personal protective equipment needed. Each step in the SOP should provide all the information needed to accomplish the task safely.

If you do not have annual training, the use of SOPs may instill a sense of confidence and refresh workers' memories for the task at hand.

Any updated procedure should be reflected in the SOP immediately to ensure its effective application. A checklist is one form of an SOP. A checklist is very effective and attempts to ensure that every step is followed.

SOPs are only useful when they are up-to-date and readily accessible at the actual job or task site. Since we can now store SOPs online, revision and modification, based on workers/supervisors' suggestions, are much simpler. Figure 18.1 depicts a typical forklift and Table 18.1 depicts an SOP for a forklift. Using the format from this example, develop your own SOPs for procedures, jobs, tasks, or equipment.

An SOP is only one accident-prevention technique or component of any safety and health initiative. There are specific jobs or tasks that lend themselves well to this approach. Make sure that you use SOPs when they benefit your type of work the most and not as a cure-all for all your accidents and injuries. Use it as one of the many tools for accident prevention.

FIGURE 18.1   Forklift.

TABLE 18.1
An SOP for a Forklift

What To Do	How To Do It	Key Points
Perform a pre-start up inspection	Walk around the vehicle checking overall conditions of:	
	1. Tires	1. Check tires to ensure adequate tread, no cuts/missing chunks, all tire bolts are present and are tight.
	2. Fluid leaks	2. Check hydraulic hose fittings for evidence of fluid leak. Look beneath vehicle for fluid on the floor.
	3. Overhead guard	3. Check for missing bolts, bent frame.
	4. Lifting forks and load backrest	4. Check the lifting forks and backrest for damage. Check that the lifting fork's width adjustment lock pins are in good condition and are working smoothly.
	5. Preventive maintenance (P/M) sticker	5. If P/M expiration date is not valid, do not operate vehicle.
Perform operating controls inspection	1. Sit in driver's seat and operate controls	1. Check to ensure emergency brake is engaged and gear shift is in park or neutral.
	2. Adjust seat for effective operation and comfort	2. Seat should be adjusted to allow foot brake pedal to be depressed without reaching with foot.

(*continued*)

TABLE 18.1 (continued)
An SOP for a Forklift

What To Do	How To Do It	Key Points
Perform operating controls inspection (continued)	3. Fasten seat belt	3. Seat belt should be snugly across hips.
	4. Turn on power to vehicle	4. Check 360° around vehicle to ensure no one is standing near vehicle.
	5. Turn on headlights	5. Headlights must be bright and in position to ensure being seen by other vehicle operators or pedestrians.
	6. Depress brake foot pedal	6. The pedal must be firm and brake lights must function and be bright.
	7. Depress horn button	7. Horn should function easily and be loud.
	8. Elevate lifting forks and tilt by pulling back on control levers	8. Hydraulic controls shall operate smoothly.
	9. Lower lifting forks to 2–4 in. above floor	9. Keep lifting forks 2–4 in. above surface when in motion.
	10. Report any safety check failure to supervisor immediately for repair	10. Do not operate if any safety check fails. Ensure the vehicle is not operated until repaired.
Operating procedure— traveling to designation	1. Depress foot brake pedal	1. Check 360° around vehicle to ensure no one is standing near vehicle. Keep lifting forks 2–4 in. above surface when in motion.
	2. Release parking brake	2. —
	3. Select direction of travel	3. Engage gear drive. Check travel direction to ensure path is clear of pedestrians or other vehicles.
	4. Remove foot from brake pedal	4. Remove foot slowly.
	5. Depress accelerator pedal	5. Depress accelerator pedal slowly to avoid quick, jerky start. Keep lifting forks 2–4 in. above surface when in motion.
	6. Obey safety rules and regulations	6. • Travel at speeds which allow vehicle to be under control at all times under any condition. • Travel single-file keeping to the right. • Pedestrians have right of way. • Emergency vehicles have right of way at all times. • Use lights and horn when necessary. • Allow at least 15 ft or three vehicle lengths between you and person in vehicle (PIV) in front.

TABLE 18.1 (continued)
An SOP for a Forklift

What To Do	How To Do It	Key Points
Operating procedure— traveling to designation (continued)		• Do not pass other vehicles traveling in the same direction at intersections or blind spots, or narrow passages. • Ensure there is adequate overhead clearance. • Do not travel over objects. • Avoid sudden stops, except in emergencies. • Stop at all stop signs, blind corners, or when entering intersecting aisle and look for pedestrians and vehicle traffic.
Operating procedure— material pick-up	1. Approach material slowly 2. Stop vehicle, depress foot brake pedal 3. Shift to park or neutral 4. Engage parking brake 5. Unfasten seat belt and dismount, set forks for maximum load width 6. Remount and fasten seat belt 7. Depress foot brake pedal 8. Release parking brake 9. Select direction of travel 10. Remove foot from brake pedal 11. Depress accelerator pedal 12. Approach the load with lifting forks level 13. Penetrate forks to back of pallet 14. Depress foot brake pedal bringing vehicle to a stop 15. Raise the lifting forks until pallet is 2–4 in. above walking surface	1. Reduce speed to avoid sudden stop. 2. Apply slow, steady pressure until vehicle stops. 3. Never select direction while in motion. 4. If on an incline block wheels. 5. Know the vehicle's capacities and load weights. 6. Seat should be adjusted to allow foot brake pedal to be depressed without over extending. 7. Check 360° around vehicle to ensure no one is standing near vehicle. 8. — 9. Engage gear drive. Check travel direction to ensure path is clear. 10. Remove foot slowly. 11. Depress accelerator pedal slowly to avoid quick, jerky start. 12. Lifting forks should be parallel with walking surface. 13. The pallet should be set against the backrest. 14. Apply slow steady pressure until vehicle stops. 15. Never raise forks while in motion.

(continued)

TABLE 18.1 (continued)
An SOP for a Forklift

What To Do	How To Do It	Key Points
Operating procedure— material pick-up (continued)	16. Tilt the lifting forks back slightly	16. Tilt the forks backward slightly to prevent the load from falling forward.
	17. Select direction of travel	17. Engage gear drive. Check travel direction to ensure path is clear of pedestrians or other vehicles.
	18. Remove foot from brake pedal	18. Remove foot slowly.
	19. Depress accelerator pedal	19. Depress accelerator pedal slowly to avoid quick, jerky start.
	20. Travel carefully to designation, obeying rules and regulations	20. Travel at speeds which allow vehicle to be under control at all times under any condition. • Travel single-file keeping to the right. • Pedestrians have right of way. • Emergency vehicles have right of way at all times. • Use lights and horn when necessary. • Allow at least 15 ft or three vehicle lengths between you and PIV in front. • Do not pass other vehicles traveling in the same direction at intersections or blind spots, or narrow passages. • Ensure there is adequate overhead clearance. • Do not travel over objects. • Avoid sudden stops, except in emergencies. • Stop at all stops sign, blind corners, or when entering intersecting aisle and look for pedestrians and vehicle traffic. • Always drive with load facing up hill. • Drive backwards when view is obstructed by large loads.
Operating procedure— material drop-off	1. Depress foot brake pedal bringing vehicle to a stop	1. Apply slow steady pressure until vehicle stops.
	2. Shift to park or neutral	2. Never select reverse while in motion.
	3. Tilt forks forward until parallel with walking surface	3. Never lower forks while in motion.

TABLE 18.1 (continued)
An SOP for a Forklift

What To Do	How To Do It	Key Points
Operating procedure— material drop-off (continued)	4. Lower lifting forks until pallet bottom is resting on surface and forks no longer support load	4. Lower load slowly to prevent sudden drop.
	5. Engage reverse drive	5. Check 360° around vehicle to ensure no one is standing near vehicle.
	6. Remove foot from brake pedal	6. Engage gear drive. Check travel direction to ensure path is clear of pedestrians or other vehicles.
	7. Depress accelerator pedal	7. Depress accelerator pedal slowly to avoid quick, jerky start.
	8. Travel enough distance until lifting forks can clear pallet	8. Travel in reverse until there is enough distance between the end of lifting forks and pallet.
	9. Depress foot brake pedal bringing vehicle to a stop	9. Apply slow steady pressure until vehicle stops.
	10. Engage forward drive	10. Check 360° around vehicle to ensure no one is standing near vehicle. Never select directional change while in motion.
	11. Remove foot from brake pedal	11. Check travel direction to ensure path is clear of pedestrians or other vehicles.
	12. Depress accelerator pedal	12. Depress accelerator pedal slowly to avoid quick, jerky start.
	13. Travel carefully to designation obeying rules and regulations	13. • Travel at speeds which allow vehicle to be under control at all times under any condition. • Travel single-file keeping to the right. • Pedestrians have right of way. • Emergency vehicles have right of way at all times. • Use lights and horn when necessary. • Allow at least 15 ft or three vehicle lengths between you and PIV in front. • Do not pass other vehicles traveling in the same direction at intersections or blind spots, or narrow passages.

(continued)

TABLE 18.1 (continued)
An SOP for a Forklift

What To Do	How To Do It	Key Points
Operating procedure— material drop-off (continued)		• Ensure there is adequate overhead clearance. • Do not travel over objects. • Avoid sudden stops, except in emergencies. • Stop at all stop signs, blind corners, or when entering intersecting aisle and look for pedestrians and vehicle traffic.
Operation procedure—shut down	1. Depress foot brake pedal bringing vehicle to a stop	1. Apply slow steady pressure until vehicle stops.
	2. Shift to park or neutral	2. Never select directional changes while in motion.
	3. Engage parking park	3. If on an incline, block wheels.
	4. Lower lifting forks slowly until resting on walking surface	4. Never lower forks while in motion.
	5. Turn off power	5. Remove key.
	6. Release seat belt	6. —
	7. Dismount	7. Wheels must be blocked if parked on an incline.
	8. Perform walk-around inspection, noting damage or operational problems	8. Report all operation problems to supervisor for repair.
	9. Remove all trash	

Source: From Reese, C.D. *Accident/Incident Prevention Techniques*, Taylor & Francis, New York, 2001.

18.4.1 COMPONENTS OF AN SOP

SOPs should provide clear instructions for safely conducting activities involved in each covered process, consistent work activity using the appropriate manufacturer's guidelines and instructions, other pertinent safety resources information, and expertise of the individual with specific safety knowledge to address at least the following elements:

1. Steps for each operating phase
   a. Initial startup
   b. Normal operations
   c. Temporary operations
   d. Emergency shutdown including the conditions under which emergency shutdown is required, and the assignment of shutdown responsibility to

qualified operators to ensure that emergency shutdown is executed in a safe and timely manner
 e. Emergency operations
 f. Normal shutdown
 g. Startup following a turnaround, or after an emergency shutdown
2. Operating limits
 a. Consequences of deviation
 b. Steps required to correct or avoid deviation
3. Safety and health considerations
 a. Properties of, and hazards presented by, the chemicals used in the process or hazards involved in the task
 b. Precautions necessary to prevent exposure, including engineering controls, administrative controls, and personal protective equipment
 c. Control measures to be taken if physical contact or airborne exposure occurs
 d. Quality control for raw materials, control of hazardous chemical inventory levels, and any other special or unique hazards
4. Safety systems and their functions

Safe or standard operating procedures should be readily accessible to employees who work with or maintain a process or operation. The SOPs should be reviewed as often as necessary to ensure that they reflect current operating practice, including any updations in technology, equipment, and facilities.

The employer should develop and implement safe work practices to provide for the control of hazards during operations such as equipment/machine operation, lockout/tagout; confined space entry; opening process equipment or piping; and control over an entrance into a facility by maintenance, contractor, laboratory, or other support personnel. These safe work practices must apply to employees and contractor employees.

18.4.2 Guidelines for Writing an SOP

SOPs are often poorly written because little thought or effort is made to do it right. At times they are mandated as a quick fix for a perceived problem. An organized and thoughtful approach will yield SOPs which are more practical. Here are some guidelines for writing an SOP:

- Decide what SOPs must be written based on a review of organizational functions.
- Check for any existing SOP that can be revised or updated.
- Gather information on the procedure from reference sources and knowledgeable employees.
- When possible, contact other agencies performing similar functions to see if they have an SOP and use it as a guide.
- Select a suitable format for the SOP to be written (e.g., Table 18.1).
- Assemble blank forms and any other documents to be referred to in the SOP.

- Write a draft of the SOP. Include copies of any blank forms referred to in the SOP.
- Review or have a fellow employee review the draft SOP for technical adequacy.
- Request that the draft SOP be reviewed for administrative adequacy by the supervisor or the person in charge.
- Incorporate any changes indicated by the reviews into a final draft.
- Date, sign, assign a file number, and distribute the new or revised SOP. The final copy should be signed both by the official responsible for preparing the SOP and by the official's supervisor or the official in charge. Copies should be provided to supervisors and officers in charge of the immediate organization, and should be posted in an SOP file for ready reference.

The most common problems and errors found in SOPs are summarized in the following list:

- Assigning responsibilities for carrying out a procedure rather than listing out methods. Regulations are the place for delineating responsibilities, not SOPs.
- Failure to clearly state specific responsibilities in the procedure. The "who" is as important as the "what."
- Inclusion of steps or procedures performed by persons outside the organization. This information has no place in an SOP because it involves actions which are beyond the direct control of the organization. Include only those steps that are carried out by the employees in the immediate organization; all else is irrelevant.
- Vagueness and imprecision. What if the reader cannot figure out exactly who (job description) is required to carry out a step in the procedure, and furthermore cannot determine precisely how it is to be carried out? Obviously, then, the SOP has failed in its primary objective, communication. This is why the prime function of the reviewer is to check whether the writer has conveyed his message clearly and unequivocally.

18.4.3 How SOPs Work

Safe or standard operating procedures describe tasks to be performed, data to be recorded, operating conditions to be maintained, samples to be collected, and safety and health precautions to be taken. The procedures need to be technically accurate, understandable to employees, and revised periodically to ensure that they reflect current operations. Operating procedures should be reviewed by engineering staff and operating personnel to ensure that they are accurate and provide practical instructions on how to actually carry out job duties safely.

Operating procedures will include specific instructions or details on what steps are to be taken or followed in carrying out the stated procedures. These operating instructions for each procedure should include the applicable safety precautions and should contain appropriate information on safety implications. For example, the operating procedures addressing operating parameters will contain operating instructions about pressure limits, temperature ranges, flow rates, what to do when an upset condition occurs, what alarms and instruments are pertinent if an upset condition occurs, and such other subjects. In some cases, different parameters will be required from those of normal operation. These operating instructions need to clearly indicate the distinctions between startup and normal operations such as the appropriate allowances for driving while fully loaded.

Operating procedures and instructions are important for training operating personnel. The operating procedures are often viewed as the SOPs for operations. Operators and operating staff, in general, need to have a full understanding of operating procedures. If workers are not fluent in English then procedures and instructions need to be prepared in a second language understood by the workers. In addition, operating procedures need to be changed when there is a change in the process as a result of the management of change procedures. The consequences of operating procedure changes the need to be fully evaluated and the information conveyed to the personnel. For example, mechanical changes to the process made by the maintenance department (like changing a valve from steel to brass or other subtle changes) need to be evaluated to determine if operating procedures and practices also need to be changed.

All management change actions must be coordinated and integrated with current operating procedures and operating personnel must be oriented to the changes in procedures before the change is made. When the process is shut down to make a change, the operating procedures must be updated before startup of the process.

Operating personnel must be trained to handle upset conditions as well as emergencies such as pump seal failures and pipeline ruptures. Communication between operating personnel and workers performing work within the operating or production area, such as nonroutine tasks, must also be maintained. The hazards of the tasks are to be conveyed to operating personnel in accordance with established procedures and to those performing the actual tasks.

REFERENCE

Reese, C.D. *Accident/Incident Prevention Techniques*. New York: Taylor & Francis, 2001.

19 Material Handling

The one constant in the goods and materials sectors is the lifting and handling of merchandise of all types.

Material handling is a common task in each sector of industry. It is very common in the industry sectors that store, receive, and sell goods and materials such as the retail, wholesale, and warehousing sectors of the service industry.

The handling of all types of materials may manifest itself in the individual worker's effort to lift or move materials using large industrial cranes. No matter which procedure is used, there are hazards and safety concerns that need to be addressed. Almost every industrial sector has to address material handling issues, especially workplaces moving materials in and products out on a set schedule. Yet improper handling and storage of materials can result in grave injuries. Materials may be anything from boxes, parts, equipment, steel beams, aircraft engines, or manufactured homes.

The efficient handling and storage of materials is vital to the function of industry. Material handling operations provide for the continuous flow of raw materials, parts, and products throughout the workplace and assure that materials and products are there when they are needed. Yet the improper handling and storage of materials can cause serious injuries. In most industry sectors, around 20%–25% of all injuries are caused by material handling.

The proper and safe handling of a wide variety of materials must be done in compliance with existing Occupational Safety and Health Administration (OSHA) regulatory guidelines for the equipment used, the methods or procedures followed, and the appropriate storage of each type of material.

The guidelines provided in this chapter should help prevent some of the material handling accidents that are occurring in the goods and material sectors.

19.1   HAZARDS INVOLVED

Injuries faced by those performing material handling tasks may be something as simple as overexertion, which results in sprains or strains, to simple cuts and lacerations from sharp edges, or contact with moving parts on equipment. The pinch (nip) points or shear weight of items being handled can result in bruises, contusions, crushing, fractures, and amputations. The larger the objects, the larger the equipment being used, and the faster the movement of materials, the greater the risk for multiple injuries, suffocation, or, worst of all, death (Figure 19.1).

Many of the materials being handled by workers include hazardous chemicals and have the potential for causing injuries and illnesses. Chemicals can cause fires or explosion hazards and can result in burns or concussion injuries. Others may present the potential for contact, ingestion, or inhalation exposures, which may cause allergic reactions or toxic (poisonous) effects in workers, when such materials are mishandled and/or spilled. It certainly seems safe to say that all of these scenarios have transpired at one time or another to workers handling hazardous chemicals.

Certainly, if some materials are too heavy and when lifting becomes repetitious, the potential for overexertion will likely result in sprains and strains. But, materials that are improperly stored or handled have the potential to shift due to their weight, shape, or potential to flow. For example, sand being moved and stored at an unusual

FIGURE 19.1   Handling goods and materials is the primary hazard faced by workers.

angle of repose may engulf a worker. This is particularly a problem around stockpiles, surge bins, or excavations. When material shifts, it may physically strike a worker, pinning him/her between a stationary object and the moving materials.

When using the wide variety of equipment available to move or handle the different types of materials that exist in the workplace, the unevenness, unsecured loads, and extreme weight of the loads being lifted or moved can cause equipment to malfunction, collapse, or, at least, function erratically. The load can potentially fall and strike a worker, swing into a worker, be caught under a piece of equipment, or under a load, which cannot be controlled. This is why it is important to regularly inspect and maintain equipment used for handling material. If a sling (wire rope, steel alloy chain, or webbed sling) fails, a crane boom collapses, or the brakes fail on a forklift or other vehicle, the end results can be disastrous.

The use of equipment to handle materials is controlled by preestablished lifting or load limits and restrictions on the supporting capacity of storage (shelving) units which can never be exceeded if safety is a primary focus. It must be ensured that the appropriate equipment is used for the job, and that it is properly used by the operator. If an operator inadvertently contacts an electrical conductor, for example, electrocution is a real possibility.

Workers do not expect to be working under a load, have a load fall on them from above, or to be run over by a piece of material-handling equipment. These hazards are preventable using fundamental safety precautions.

The last hazard is derived from the myriad of chemicals handled or stored within the workplace. Not only do they present the potential to cause physical harm (as noted earlier when chemical containers shift, roll, or strike a worker injuring or killing that worker), they also pose another type of hazard. They can potentially be toxic (poisonous) or cause burns if mishandled, spilled, or not properly controlled. Some chemicals may also create an explosion or fire.

As can be seen, the movement, stacking, and storage of materials pose many hazards within the workplace. The philosophical approach to these hazards must be that they are identifiable, preventable, and the accidents, that result in injuries, illnesses, and deaths from improper material handling, can significantly be reduced.

19.2 SAFE HANDLING

The efficient handling and storing of materials are vital to industries. In addition to raw materials, these operations provide a continuous flow of parts and assemblies through the workplace and ensure that materials are available when needed. Unfortunately, the improper handling and storing of materials often result in serious injuries.

In addition to training and education, applying general safety principles—such as proper work practices, equipment, and controls—can help reduce workplace accidents involving the moving, handling, and storing of materials. Whether moving materials manually or mechanically, employees should know and understand the potential hazards associated with the task at hand and how to control their workplaces to minimize the danger.

Because numerous injuries can result from improperly handling and storing materials, workers should also be aware of accidents that may result from the unsafe or improper handling of equipment as well as from improper work practices. In addition, workers should be able to recognize the methods for eliminating or at least minimizing the occurrence of such accidents. Employers and employees should examine their workplaces to detect any unsafe or unhealthful conditions, practices, or equipment and take corrective action.

Workers frequently cite the weight and bulkiness of objects that they lift as major contributing factors to their injuries. In 1999, for example, more than 420,000 workplace accidents resulted in back injuries. Bending, followed by twisting and turning, were the more commonly cited movements that caused back injuries (Figure 19.2).

Other hazards include falling objects, improperly stacked materials, and various types of equipment. Employees should be made aware of potential injuries when manually moving materials, including the following:

- Strains and sprains from lifting loads improperly or from carrying loads that are either too large or too heavy
- Fractures and bruises caused by being struck by materials or by being caught in pinch points
- Cuts and bruises caused by falling materials that have been improperly stored or by incorrectly cutting ties or other securing devices

19.2.1 MANUAL HANDLING SAFETY

When moving materials manually, workers should attach handles or holders to loads. In addition, workers should always wear appropriate personal protective equipment

FIGURE 19.2   Material handling tasks result in strain being placed on the back.

and use proper lifting techniques. To prevent injury from oversize loads, workers should seek help when the following conditions exist:

- When a load is so bulky that employees cannot properly grasp or lift it
- When employees cannot see around or over a load (Figure 19.3)
- When employees cannot safely handle a load

Using the following personal protective equipment prevents needless injuries when manually moving materials:

- Hand and forearm protection, such as gloves, for loads with sharp or rough edges
- Eye protection
- Steel-toed safety shoes or boots
- Metal, fiber, or plastic metatarsal guards to protect the instep area from impact or compression

Employees should use blocking materials to manage loads safely. Workers should also be cautious when placing blocks under a raised load to ensure that the load is not released before removing their hands from under the load. Blocking materials should be large and strong enough to support the load safely. In addition to materials with cracks, workers should not use materials with rounded corners, splintered pieces, or dry rot for blocking.

FIGURE 19.3 The inability to see around a load creates a hazard.

19.2.2 MECHANICAL HANDLING EQUIPMENT

Using mechanical equipment to move and store materials increases the potential for employee injuries. Workers must be aware of both manual handling safety concerns and safe equipment operating techniques. Employees should avoid over-loading equipment when moving materials mechanically by choosing the equipment based on the weight, size, and shape of the material being moved. All materials-handling equipment have rated capacities that determine the maximum weight the equipment can safely handle and the conditions under which it can handle that weight. Employers must ensure that the equipment's rated capacity is displayed on each piece of equipment and is not exceeded except for load testing (Figure 19.4).

Although workers may be knowledgeable about powered equipment, they should take precautions when stacking and storing material. When lifting items with a powered industrial truck (forklift), workers must do the following:

- Center the load on the forks as close to the mast as possible to minimize the potential for the truck tipping or the load from falling.
- Avoid overloading a lift truck because it impairs control and causes tipping over.
- Do not place extra weight on the rear of a counterbalanced forklift to allow an overload.
- Adjust the load to the lowest position when traveling.
- Follow the forklift manufacturer's operational requirements.
- Pile and cross-tier all stacked loads correctly when possible.

Chapter 20 provides more detailed information on the safe use of forklifts.

FIGURE 19.4   Care must be taken by operator to not overload stockpickers.

19.2.3   SAFE STORAGE OF MATERIALS

Stored materials must not create a hazard for employees. Employers should inform workers of such factors as the materials' height and weight, accessibility of the stored materials, and the condition of the containers where the materials are being stored when stacking and piling materials. To prevent creating hazards when storing materials, employers must do the following:

- Keep storage areas free from accumulated materials that cause tripping, fires, or explosions, or that may contribute to the harboring of rats and other pests.
- Place stored materials inside buildings that are under construction and at least 6 ft from hoist ways, or inside floor openings and at least 10 ft away from exterior walls.
- Separate noncompatible material.
- Equip employees who work on stored grain in silos, hoppers, or tanks, with lifelines and safety harnesses.
- In addition, workers should consider placing bound material on racks, and secure it by stacking, blocking, or interlocking to prevent it from sliding, falling, or collapsing.

19.2.4   SAFE STACKING OF MATERIALS

Stacking materials can be dangerous if workers do not follow safety guidelines. Falling materials and collapsing loads can crush or pin workers, causing injuries or death. To help prevent injuries when stacking materials, workers must do the following:

- Stack lumber no more than 16 ft high if it is handled manually, and no more than 20 ft if using a forklift.
- Remove all nails from used lumber before stacking.
- Stack and level lumber on solidly supported bracing.
- Ensure that stacks are stable and self-supporting.
- Do not store pipes and bars in racks that face main aisles to avoid creating a hazard to passersby when removing supplies.
- Stack bags and bundles in interlocking rows to keep them secure (Figure 19.5).
- Stack bagged material by stepping back the layers and cross-keying the bags at least every 10 layers (to remove bags from the stack, start from the top row first).

During materials stacking activities, workers must also do the following:

- Store baled paper and rags inside a building no closer than 18 in. to the walls, partitions, or sprinkler heads.
- Band boxed materials or secure them with cross-ties or shrink plastic fiber.
- Stack drums, barrels, and kegs symmetrically.

FIGURE 19.5   Proper stacking of bags.

- Block the bottom tiers of drums, barrels, and kegs to keep them from rolling if stored on their sides.
- Place planks, sheets of plywood dunnage, or pallets between each tier of drums, barrels, and kegs to make a firm, flat, stacking surface when stacking on end.
- Check the bottom tier of drums, barrels, and kegs on each side to prevent shifting in either direction when stacking two or more tiers high.
- Stack and block poles as well as structural steel, bar stock, and other cylindrical materials to prevent spreading or tilting unless they are in racks.

In addition, the following recommendations should be followed to make this easier for workers:

- Paint walls or posts with stripes to indicate maximum stacking heights for quick reference.
- Observe height limitations when stacking materials.
- Consider the need for availability of the material.
- Stack loose bricks no more than 7 ft in height. (When these stacks reach a height of 4 ft, taper them back 2 in. for every foot of height above the 4 ft level. When masonry blocks are stacked higher than 6 ft, taper the stacks back one-half block for each tier above the 6 ft level.)

19.2.5   HOUSEKEEPING

The importance of preventing the type of accidents that can be caused by poor housekeeping practices is an integral part of good material handling practices.

The typical accidents that frequently result from inadequate housekeeping are as follows:

- Tripping over loose objects on floors, stairs, and platforms
- Slipping on wet, greasy, or dirty floors
- Bumping against projecting or misplaced materials
- Puncturing or scratching hands or other parts of the body on protruding nails, hooks, or rods
- Injuries from falling objects

Employees tend to take housekeeping for granted and may sometimes be careless. Housekeeping is the one area of accident prevention in which all employees (blue-collar, white-collar, administrative, etc.) must share the responsibility. The emphasis on housekeeping should be at

- Work areas: Avoid unnecessary clutter (Figure 19.6).
- Machines and equipment: Avoid crowding; provide racks or containers for tools, jigs, and fixtures.
- Aisles: Keep free of material, finished parts, and scrap.
- Floors: Make sure they are vacuumed and scrubbed regularly; spills should be cleaned up immediately.
- Walls and ceilings: See to it that they are scrubbed and painted when necessary; clothing and supplies should be hung on racks; clutter should be confined to a bulletin board.

FIGURE 19.6 Poor housekeeping is a primary cause of accidents.

- Storage facilities: Follow appropriate storage procedures.
- Employee facilities: Keep personal belongings in lockers; washrooms should be cleaned regularly.

Housekeeping inspections should occur regularly. It should be easy to recognize the benefits of good housekeeping since it

- Reduces operating costs: Once a housekeeping system has been established, less time and effort are required to keep the work area clean.
- Increases production: Delays and interference from excess materials, loose tools, etc., are avoided.
- Improves production control: Material and parts do not get lost or mixed up.
- Conserves materials and parts: Unused materials are easily and quickly removed to the proper place.
- Saves production time: There is no need to search for tools, parts, etc.
- Lowers accident rates since open aisles permit faster traffic with fewer collisions; clean floors mean less slipping, tripping, and falling; reduction of object hazards results in fewer injuries.
- Reduces fire hazards: Fires result from, or are spread by, poor housekeeping conditions.

Some common sense tips for good housekeeping are as follows:

- Give your immediate work area a good cleaning at the end of each day. Sweep up rubbish and put tools away.
- Dispose of combustible rubbish in fire-resistant receptacles.
- Dispose of oily rags in closed metal containers. Maintain regular disposal.
- Clean oily deposits from walls, ceilings, exhaust ducts, and mechanical equipment periodically.
- Remove combustible lint and dust from ledges, beams, and equipment as it accumulates.
- Remove oily metal chips and rubbish to outside storage locations on schedule.
- Keep aisles and passageways clear of stock and other obstructions at all times.
- Do not block valves, hose stations, fire extinguishers, or fire exits.
- Keep packing material in metal-lined bins with self-closing covers.
- Follow the clean-out schedules for electric motors, switch enclosures, contacts, etc.
- Keep solvents, thinners, etc., in approved safety containers.
- Avoid drippings—clean up spills promptly.
- Keep spray residues in covered containers and remove at the end of each day.
- Limit flammable liquids, cement, paint, etc., to only one day's supply at the working area.
- Empty drip pans and replace absorbent compounds frequently.

- Avoid overlubrication.
- Keep a minimum supply of lubricants in the work area.

19.3   MATERIAL-HANDLING EQUIPMENT

Probably few facets of safety and health have such a wide variety of equipment with which to be concerned as material handling. Some equipment are as simple as a handcart or dolly and some as complex as industrial robots. Examples of nonpowered equipment are dollies, hand trucks, carts, dockboards, and ramps.

A mechanical advantage can be enhanced by using powered equipment for smaller items not requiring industrial hoists, conveyors, forklifts, cranes, or heavy-duty trucks. Some of these types of available handling devices are operated by hydraulic, compressed air, or electrical energy.

To reduce the number of accidents associated with workplace equipment, employers must train employees in the proper use and limitations of the equipment they operate. In addition to powered industrial trucks, this includes knowing how to safely and effectively use equipment such as conveyors, cranes, and slings. The following are some commonly used material-handling equipment.

19.3.1   CONVEYORS

When using conveyors, workers may get their hands caught in nip points where the conveyor medium runs near the frame or over support members or rollers. Workers also may be struck by material falling off the conveyor, or they may get caught in the conveyor and drawn into the conveyor path as a result (Figure 19.7). To prevent or

FIGURE 19.7   Conveyors are used to decrease manual handling.

reduce the severity of an injury, employers must take the following precautions to protect workers:

- Install an emergency button or pull cord designed to stop the conveyor at the employee's workstation.
- Install emergency stop cables that extend the entire length of continuously accessible conveyor belts so that the cables can be accessed from any location along the conveyor.
- Design the emergency stop switch so that it must be reset before the conveyor can be restarted.
- Ensure that appropriate personnel inspect the conveyor and clear the stoppage before restarting a conveyor that has stopped due to an overload.
- Prohibit employees from riding on a materials-handling conveyor.
- Provide guards where conveyors pass over work areas or aisles to keep employees from being struck by falling material. (If the crossover is low enough for workers to run into it, mark the guard with a warning sign or paint it a bright color to protect employees.)
- Cover screw conveyors completely except at loading and discharging points. (At those points, guards must protect employees against contacting the moving screw. The guards are movable, and they must be interlocked to prevent conveyor movement when the guards are not in place.)

19.3.2 CRANES

Employers must permit only thoroughly trained and competent workers to operate cranes. Operators should know what they are lifting and what it weighs. For example, the rated capacity of mobile cranes varies with the length of the boom and the boom radius. When a crane has a telescoping boom, a load may be safe to lift at a short boom length or a short boom radius, but may overload the crane when the boom is extended and the radius increases. To reduce the severity of an injury, employers must take the following precautions:

- Equip all cranes that have adjustable booms with boom angle indicators.
- Provide cranes with telescoping booms with some means to determine boom lengths unless the load rating is independent of the boom length.
- Post load rating charts in the cab of cab-operated cranes. (All cranes do not have uniform capacities for the same boom length and radius in all directions around the chassis of the vehicle.)
- Require that workers always check the crane's load chart to ensure that the crane will not be overloaded by operating conditions.
- Instruct workers to plan lifts before starting them to ensure that they are safe.
- Tell workers to take additional precautions and exercise extra care when operating around power lines.
- Teach workers that outriggers on mobile cranes must rest on firm ground, on timbers, or be sufficiently cribbed to spread the weight of the crane and

the load over a large enough area. (Some mobile cranes cannot operate with outriggers in the traveling position.)

- Direct workers to always keep hoisting chains and ropes free of kinks or twists and never wrapped around a load.
- Train workers to attach loads to the load hook by slings, fixtures, and other devices that have the capacity to support the load on the hook.
- Instruct workers to pad sharp edges of loads to prevent cutting slings.
- Teach workers to maintain proper sling angles so that slings are not overloaded.
- Ensure that all cranes are inspected frequently by authorized persons, the methods of inspecting the crane, and what can render the crane unserviceable. Crane activity, the severity of use, and environmental conditions should determine inspection schedules.

Ensure that the critical parts of a crane—such as crane operating mechanisms, hooks, air, or hydraulic system components and other load-carrying components—are inspected daily for any maladjustment, deterioration, leakage, deformation, or other damage.

19.3.3 SLINGS

Employers must designate a competent person to conduct inspections of slings before and during use, especially when service conditions warrant. In addition, workers must observe the following precautions when working with slings:

- Remove immediately damaged or defective slings from service.
- Do not shorten slings with knots or bolts or other makeshift devices.
- Do not kink sling legs.
- Do not load slings beyond their rated capacity.
- Keep suspended loads clear of all obstructions.
- Remain clear of loads about to be lifted and suspended.
- Do not engage in shock loading.
- Avoid sudden crane acceleration and deceleration when moving suspended loads.

19.3.4 FORKLIFTS

Workers who handle and store materials often use fork trucks, platform lift trucks, motorized hand trucks, and other specialized industrial trucks powered by electrical motors or internal combustion engines. Employers must make these workers aware of the safety requirements pertaining to the design, maintenance, and use of these trucks. For more details on these requirements refer Chapter 20.

19.4   TRAINING AND EDUCATION

OSHA recommends that employers establish a formal training program to teach workers how to recognize and avoid materials-handling hazards. Instructors should be well versed in safety engineering and materials handling and storing. The training should reduce workplace hazards by emphasizing the following factors:

- Dangers of lifting without proper training
- Avoidance of unnecessary physical stress and strain
- Awareness of what a worker can comfortably handle without undue strain
- Proper use of equipment
- Recognition of potential hazards and how to prevent or correct them

The prevention of some injuries receives special emphasis. Because of the high incidence of back injuries, both supervisors and employees should demonstrate and practice safe manual lifting techniques. Training programs on proper lifting techniques should cover the following topics:

- Health risks of improper lifting, citing organizational case histories, versus the benefits of proper lifting
- Basic anatomy of the spine, muscles, and joints of the trunk, and the contributions of intra-abdominal pressure while lifting
- Body strengths and weaknesses—determining one's own lifting capacity
- Physical factors that might contribute to an accident and how to avoid the unexpected
- Safe postures for lifting and timing for smooth, easy lifting
- Aids such as stages, platforms, or steps, trestles, shoulder pads, handles, and wheels
- Body responses—warning signals—to be aware of when lifting

To have an effective safety and health program covering materials handling and storing, managers must actively participate in its development. First line supervisors must be convinced of the importance of controlling hazards associated with materials handling and storing and must be accountable for employee training. An ongoing safety and health management system can motivate employees to continue using necessary protective gear and observing proper job procedures. Instituting such a program, along with providing the correct materials-handling equipment, can enhance worker safety and health in the area of materials handling and storing.

19.5   MATERIAL HANDLING REGULATIONS

Specific regulations, applicable to material handling within General Industry (29 CFR 1910) Standards are as follows:

Subpart H—Hazardous materials

   1910.101—Compressed gases
   1910.102—Acetylene
   1910.103—Hydrogen
   1910.104—Oxygen

1910.105—Nitrous oxide
1910.106—Flammable and combustible liquids
1910.107—Storage and handling of anhydrous ammonia
1910.109—Explosives and blasting agents
1910.110—Storage and handling of liquefied petroleum gases

Subpart J—General environmental controls

1910.144—Safety color code for marking physical hazards
1910.145—Specifications for accident prevention tags

Subpart N—Material handling and storage

1910.176—Handling materials—general
1910.178—Powered industrial trucks
1910.179—Overhead and gantry crane
1910.180—Crawler, locomotion, and truck cranes
1910.181—Derricks
1910.183—Helicopters
1910.184—Slings

Subpart Q—Welding, cutting, and brazing

1910.253—Oxygen-fuel gas welding and cutting

Although other 29 CFR 1910 regulations are tangential, related to material handling, and apply to most workplaces, the previous list includes the ones most applicable to material handling.

19.6 SUMMARY

Employers can reduce injuries resulting form handling and storing material by using some basic safety procedures such as adopting sound ergonomic practices, taking general fire safety precautions, and keeping aisles and passageways clear.

20 Powered Industrial Trucks (aka Forklifts)

Example of a forklift.

20.1 FORKLIFTS

Forklifts (powered industrial trucks) are among the most useful and important material handling vehicles within the workplace or jobsite. In recent years, we have become very aware that the misuse of this type of lifting vehicle has resulted in many injuries and deaths. Thus, special precautions and driver training are of the utmost importance in the safe use of powered industrial trucks.

20.2 INCIDENCE OF LIFT-TRUCK INJURIES

Every year, it is estimated that more than 95,000 powered lift-truck-related injuries and 100 deaths (Table 20.1) occur in U.S. industry (OSHA, 1999). Injuries involve employees being struck by lift trucks or falling while standing/working from elevated pallets and tines. Many employees are injured when lift trucks are inadvertently driven off loading docks or when the lift falls between a dock and an unchecked trailer. For each employee injured, there are probably numerous incidents that are unnoticed or unreported to supervisors. All mishaps, no matter how small, are costly. Most incidents also involve property damage. Damage to overhead sprinklers, racking, pipes, walls, machinery, and various other equipment occurs all too often. In fact, millions of dollars are lost in damaged equipment, destroyed

TABLE 20.1

Classification of Forklift Fatalities, 1991–1992

How Accident Occurred	Number	Percent
Forklift overturned	41	24
Forklift struck something, or ran off dock	13	8
Worker pinned between objects	19	11
Worker struck by material	29	17
Worker struck by forklift	24	14
Worker fell from forklift	24	14
Worker died during forklift repair	10	6
Other accidents	10	6
Total	170	100

Source: From Bureau of Labor Statistics, *Fatal Workplace Injuries in 1992, A Collection of Data and Analysis*, Report 870, April 1994.

products, or missed shipments. Unfortunately, a majority of employee injuries and property damage can be attributed to lack of procedures, insufficient or inadequate training, and lack of safety-rule enforcement.

20.2.1 UNSAFE ACTS AND CONDITIONS

Some examples of the unsafe acts and conditions that occur during the use of powered industrial trucks are as follows:

20.2.1.1 Unsafe Acts

- Inadequately trained maintenance personnel, inspectors, and operators
- Wrong truck selected for the job (too big, too small, wrong for hazardous location)
- Hurrying, taking shortcuts, not paying attention, fatigue, boredom, or not following the rules
- Overloading trucks
- Improper selection and installation of dockboards and bridge plates

20.2.1.2 Unsafe Conditions

- Gouges or large chunks missing from solid tires
- Blind corners
- Leaky connectors and hydraulic cylinders
- Too much free play in the steering
- Unsafe refueling or recharging practices

20.3 HAZARDS AND EFFECTS

Many hazards associated with the operation of powered industrial trucks are the result of common operator mistakes. For instance, collisions between trucks and stationary objects often occur while trucks are backing up—usually while turning and maneuvering. Unless care is exercised, operators can cause damage to overhead fixtures (e.g., sprinklers, piping, electrical conduits) while traveling and maneuvering under them.

Accidents often occur when an operator leaves a truck so that it obstructs a passageway and an unauthorized (untrained) worker tries to move it. Other common hazards include carrying unstable loads, tipping over trucks, dropping loads on operators or others, running into or over others, and pinning others between the truck and fixed objects.

Unauthorized passengers are often seriously injured from falling off trucks. Unless space is provided, do not allow passengers to ride on the trucks. Dangerous misuse of trucks includes bumping skids, moving piles of material out of the way, moving heavy objects by means of makeshift connections, and pushing other trucks. All these activities can cause accidents or injuries; they also indicate poor operator training.

Factors that can influence stability (resistance to overturning) must be considered. These include the following:

- Weight, weight distribution, wheel base, tire tread, truck speed, and mast defection under load
- Improper operation, faulty maintenance, and poor housekeeping
- Ground and floor conditions, grade, speed, and judgment of the operator

20.4   PREVENTION OVERVIEW

Whether the operator is new to the job or experienced, he/she should visually check forklift trucks every day. Good prevention consists mainly of proper maintenance, trained operators, and adherence to established safety procedures. Special attention should be given to the following areas:

- Proper truck selection (size, load-carrying capacity, hazardous locations)
- Condition and inflation of pressure lines
- Battery, lights, and warning devices
- Controls, including lift and tilt system and limit switches
- Brakes and steering mechanisms
- Fuel system

20.5   TYPES OF POWERED INDUSTRIAL TRUCKS

These general requirements for powered industrial trucks contain safety requirements related to fire protection, design, maintenance, and the use of fork trucks, tractors, platform lift trucks, motorized hand trucks, and other specialized industrial trucks powered by electric motors or internal combustion engines. These requirements do not apply to compressed air or nonflammable compressed gas-operated industrial trucks, nor to farm vehicles, or to vehicles intended primarily for earth moving or over-the-road hauling.

Approved powered industrial trucks should bear a label or some other identifying mark indicating approval by the testing laboratory. Modifications and additions which

affect capacity and safe operation of these trucks should not be performed by the user without manufacturers' prior written approval.

The terms "approved truck" or "approved industrial trucks" as used in this chapter mean a truck that is listed or approved for fire safety purposes for the intended use by a nationally recognized testing laboratory, using nationally recognized testing standards.

There are four different sources of energy to power forklifts: diesel, electric, gasoline, and liquid petroleum gas with combinations and safeguards. Atmospheres or locations that pose specific hazards need to be determined before selecting a forklift. These hazards are explosive gases, flammable gases, combustible dust, or ignitable fibers. For more details, see NFPA No. 505-1969, Powered Industrial Trucks.

20.6 PROTECTIVE DEVICES

The use of protective devices is an important factor in safe forklift operation. Safety specialists can assist supervisors in determining what protective devices are necessary. Although forklifts need not be equipped alike, there are some similarities such as lights. Also, manufacturers are required by federal standards to equip forklifts with certain mandatory features such as backup alarms. When a truck is about to reverse it should sound a warning. Some other protective devices include the following:

- Overhead protection to guard the operator from falling objects
- Wheel plates to protect the operator from objects picked up and thrown by tires
- On-board fire extinguishers
- Horns to warn others when the truck is moving forward

Other protection devices that might be seen in the work area, which are specifically designed for the operator, include the following:

- Signs—such as stop, caution, danger, and speed limits—to inform operators of conditions
- Gloves and safety shoes
- Eyewash stations
- Concave mirrors
- Eye protection devices
- Hardhats to protect operators when there is an overhead hazard

20.7 WORK PRACTICES

20.7.1 SELECTION AND INSPECTION OF TRUCKS

Industrial trucks should be examined before being put into service, and should not be utilized if there is any defect. Examinations should be done at least daily. Where trucks are used on a round-the-clock basis, they should be examined after each shift. Figure 20.1 shows the major component of a standard forklift.

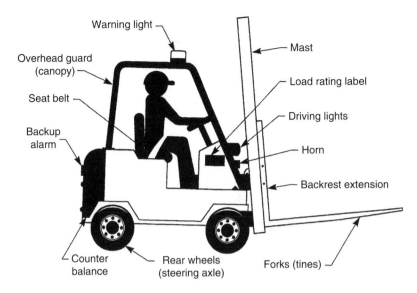

FIGURE 20.1   Components of a forklift. (Courtesy of Department of Energy.)

The proper truck (size, load capacity, and use) must be selected and inspected to ensure that all controls and other safety features are working properly. All powered industrial truck operators must check the vehicle, which they are operating at the start of each shift. If the vehicle is found to be unsafe, it must be reported to the manager immediately. No powered industrial truck should be operated in an unsafe condition. It is a good idea to use and maintain a daily preshift checklist to monitor the condition of powered industrial truck (forklifts). The operating condition of a forklift may change throughout the day and between shifts. An inspection identifies potential hazards both before operation and at the end of use of the powered industrial truck. Attention should be given to the proper functioning of tires, horns, lights, battery, controller, brakes, steering mechanism, and the lift system of fork lifts (fork chains, cables, and limit switches). Special attention should be given to the following:

- Before initial use, all new, altered, modified, or extensively repaired fork-lifts should be inspected by a qualified inspector to assure compliance with the provisions of the manufacturers' instructions.
- Brakes, steering mechanisms, control mechanisms, warning devices, lights, governors, lift overload devices, guards, and other safety devices should be inspected regularly and maintained in a safe operating condition.
- All parts of the lift and tilt mechanisms and frame members must be carefully and regularly inspected and maintained in a safe operating condition.
- Special trucks or devices, designed and approved for operation in hazardous areas, should receive special attention to ensure that the original, approved safe operating features are maintained.

- Fuel systems should be checked for leaks and condition of parts. Special consideration must be given in the case of a leak in the fuel system. Action should be taken to prevent the use of the truck until the leak has been corrected.
- All hydraulic systems must be regularly inspected and maintained properly. Tilt cylinders, valves, and other similar parts should be checked to assure that "drift" has not developed to the extent that it would create a hazard.
- Capacity, operation, and maintenance-instruction plates, tags, and decals must be maintained in a legible condition.
- Batteries, motors, controllers, limit switches, protective devices, electrical conductors, and connections should be inspected and maintained properly. Special attention must be paid to the condition of electrical insulation.
- Inspect the mast for broken or cracked weld points and any other obvious damage.
- Make sure roller tracks are greased and that chains are free to travel.
- Make sure the forks are equally spaced and free from cracks along the blade and at the heels.
- Check tires for excessive wear, splitting, or missing tire material as well as inflation levels.
- If a powered industrial truck (forklift) is powered by propane, inspect the tank for cracks, broken weld points, and other damages. Make sure all valves, nozzles, and hoses are secure and do not leak.

Once the inspection and maintenance has been completed, the operator should then get in the seat to check the following:

- Brakes
- Oil pressure gauge, water temperate gauge
- Steering (the wheel should turn correctly in both directions)
- Operation of the headlights, taillights, and warning lights
- Clutch
- Backup alarm

20.7.2 MAINTENANCE AND REPAIR OF TRUCKS

It is required that trained and authorized personnel maintain and inspect the powered (forklift) industrial trucks. All work should be done in accordance with the manufacturer's specifications. Because of everyday use of these vehicles, it is particularly important for personnel to follow the maintenance, lubrication, and inspection schedules. Special attention should be given to forklift control and lifting features, such as brakes, steering, lift apparatus, overload devices, and tilt mechanism.

Any power-operated industrial truck not in safe operating condition should be removed from service. All repairs must be made by authorized personnel. No repairs should be made in Class I, II, or III locations. Repairs to the fuel and ignition systems, which involve fire hazards, must be conducted only in locations designated for such repairs.

20.7.3 CHANGING AND CHARGING STORAGE BATTERIES

Workplaces using electrically powered industrial trucks will have battery-charging areas somewhere in the plant. In many cases, depending on the number of electrically powered industrial trucks, there will be more than one changing and charging area. This section only applies to storage battery changing and charging areas associated with powered industrial trucks. It does not apply to areas where other batteries, such as those used in motor vehicles (cars or trucks), are charged, although some of the same hazardous conditions may exist. Some of the requirements specified in the regulation include the following:

- Make sure batteries are checked for cracks or holes, security sealed cells, frayed cables, broken insulation, tight connections, and clogged vent caps.
- Battery-charging installations should be located in areas designated for that purpose.
- Facilities must be provided for flushing and neutralizing spilled electrolyte, fire protection, protecting charging apparatus from damage by trucks, and adequate ventilation for dispersal of air contaminants from gassing batteries.
- A conveyor, overhead hoist, or equivalent material handling equipment should be provided for handling batteries.
- Smoking should be prohibited in the charging area.
- Precautions should be taken to prevent open flames, sparks, or electric arcs in battery-charging areas.

20.7.4 RATED CAPACITY

Rated capacity is the maximum weight that a powered industrial truck can transport and stack at a specified load center and for a specified load elevation. When originally purchased, this is usually the maximum weight, expressed in kilograms (pounds) of a 1200 mm (48 in.) homogenous cube (600 mm load center) that a truck can transport and stack to a height established by the manufacturer. Industrial trucks should not be used or tested above their special rated capacity (see ANSI/ASME B56.1).

20.7.5 LOAD TESTING

Forklifts should be load tested and inspected by a qualified inspector when assigned to service, and thereafter at 12 month intervals. Load test records should be kept on file and readily available to appointed personnel. The load tests required must not exceed the rated capacity of the equipment. Test weights should be accurate to within 5% plus 0% of stipulated values. Load slippage for this equipment must not be greater than a maximum of 3 in. vertically and 1 in. horizontally at the cylinder during a static test period of at least 10 min. If a test has not been completed by the end of the required period, the equipment should be down rated as follows:

- Thirty calendar days after the end of the period, the equipment should be down rated to 75% of the rated capacity.
- Sixty calendar days after the end of the period, the equipment should be down rated to 50% of the rated capacity.
- Ninety calendar days after the end of the period, the equipment should be taken out of service until the required inspection has been completed.

20.7.6 INDUSTRIAL TRUCK NAMEPLATE

Every forklift (powered industrial truck) should have appended to it a durable, corrosion-resistant nameplate with the model or serial number and appropriate weight of the truck legibly inscribed. The serial number should also be stamped on the frame of the truck. The truck must be accepted by a recognized national testing laboratory and the nameplate should be marked. The truck should meet all other nameplate requirements of ANSI/ASME B56.1.

Every removable attachment (excluding fork extensions) must have installed a durable corrosion-resistant nameplate with the following information legibly and permanently inscribed:

- Serial number.
- Weight of attachment.
- Rated capacity of attachment.
- The following instructions (or equivalent): "Capacity of truck and attachments combination may be less than capacity shown on attachment—consult truck nameplate."

20.8 SAFETY TIPS FOR OPERATING POWERED INDUSTRIAL TRUCKS

20.8.1 SAFE OPERATIONS

Operators must follow all safety rules related to speed, parking, fueling, loading, and moving loads. While the forklift is in operation keep the forks low with the mast tilted slightly back. Too tall or "top-heavy" loads can change the forklift's center of gravity and cause it to tip over. Follow safe speed limits. Loaded forklifts should travel at low speeds. Without loads, forklifts are not weighted and are especially unstable. Avoid sharp turns. Forklifts can turn over if turns are made too fast. When parking on a hill, always chock the forklift's wheels, lower the tines, and set the parking brake. Also, to avoid tipping, always carry loads up a grade and back down ramps. Never turn on grades. Keep safe visibility. If a load blocks forward vision, drive backward. Always use the horn at intersections. Be cautious around uneven surfaces; chuckholes and other uneven ground can cause forklifts to tip. The following are some general safety rules for operating a powered industrial truck:

- Only drivers authorized by the company and trained in the safe operation of forklift trucks or pickers should be permitted to operate such vehicles. Drivers may not operate trucks other than those for which they are authorized.
- Drivers must check the vehicle at least once per day and if it is found to be unsafe, the matter should be reported immediately to the manager or mechanic, and the vehicle should not be used again until it has been made safe.
- No person should be allowed to stand or pass under the elevated portion of any truck, whether loaded or empty.
- Unauthorized personnel should not be permitted to ride on powered industrial trucks. A safe place to ride should be provided where riding of trucks is authorized.
- When a powered industrial truck is left unattended, load-engaging means should be fully lowered, controls must be neutralized, power must be shut off, and brakes set. Wheels should be blocked if the truck is parked on an incline. A powered industrial truck is "unattended" when the operator is 25 ft or more away from the vehicle which remains in operator's view, or whenever the operator leaves the vehicle and it is not in his view.
- When the operator dismounts and is within 25 ft of the truck still in his/her view, the load-engaging means should be fully lowered, control neutralized, and the brakes set to prevent movement (see Figure 20.2).
- The vehicle should not exceed the authorized or safe speed, must always maintain a safe distance from other vehicles, and must observe all established traffic regulations. For trucks traveling in the same direction, a safe distance may be considered to be approximately three truck lengths or, preferably, a time lapse of 3 s before passing the same point. Exercise extreme care when turning corners. Sound horn at blind corners.
- Employees should not place any part of their bodies outside the running lines of the forklift truck or between mast uprights or other parts of the truck where shear or crushing hazards exist.

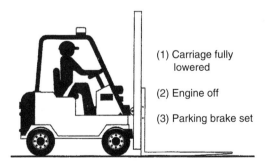

(1) Carriage fully lowered

(2) Engine off

(3) Parking brake set

FIGURE 20.2 Properly set forklift. (Courtesy of the U.S. Department of Energy.)

- The width of one tire on the forklift should be the minimum distance maintained by the truck from the edge while it is on any elevated dock, platform, or freight car.
- Stunt driving and horseplay are prohibited.
- Trucks should not be loaded in excess of their rated capacity.
- Extreme care must be taken when lifting loads and loaded vehicles should not be moved until the load is safe and secure.
- Extreme care should be taken when tilting loads. Elevated loads should not be tilted forward except when the load is being deposited onto a storage rack or equivalent. When stacking or tiering, backward tilts should be limited to that which is necessary to stabilize the load.
- Operators must look in the direction of travel and should not move a vehicle until certain that all persons are in the clear.
- Vehicles should not be operated on floors, sidewalk doors, or platforms that will not safely support the vehicle, empty or loaded. Any damage to forklift trucks and/or structures must be reported immediately to the manager. Additionally, doors adjacent to the path of vehicles should be marked and secured where possible.
- The forks should always be carried as low as possible, consistent with safe operation.
- Special precautions must be taken in the securing and handling of loads by trucks equipped with attachments, and during the operation of these trucks after the loads have been removed.
- Vehicles should not be driven in and out of highway trucks and trailers at unloading docks, until such trucks are securely blocked and brakes set.
- No truck should operate with a leak in the fuel system.
- The load-engaging device must be placed in such a manner that the load will be securely held or supported.
- No smoking is permitted while operating or refueling forklifts.
- A fire extinguisher must be installed on the forklift and should be maintained in a serviceable condition.
- The operating area should be kept free of water, snow, ice, oil, and debris that could cause the operator's hands and feet to slip from the controls.

20.8.2 PICKING UP AND MOVING LOADS

It is important to know how much a load weighs before trying to move it. If the weight of the load is not clearly marked, try a simple test to see if it is safe to move. Lift the load an inch or two. Powered industrial trucks should feel stable and the rear wheels should be in firm contact with the floor. If everything is operating properly and steering seems normal begin to move the load. If the forklift struggles, set the load down and check with the supervisor before continuing. Operators need to practice picking up loads in various locations and in whatever situation they are expected to work.

All loads should be squared up on the center of the load and approached straight on with forks in traveling position. Stop when the tips of the forks are about a foot away from the load. Level the forks and slowly drive forward until the load rests against the backrest. Lift the load high enough to clear whatever is under it. Look in all directions to make sure the travel path is clear, and back out. Carefully tilt the mast back to stabilize the load.

20.8.3 TRAVELING WITH A LOAD

The nature of the terrain, the surface upon which the truck is to operate, is a very important factor in the stability of load-truck system. The designated person should assure that a proper truck has been selected to operate on the surface available. In general, small, three-wheeled trucks are to be operated on smooth, hard surfaces only and are not suitable for outdoor work. The operator should assure that the load is well secured and properly balanced before it is lifted. The lift must be done slowly with no sudden acceleration of the load nor should it contact any obstruction. Here are some requirements for traveling in powered industrial trucks. Some of these requirements include the following:

- All traffic regulations must be observed, including authorized plant speed limits.
- The driver should be required to slow down and sound the horn at cross aisles and other locations where vision is obstructed. If the load being carried obstructs the forward view, the driver is required to travel with the load trailing.
- Railroad tracks should be crossed diagonally whenever possible. Parking closer than 8 ft from the center of railroad tracks is prohibited.
- When ascending or descending grades in excess of 10%, loaded trucks should be driven with the load upgrade.
- Always travel with a load tilted slightly back for added stability.
- Travel with the load at the proper height. A stable clearance height is 4–6 in. at the tips and 2 in. at the heels to clear most uneven surfaces and avoid debris (Figure 20.3).
- Dockboards and bridgeplates should be properly secured, before they are driven over.
- Dockboards and bridgeplates should be driven over carefully and slowly and their rated capacity never exceeded.
- Turning a powered industrial truck will require a little more concentration than driving a car.
- Because it steers from the rear, the forklift handles very differently from a car and other roadway vehicles. The back end of the forklift swings wide and can injure coworkers and damage products or equipment.
- Once the load has been picked up never make a turn at normal speed. Always slow down to maintain balance.

1. Always ensure the load is against the backrest. Drive a loaded forklift with the load on the uphill side. Back down.

2. Always drive an unloaded forklift with the forks on the downhill side. Drive down forward and back up.

3. Never turn a forklift sideways on a ramp.

FIGURE 20.3   Safe traveling for forklift trucks. (Courtesy of the U.S. Department of Energy.)

20.8.4 STACKING AND UNSTACKING LOADS

The use of powered industrial trucks to stack products and increase storage capacity is frequently undertaken. When stacking or unstacking a product, keep in mind that the higher the load is positioned the less stable the truck becomes. Lifting a load from a stack is similar to lifting a load from the floor:

- Approach the load slowly and squarely with the forks in the traveling position.
- Stop about a foot from the load and raise the mast so the forks are at the correct height.
- Level the forks and drive forward until the load is flush against the backrest.
- Lift it high enough to clear the bottom load, look in all directions, and slowly back straight out.
- Once the top of the stack has been cleared, stop and lower the mast to the traveling position. Tilt the forks back and proceed.
- To stack one load on top of another stop about a foot away from the loading area and lift the fork tips enough to clear the top of the stack.
- Slowly move forward until the load is square over the top.

- Level the forks and lower the mast until the load is no longer supported by the fork.
- Look over both shoulders and slowly back straight out.
- Never lift a load while moving.

20.8.5 STANDARD SIGNALS

Standard hand signals for use should be as specified in the latest edition of the ANSI (American Nation Standards Institute) standards regarding powered industrial trucks. The operator should recognize signals from the designated signaler with the only exception being a "Stop" signal, which should be obeyed no matter who gives it. These signals are those provided by the American Society of Mechanical Engineers (ASME).

20.8.6 SAFETY GUARDS

All high-lift rider trucks should be fitted with overhead guards, where overhead lifting is performed, unless operating conditions do not permit. In such cases where high-lift rider trucks must enter, as with, truck trailers when the overhead guard will not permit their entry, the guard may be removed or a powered industrial truck without a guard may be used. If a powered industrial fork truck carries a load that presents a hazard of falling back onto the operator, it should be equipped with a vertical load backrest extension.

20.8.7 TRUCKS AND RAILROAD CARS

In receiving and shipping areas, forklifts (powered industrial trucks) are often used to load and unload materials from trucks and railroad cars. The brakes of highway trucks should be set and wheel chocks placed under the rear wheels to prevent trucks from rolling while they are boarded with powered industrial trucks.

Wheel stops or other positive protection should be provided to prevent railroad cars from moving during loading or unloading operations. Fixed jacks may be necessary to support a semitrailer and prevent unending movement during the loading or unloading when the trailer is not coupled to a tractor.

20.8.8 COWORKER SAFETY

Never carry hitchhikers—they can easily fall off and become injured. If coworkers are on a safety platform, always ensure that the platform is securely attached to the forklift and personnel are wearing proper personal protective equipment (e.g., hardhats and safety harness). Never travel with coworkers on the platform. Watch out for overhead obstructions.

20.8.9 PEDESTRIAN SAFETY

Pedestrians working nearby should be sure to keep a safe distance from forklifts. That means staying clear of the forklift's turning radius and making sure the driver knows where you are.

20.8.10 Conduct of the Operator

The operator's driving skill, attitude, adherence to safety rules, and conduct will play an important role in forklift safety. The operator should

- Not engage in any practice, which will divert attention while operating the forklift.
- Not operate the forklift when physically or mentally incapacitated.
- Before operation of electric powered machines, check location of the battery plug for quick disconnection in case of a short circuit.
- Avoid sudden stops.
- Face in the direction of travel, except as follows:
  - For better vision with large loads, operate the truck in reverse gear.
  - Do not descend ramps with the load in front.
- Watch blind corners, stop at all intersections and doorways and sound the horn.
- Operate at safe speeds: in-plant buildings—5 miles/h; on roads—15 miles/h maximum.
- Go slow around curves.
- Use low gear for the slowest speed control when descending ramps.
- Know the rated capacity of the truck and stay within it.
- Consider both truck and load weight.
- Watch overhead clearance; if in doubt, measure.
- Keep clear of the edge of the loading dock.
- Watch rear-end swing.
- Before handling, assure that stacks and loads are stable. Block and lash them if necessary.
- Always spread the forks to suit the load width.
- Lower and raise the load slowly. Make smooth gradual stops.
- Lift and lower loads only while the vehicle is stopped.
- Use special care when high-tiering. Return the lift to a vertical position before lowering load.
- Lift, lower, and carry loads with the upright vertical tilted back, never forward.
- To avoid personal injury, keep arms and legs inside the operator's area of the machine.
- Never travel with forks raised to unnecessary heights. Approximately 4–6 in. above floor level is adequate.
- When loading trucks or trailers, see that the wheels are chocked and the brakes set.
- Operate in front end of the semitrailer only if the tractor is attached, or adequate trailer (railroad) jacks are in place.
- Inspect floors on trucks, boxcars, unfamiliar ramps, or platforms before start of operation.
- Be sure bridge plates into trucks or freight cars are sufficiently wide, strong, and secure.

- Never butt loads with forks or rear end of truck.
- Fork trucks should not be used as tow trucks, unless a towing hitch is supplied by the manufacturer. They are built for lifting only. Use tow bars rather than cable for towing.
- Stop engine before refueling.
- Use only approved explosion-proof lights to check gas tank and battery water levels.
- Smoking is prohibited during this operation.
- Place forks flat on the floor when truck is parked.
- Turn switch key off when leaving the machine.
- Always set brakes before leaving the truck.
- Report evidence of faulty truck performance.
- When alighting from truck, step down—do not jump.
- Report all accidents promptly to your supervisor.

Operators who are properly trained are expected to adhere to all of the previously iterated requirements for operator conduct and safe work practices when using powered industrial trucks.

20.9 TRAINING OF OPERATORS

From March 1, 1999, employers who use powered industrial trucks (forklifts) in the general industry, construction, or maritime industries must comply with Occupational Safety and Health Administration's (OSHA) new forklift training standards, 29 CFR 1910.178(l), 29 CFR 1915.120, and new 1926.602(d) which are identical to 1910.178(l) and CFR 1917 and 1918, which includes the training requirements by reference to 178(l).

The useful forklift looks easy to operate and most workers think they can. The forklift does not appear as dangerous or formidable as large powered lift trucks or other types of industrial vehicles. But, about 100 workers are killed every year in incidents related to forklift operation and nearly 95,000 suffer injuries every year that result in lost workdays. Approximately 30% of these incidents are, at least in part, caused by inadequate training.

Federal regulations on training of all forklift operators are in 29 CFR 1910.178. These regulations require that only trained and authorized persons should be permitted to operate a powered industrial truck, the regulatory definition of "forklift." This includes all employees who may use a forklift, even if it is only a casual or occasional part of their job duties.

20.9.1 GENERAL TRAINING REQUIREMENTS

The employer must ensure that each powered industrial truck operator is competent to operate a powered industrial truck safely, as demonstrated by the successful completion of training and evaluation. Before permitting an employee to operate a powered industrial truck (except for training purposes), the employer should ensure that each operator has successfully completed the required training.

While implementing training, trainees may operate a powered industrial truck under the direct supervision of persons who have the knowledge, training, and experience to train operators and evaluate their competency, and where such operation does not endanger the trainee or other employees.

All training is to consist of a combination of formal instruction (e.g., lecture, discussion, interactive computer learning, video tape, or written material), practical training (demonstrations performed by the trainer and practical exercises performed by the trainee), and evaluation of the operator's performance in the workplace.

The employer should ensure that all operator training and evaluation be conducted by persons who have the knowledge, training, and experience to train powered industrial truck operators and evaluate their competencies.

20.9.2 TRAINING PROGRAM CONTENT

Powered industrial truck operators must receive initial training in the following topics, when the exception of topics that the employer can demonstrate is not applicable to safe operation of the truck in his/her workplace. The topics are as follows:

- Operating instructions, warnings, and precautions for the types of trucks the operator will be authorized to operate
- Differences between the truck and the automobile
- Truck controls and instrumentation: where they are located, what they do, and how they work
- Engine or motor operation
- Steering and maneuvering
- Visibility (including restrictions due to loading)
- Fork and attachment adaptation, operation, and use limitations
- Vehicle capacity
- Vehicle stability
- Any vehicle inspection and maintenance that the operator will be required to perform
- Refueling and/or charging and recharging of batteries
- Operating limitations
- Any other operating instructions, warnings, or precautions listed in the operators' manual for the types of vehicles that the employee is being trained to operate

The training must also consist of specific workplace-related hazards and topics such as the following:

- Surface conditions where the vehicle will be operated
- Composition of loads to be carried and load stability
- Load manipulation, stacking, and unstacking
- Pedestrian traffic in areas where the vehicle will be operated
- Narrow aisles and other restricted places where the vehicle will be operated
- Hazardous (classified) locations where the vehicle will be operated

- Ramps and other sloped surfaces that could affect the vehicle's stability
- Closed environments and other areas where insufficient ventilation or poor vehicle maintenance could cause a buildup of carbon monoxide or diesel exhaust
- Other unique or potentially hazardous environmental conditions in the workplace that could affect safe operation

20.9.3 REFRESHER TRAINING AND EVALUATION

Refresher training, including an evaluation of the effectiveness of that training, should be conducted to ensure that the operator has the knowledge and skills needed to operate the powered industrial truck safely. Refresher training in relevant topics should be provided to the operator when

- The operator has been observed to operate the vehicle in an unsafe manner.
- The operator has been involved in an accident or near-miss incident.
- The operator has received an evaluation that reveals that the operator is not operating the truck safely.
- The operator is assigned to drive a different type of truck.
- A condition in the workplace changes in a manner that could affect safe operation of the truck.

20.9.4 REEVALUATION

An evaluation of each powered industrial truck operator's performance should be conducted at least once every 3 years.

20.9.5 AVOIDANCE OF DUPLICATIVE TRAINING

If an operator has prior training in the previously specified topics, and such training is appropriate to the truck and present working conditions encountered, and the operator has been evaluated and found competent to operate the truck safely, then additional training in that topic is not required.

20.9.6 CERTIFICATION

The employer should certify that each operator has been trained and evaluated as required. Certification should include the name of the operator, the date of the training, the date of the evaluation, and the identity of the persons performing the training and evaluation.

20.9.7 IN-HOUSE TRAINING DEVELOPMENT

Training programs should be tailored to employees' work situations. Employees benefit more from training that simulates their daily processes, rather than from watching "canned" programs that are not applicable to their specific operations. Training programs should be devised so that employees can demonstrate the knowledge and skills required for their job.

20.9.8 DRIVING SKILL EVALUATIONS

A key dimension of operator training is driver certification. Operators should be required to demonstrate their skills. Adequate completion of skills tests demonstrates both that the operator knows and understands the unit's functional features, and is familiar with overall departmental safety rules and can identify specific safety factors at a dock and battery recharge station. He/she must also demonstrate overall driving skills. Testing can be administered on the job during the employee's normal workday. A written record of the evaluation of the operator must be made and retained.

20.10 APPLICABLE STANDARDS AND REGULATIONS

The following is a list of the applicable standards relevant to powered industrial trucks from varied official organizations (Figure 20.4).

20.10.1 FORKLIFT TRUCKS (POWERED INDUSTRIAL TRUCKS) (29 CFR 1910.178)

The ASME defines a powered industrial truck as a mobile, power-propelled truck used to carry, push, pull, lift, stack, or tier materials. Powered industrial trucks are also commonly known as forklifts, pallet trucks, rider trucks, forktrucks, or lifttrucks. Each year, tens of thousands of forklift-related injuries occur in U.S. workplaces. Injuries

Applicable forklift standards		
Organization	Standard	Title
OSHA	29 CFR 1910.178	Powered industrial trucks
OSHA	29 CFR 1910.1000	Air contaminants
OSHA	29 CFR 1926.602	Material handling equipment
ANSI	B56.1–1988	American national standard for powered industrial trucks
NFPA	NFPA No. 30–1969	NFPA flammable and combustible liquids code
NFPA	NFPA No. 58–1969	NFPA storage and handling of liquefied petroleum gases
NFPA	NFPA No. 505–1969	Powered industrial trucks
UL	583	Standard for safety for electric or battery-powered industrial trucks
UL	558	Standard for safety for internal combustion or engine-powered industrial trucks
ANSI/NFPA	30–1987	Flammable and combustible liquid code
ANSI/NFPA	58–1986	Storage and handling of liquefied petroleum gases
ANSI/NFPA	505–1987	Fire safety standard for powered industrial trucks—type designations, areas of use, maintenance, and operation

OSHA = Occupational Safety and Health Administration
ANSI = American National Standards Institute
NFPA = National Fire Protection Association
UL = Underwriters Laboratory

FIGURE 20.4 Applicable forklift standards.

usually involve employees being struck by lift trucks or falling while standing or working from elevated pallets and tines. Many employees are injured when lift trucks are inadvertently driven off loading docks or when the lift falls between a dock and an unchocked trailer. Most incidents also involve property damage, including damage to overhead sprinklers, racking, pipes, walls, machinery, and other equipment. Unfortunately, a majority of employee injuries and property damage can be attributed to lack of procedures, insufficient or inadequate training, and lack of safety-rule enforcement.

If, at any time, a powered industrial truck is found to be in need of repair, defective, or in any way unsafe, the truck is to be taken out of service until it has been restored to safe operating condition.

High-lift rider trucks should be equipped with substantial overhead guards unless operating conditions do not permit. Fork trucks are to be equipped with vertical-load backrest extensions when the types of loads present a hazard to the operators. Each industrial truck is to have a warning alarm, whistle, gong, or other device which can be clearly heard above the normal noise in the areas where operated. The brakes of trucks are to be set and wheel chocks placed under the rear wheels to prevent the movement of trucks, trailers, or railroad cars while loading or unloading.

Only trained and authorized operators are permitted to operate a powered industrial truck. Methods are to be devised to train operators in the safe operation of powered industrial trucks.

20.11   FORKLIFT CHECKLIST

The checklist in Figure 20.5 is to be used to assure the safe operation of a forklift in handling materials.

Powered industrial truck safety checklist

These checks must be made at the start of each work shift:

Date_____

Visual Checks	Operational Checks
_____ Tires, wheels, rims	_____ Seat Belt
—in good condition, not excessively worn, no punctures	—if it is there, use it!
_____ Hydraulics	_____ Horn
—no fluid leaks, piston free of debris	
_____ Forks and Carriage	_____ Brakes
—no cracks or bends, any attachments securely fastened	—brings machine to complete stop, parking brake holds machine in fixed position
_____ Overhead Guard	_____ Accelerator
—no cracks, nothing stacked on top or impeding vision	—even acceleration, does not stick

---

FIGURE 20.5   Forklift safety checklist.

(*continued*)

_____ Mast Chains   _____ Steering
—clean, links intact, no excessive slack   —responsive, minimal looseness

_____ ID Plate   _____ Mast Lift
—visible and legible, lifting capacity enough to   _____ Mast Tilt
perform the designated task?   _____ Mast Shift

_____ Battery
—charged and in good condition, caps secure   _____ Backup Alarm Lights
_____ Propane Tank   — if present are they operational
—no leaks in connections, tank secured to forklift
in designated position

Failed Checks/Areas in Need of Service: _____

_____

_____

Name_____ Date _____

FIGURE 20.5 (continued)

20.12 SUMMARY

Although forklifts are the most universally common powered material vehicles, they are also one of the more dangerous. The usefulness and size make it look easy to operate. Accident facts over a period of years indicated that in most cases accidents were caused by inexperienced and usually untrained operators. With these facts in mind OSHA now requires that operators be trained in a formal manner and certified to operate the specific forklift that they are expected to use. Also, each operator should be able to answer in the affirmative all the questions in Figure 20.6 to insure safe operations of a forklift (Figure 20.6).

Operator safe operation questionnaire

Operators should answer the following questions yes to operate a forklift safely:

- When operate or work near forklifts, do you take these steps to protect yourself?
- When you operate a forklift have been trained and licensed?
- Do you use seat belts if they are available?
- Do you report to your supervisor any damage or problems that occur to a forklift during your shift?
- Do you remember not jump from an overturning, sit-down type forklift to stay with the truck, holding on firmly and leaning in the opposite direction of the overturn?
- Do you remember to exit from a stand-up type forklift with rear-entry access by stepping backward if a lateral tipover occurs?
- Do you use extreme caution on grades or ramps?
- On grades, do you tilt the load back and raise it only as far as needed to clear the road surface?

FIGURE 20.6 Operator safe operation questionnaire.

- Do you not raise or lower the forks while the forklift is moving?
- Do you not handle loads that are heavier than the weight capacity of the forklift?
- Do you operate the forklift at a speed that will permit it to be stopped safely?
- Do you slow down and sound the horn at cross aisles and other locations where vision is obstructed?
- Do you look toward the travel path and keep a clear view of it?
- Do you not allow passengers to ride on forklift trucks unless a seat is provided?
- When dismounting from a forklift, do you set the parking brake, lower the forks or lifting carriage, and neutralize the controls?
- Do you not drive up to anyone standing in front of a bench or other fixed object?
- Do you not use a forklift to elevate workers who are standing on the forks?
- Do you elevate a worker on a platform only when the vehicle is directly below the work area?
- Whenever a truck is used to elevate personnel, do you secure the elevating platform to the lifting carriage or forks of the forklift?
- Do you use a restraining means such as rails, chains, or a body belt with a lanyard or deceleration device for the workers on the platform?
- Do you not drive to another location with the work platform elevated?

FIGURE 20.6 (continued)

REFERENCE

Bureau of Labor Statistics. *Fatal Workplace Injuries in 1992, A Collection of Data and Analysis*, Report 870, Washington, April 1994.

U.S. Department of Labor. Occupational Safety and Health Administration, *Forklift Data*, Washington, 1999. http://www.osha.gov

American National Standard Institute/American Society of Mechanical Engineers, *Consewsys Standard 1356.1-2004: Safety Standard for Low Lift and High Lift Trucks*, New York, 2004.

21 Workplace Security and Violence

Security in a mall is much more visible as warning to those who endanger others.

21.1 WORKPLACE SECURITY AND VIOLENCE

There is a very close alliance between security and violence. Steps taken to provide security are often the same as those taken to prevent violence. If security were to be adequate, the risk of violent acts would be reduced. If violent acts are prevented the security is a success. Thus, this chapter provides a blended approach to security and safety. Many members of the workforce in the goods and material services sectors are in constant contact with the public and are very visible to the public.

21.1.1 BACKGROUND

Workplace violence is a serious safety and health hazard in many workplaces. According to the Bureau of Labor Statistics (BLS), homicide is the second leading cause of death to American workers, claiming the lives of 912 workers in 1996 and accounting for 15% of the 6112 fatal work injuries in the United States (BLS, 1997). Violent incidents at work also resulted in 20,438 lost workday cases in 1994 (BLS, 1996). Violence inflicted upon employees may come from many sources, including customers, robbers, muggers, relations, acquaintances, and coworkers to mention a few.

Although workplace violence may appear to be random, many incidents can be anticipated and avoided and security and preventive measures can be taken. Even

311

where a potentially violent incident occurs, a timely and appropriate response can prevent the situation from escalating and resulting in injury or death.

21.1.2 HIGH-RISK ESTABLISHMENTS

From 1980 to 1992, the overall rate of homicide was 1.6 per 100,000 workers per year in the retail industry, compared with a national average of 0.70 per 100,000 workers (NIOSH, 1996). Job-related homicides in retail trade accounted for 48% of all workplace homicides in 1996 (BLS, 1997). The wide diversity within the retail industry results in substantial variation in levels of risk of violence.

Homicides in convenience and other grocery stores, eating and drinking places, and gasoline service stations constituted the largest share of homicides in retail establishments (BLS, 1997). The most vulnerable appear to be liquor stores, gasoline service stations, jewelry stores, grocery stores, convenience stores, and eating and drinking places.

Of course, occupations such as gasoline service and garage workers, stock handlers and baggers, sales supervisors and proprietors, and sales counter clerks are at greatest risk.

21.2 SYSTEMATIC APPROACH TO PREVENTION

The basic recommendation is to address the five key areas of any safety and health program. An effective approach to preventing workplace violence and insuring security includes five key components: (1) management commitment and employee involvement, (2) worksite analysis, (3) hazard prevention and control, (4) safety and health training, and (5) evaluation. Using these basic elements, an employer can devise prevention plans that are appropriate for his/her establishment, based on the hazards and circumstances of the particular situation and address both security and violence as an integral part of doing business.

It would be best if employers were to develop a written program for workplace security and violence prevention. A written statement of policy serves as a touchstone for the many separate plans, procedures, and actions required for an effective prevention program. The extent to which the components of the program are in writing, however, is less important than how effective the program is in practice. In smaller establishments, a program can be effective without being heavily documented. As the size of a workplace or the complexity of hazard control increases, written guidance assumes more importance as a way to ensure clear communication and consistent application of policies and procedures.

21.3 MANAGEMENT COMMITMENT AND EMPLOYEE INVOLVEMENT

21.3.1 MANAGEMENT COMMITMENT

Management provides the motivation and resources to deal effectively with workplace violence. The visible commitment of management to worker safety and health is an essential precondition for its success. Management can demonstrate its commitment to violence prevention through the following actions:

- Create and disseminate a policy to managers and employees that expressly disapproves of workplace violence, verbal and nonverbal threats, and related actions.
- Take all violent and threatening incidents seriously, investigate them, and take appropriate corrective action.
- Outline a comprehensive plan for maintaining security in the workplace. Uniformed security guards can be a part of this plan (Figure 21.1).
- Assign responsibility and authority for the program to individuals or teams with appropriate training and skills. This means ensuring that all managers and employees understand their obligations.
- Provide necessary authority and resources for staff to carry out violence prevention responsibilities.
- Hold managers and employees accountable for their performance. Stating expectations means little if management does not track performance, reward it when competent, and correct it when it is not.
- Take appropriate action to ensure that managers and employees follow the administrative controls or work practices.
- Institute procedures for prompt reporting and tracking of violent incidents and breaches of security that occur in and near the establishment.
- Encourage employees to suggest ways to reduce risks and improve security, and implement appropriate recommendations from employees and others whenever possible.

FIGURE 21.1 Uniformed security personnel are a visible deterrent to crime and violence.

- Ensure that employees who report or experience workplace violence are not punished or otherwise suffer discrimination.
- Work constructively with other parties such as landlords, lessees, local police, and other public safety agencies to improve the security of the premises.

21.3.2 EMPLOYEE INVOLVEMENT

Management commitment and employee involvement are complementary elements of an effective safety and health program. To ensure an effective program, management, frontline employees, and employee representatives need to work together in the structure and operation of their violence prevention program.

Employee involvement is important for several reasons. First, frontline employees are an important source of information about the operations of the business and the environment in which the business operates. This may be particularly true for employees working in wholesale, retail, and warehousing establishments where higher level managers may not routinely be on duty. Second, inclusion of a broad range of employees in the violence prevention program has the advantage of harnessing a wider range of experience and insight than that of management alone. Third, frontline workers can be very valuable problem solvers, as their personal experience often enables them to identify practical solutions to problems and to perceive hidden impediments to proposed changes. Finally, employees who have a role in developing prevention programs are more likely to support and carry out those programs.

Employees and employee representatives can be usefully involved in nearly every aspect of a security and violence prevention program. Their involvement may include the following:

- Participate in surveys and offer suggestions about safety and security issues.
- Participate in developing and revising procedures to minimize the risk of violence in daily business operations.
- Assist in the security analysis of the establishment.
- Participate in performing routine security inspections of the establishment.
- Participate in the evaluation of prevention and control measures.
- Participate in training current and new employees.
- Share on-the-job experiences to help other employees recognize and respond to escalating agitation, assaultive behavior, or criminal intent, and discuss appropriate responses.

21.4 WORKSITE ANALYSIS

21.4.1 COMMON RISK FACTORS IN GOODS AND MATERIALS SECTOR

The National Institute for Occupational Safety and Health (NIOSH) has identified a number of factors that may increase a worker's risk for workplace assault. Some of the common risk factors that are most often mentioned in the goods and materials sector are as follows:

- Contact with the public
- Exchange of money
- Delivery of passengers, goods, or services
- Working alone or in small numbers
- Working late night or early morning hours
- Working in high-crime areas

Employees in some establishments may be exposed to multiple risk factors. The presence of a single risk factor does not necessarily indicate that the risk of violence is a problem in a workplace. The presence, however, of multiple risk factors or a history of workplace violence should alert an employer that the potential for workplace violence is increased.

Research indicates that the greatest risk of work-related homicide comes from violence inflicted by third parties such as robbers and muggers. Robbery and other crimes were the motive in 80% of workplace homicides across all industries in 1996 (BLS, 1996).

Sexual assault is another significant occupational risk in the retail industry. Indeed, the risk of sexual assault for women is equal to or greater than the risk of homicide for employees in general. Sexual assault is usually not robbery related, but may occur more often in stores with a history of robbery. These assaults occur disproportionately at night and involve a female clerk alone in a store in the great majority of cases.

The establishments that were most attractive had large amounts of cash on hand, an obstructed view of counters, poor outdoor lighting, and easy escape routes. Subsequent studies have confirmed that robbers do not choose targets randomly but, instead, consider environmental factors. The time of day also affects the likelihood of robbery. Studies have consistently found that businesses face an elevated risk of robbery during the nighttime hours.

21.4.2 WORKPLACE HAZARD ANALYSIS

A worksite hazard analysis involves a step-by-step, commonsensical look at the workplace to find existing and potential hazards for workplace violence. This entails the following steps: (1) review records and past experiences, (2) conduct an initial worksite inspection and analysis, and (3) perform periodic safety audits.

Because the hazard analysis is the foundation for determining security weaknesses and the violence prevention program, it is important to select carefully the persons for this task. The employer can delegate the responsibility to one person or a team of employees. If a large employer uses a team approach, it may wish to draw the team members from different parts of the enterprise, such as representatives from senior management, operations, employee assistance, security, occupational safety and health, legal, human resources staff, and employees or union representatives. Small establishments might assign the responsibility to a single staff member or a consultant.

21.4.3 Review of Records and Past Incidents

As a starting point for the hazard analysis, the employer would review the experience of the business over the previous 2 or 3 years. This involves collecting and examining any existing records that may shed light on the magnitude and prevalence of the risk of workplace violence or security failures. The following questions may be helpful in compiling information about past incidents:

- Has your business been robbed during the last 2–3 years? Were robberies attempted? Did injuries occur due to robberies or attempted robberies?
- Have employees been assaulted in altercations with customers?
- Have employees been victimized by other criminal acts at work (including shoplifting that became assaultive)? If yes, of what kind?
- Have employees been threatened or harassed while on duty? What was the context of those incidents?
- In each injury case, how serious were the injuries?
- In each case, was a firearm involved, discharged, or threatened to be used? Were other weapons used?
- What part of the business was the target of the robbery or other violent incident?
- At what time of day did the robbery or other incident occur?
- How many employees were on duty?
- Were the police called to your establishment in response to the incident? (When possible, obtain reports of the police investigation.)
- What tasks were the employees performing at the time of the robbery or other incident? What processes and procedures may have put employees at risk of assault? Similarly, were there factors that may have facilitated an outcome without injury or harm?
- Were preventive measures already in place and used correctly?
- Were there failures in the security system?
- How did the victim react during the incident? Did these actions affect the outcome of the incident in any way?

Employers with more than one store or business location could review the history of violence at each operation. Different experiences in those stores can provide insights into factors that can aid workplace violence. Contacting similar local businesses, community and civic groups, and local police departments is another way to learn about workplace violence incidents in the area. In addition, trade associations and industry groups often provide useful information about conditions and trends in the industry as a whole.

21.4.4 Workplace Security Analysis

The team or coordinator could conduct a thorough initial risk assessment to identify hazards, conditions, operations, and situations that could lead to violence. The initial risk assessment includes a walkthrough survey to provide the data for risk identification and the development of a comprehensive workplace violence prevention program. The assessment process includes the following:

- Analyze incidents, including the characteristics of assailants and victims. Give an account of what happened before and during the incident, and note the relevant details of the situation and its outcome.
- Identify any apparent trends in injuries or incidents relating to a particular worksite, job title, activity, or time of day or week. The team or coordinator should identify specific tasks that may be associated with increased risk.
- Identify factors that may make the risk of violence more likely, such as physical features of the building and environment, lighting deficiencies, lack of telephones and other communication devices, areas of unsecured access, and areas with known security problems.
- Evaluate the effectiveness of existing security measures. Assess whether those control measures are being properly used and whether employees have been adequately trained in their use.

A sample list of questions that illustrates a number of questions that may be helpful for the security analysis and can be altered to meet the needs of your business is as follows:

- Environmental factors
  - Do employees exchange money with the public?
  - Is the business open during evening or late-night hours?
  - Is the site located in a high-crime area?
  - Has the site experienced a robbery in the past 3 years?
  - Has the site experienced threats, harassment, or other abusive behavior in the past 3 years?
- Engineering control
  - Do employees have access to a telephone with an outside line?
  - Are emergency telephone numbers for law enforcement, fire and medical services, and an internal contact person posted adjacent to the phone?
  - Is the entrance to the building easily seen from the street and free of heavy shrub growth?
  - Is lighting bright in parking and adjacent areas?
  - Are all indoor lights working properly?
  - Are windows and views outside and inside clear of advertising or other obstructions?
  - Is the cash register in plain view of customers and police cruisers to deter robberies?
  - Is there a working drop safe or time access safe to minimize cash on hand?
  - Are security cameras and mirrors placed in locations that would deter robbers or provide greater security for employees?
  - Are there height markers on exit doors to help witnesses provide more complete descriptions of assailants?

- Are employees protected through the use of bullet-resistant enclosures in locations with a history of robberies or assaults in a high-crime area?
- Administrative/work practice controls
  - Are there emergency procedures in place to address robberies and other acts of potential violence?
  - Have employees been instructed to report suspicious persons or activities?
  - Are employees trained in emergency response procedures for robberies and other crimes that may occur on the premises?
  - Are employees trained in conflict resolution and in nonviolent response to threatening situations?
  - Is cash control a key element of the establishment's violence and robbery prevention program?
  - Does the site have a policy limiting the number of cash registers open during late-night hours?
  - Does the site have a policy to maintain less than $50 in the cash register? (This may not be possible in stores that have lottery ticket sales and payouts.)
  - Are signs posted notifying the public that limited cash, no drugs, and no other valuables are kept on the premises?
  - Do employees work with at least one other person throughout their shifts, or are other protective measures utilized when employees are working alone in locations with a history of robberies or assaults in a high-crime area?
  - Are there procedures in place to assure the safety of employees who open and close the store?

21.4.5 PERIODIC SAFETY AUDITS

Hazard analysis is an ongoing process. A good violence prevention program will institute a system of periodic safety audits to review workplace hazards and the effectiveness of the control measures that have been implemented. These audits can also evaluate the impact of other operational changes (such as new store hours, or changes in store layout) that were adopted for other reasons but may affect the risk of workplace violence. A safety audit is important in the aftermath of a violent incident or other serious event for reassessing the effectiveness of the violence prevention program.

21.5 HAZARD PREVENTION AND CONTROL

21.5.1 PREVENTION STRATEGIES

After assessing violence hazards and the effectiveness of security, the next step is to develop measures to provide security and protect employees from the identified risks of injury and violent acts. Workplace security and violence prevention and control programs include specific engineering and work practice controls to address identified hazards. The tools listed in this section are not intended to be a "one-size-fits-all" prescription. No single control will protect employees. To provide effective deterrents to violence, the employer may wish to use a combination of controls in relation to the hazards identified through the hazard analysis.

In general, a business may reduce the risk of robbery by

- Increasing the effort that the perpetrator must expend (target hardening, controlling access, and deterring offenders)
- Increasing the risks to the perpetrator (entry/exit screening, formal surveillance, and surveillance by employees and others)
- Reducing the rewards to the perpetrator (removing the target, identifying property, and removing inducements)

Other deterrents that may reduce the potential for robbery include making sure that there are security cameras, time-release safes, other 24 h businesses at the location, no easy escape routes or hiding places, and that the store is closed during late-night hours.

21.5.2 ENGINEERING CONTROLS AND WORKPLACE ADAPTATION

Engineering controls remove the hazard from the workplace or create a barrier between the worker and the hazard. The following physical changes in the workplace can help reduce violence-related risks or hazards in retail establishments:

- Improve visibility as visibility is important in preventing robbery in two respects: First, employees should be able to see their surroundings, and second, persons outside the store, including police on patrol, should be able to see into the store (Figure 21.2). Employees in the store should have an unobstructed view of the street, clear of shrubbery, trees, or any form of clutter that a criminal could use to hide. Signs located in windows should be either low or high to allow good visibility into the store. The customer service and cash register areas should be visible from outside the establishment. Shelves should be low enough to assure good visibility throughout the establishment. Convex mirrors, two-way mirrors, and an elevated vantage point can give employees a more complete view of their surroundings.
- Maintain adequate lighting within and outside the establishment to make it less appealing to a potential robber by making detection more likely. The parking area and the approach to the retail establishment should be well lit during nighttime hours of operation. Exterior illumination may need upgrading to allow employees to see what is occurring outside the store (Figure 21.3).
- Use fences and other structures to direct the flow of customer traffic to areas of greater visibility.
- Use drop safes to limit the availability of cash to robbers. Employers using drop safes can post signs stating that the amount of cash on hand is limited.

FIGURE 21.2   Roving patrols increase security visibility.

- Install video surveillance equipment and closed circuit TV (CCTV) to deter robberies by increasing the risk of identification. This may include interactive video equipment. The video recorder for the CCTV should be secure

FIGURE 21.3   Well-designed parking lots are important security measures.

and out of sight. Posting signs that surveillance equipment is in use and placing the equipment near the cash register may increase the effectiveness of the deterrence.

- Put height markers on exit doors to help witnesses provide more complete descriptions of assailants.
- Use door detectors to alert employees when persons enter the establishment.
- Control access to the establishment with door buzzers.
- Use silent and personal alarms to notify police or management in the event of a problem. To avoid angering a robber, however, an employee may need to wait until the assailant has left before triggering an alarm.
- Install physical barriers such as bullet-resistant enclosures with pass-through windows between customers and employees to protect employees from assaults and weapons in locations with a history of robberies or assaults and located in high-crime areas.

21.5.3 ADMINISTRATIVE AND WORK PRACTICE CONTROLS

Administrative and work practice controls affect the way employees perform jobs or specific tasks. The following examples illustrate work practices and administrative procedures that can help prevent incidents of workplace violence:

- Integrate violence prevention activities into daily procedures, such as checking lighting, locks, and security cameras, to help maintain worksite readiness.
- Keep a minimal amount of cash in each register (e.g., $50 or less), especially during evening and late-night hours of operation. In some businesses, transactions with large bills (over $20) can be prohibited. In situations where this is not practical because of frequent transactions in excess of $20, cash levels should be as low as is practical. Employees should not carry business receipts on their person unless it is absolutely necessary.
- Adopt proper emergency procedures for employees to use in case of a robbery or security breach.
- Establish systems of communication in the event of emergencies. Employees should have access to working telephones in each work area, and emergency telephone numbers should be posted by the phones.
- Adopt procedures for the correct use of physical barriers, such as enclosures and pass-through windows.
- Increase staffing levels at night at the establishment with a history of robbery or assaults and located in high-crime areas. It is important that clerks be clearly visible to patrons.
- Lock doors used for deliveries and disposal of garbage when not in use. Also, do not unlock delivery doors until the delivery person identifies himself or herself. Take care not to block emergency exits—doors must open from the inside without a key to allow persons to exit in case of fire or other emergency.

- Establish rules to ensure that employees can walk to garbage areas and outdoor freezers or refrigerators without increasing their risk of assault. The key is for employees to have good visibility, thereby eliminating potential hiding places for assailants near these areas. In some locations, taking trash out or going to outside freezers during daylight may be safer than doing so at night.
- Keep doors locked before business officially opens and after closing time. Establish procedures to assure the security of employees who open and close the business when staffing levels may be low. In addition, the day's business receipts may be a prime robbery target at store closing.
- Limit or restrict areas of customer access, reduce the hours of operation, or close portions of the establishment to limit risk.
- Adopt safety procedures and policies for off-site work, such as deliveries.

Administrative controls are effective only if they are followed and used properly. Regular monitoring helps ensure that employees continue to use proper work practices. Giving periodic, constructive feedback to employees helps to ensure that they understand these procedures and their importance.

21.6 POST-INCIDENT RESPONSE

Post-incident response and evaluation are important parts of an effective violence prevention program. This involves developing standard operating procedures for management and employees to follow in the aftermath of a violent incident. Such procedures may include the following:

- Assure that injured employees receive prompt and appropriate medical care. This includes providing transportation of the injured to medical care. Prompt first-aid and emergency medical treatment can minimize the harmful consequences of a violent incident.
- Report the incident to the police.
- Notify other authorities, as required by applicable laws and regulations.
- Inform management about the incident.
- Secure the premises to safeguard evidence and reduce distractions during the post-incident response process.
- Prepare an incident report immediately after the incident, noting details that might be forgotten over time. A sample violence incident report can be found in Appendix D.
- Arrange appropriate treatment for victimized employees. In addition to physical injuries, victims and witnesses may suffer psychological trauma, fear of returning to work, feelings of incompetence, guilt, power-lessness, and fear of criticism by supervisors or managers. Post-incident debriefings and counseling can reduce psychological trauma and stress among victims and witnesses. An emerging trend is to use critical incident stress management to provide a range or continuum of care tailored to the individual victim or the organization's needs.

21.7   TRAINING AND EDUCATION

Training and education ensure that all staff are aware of potential security hazards and the procedures for protecting themselves and their coworkers. Employees with different roles in the business may need different types and levels of training.

21.7.1   GENERAL TRAINING

Employees need instruction on the specific hazards associated with their job and worksite to help them minimize their risk of assault and injury. Such training would include information on potential hazards identified in the establishments, and the methods to control those hazards. Topics may include the following:

- An overview of the potential risk of assault
- Operational procedures, such as cash handling rules that are designed to reduce risk
- Proper use of security measures and engineering controls that have been adopted in the workplace
- Behavioral strategies to defuse tense situations and reduce the likelihood of a violent outcome, such as techniques of conflict resolution and aggression management
- Specific instructions on how to respond to a robbery (such as the instruction to turn over money or valuables without resistance) and how to respond to attempted shoplifting
- Emergency action procedures to be followed in the event of a robbery or violent incident

Training should be conducted by persons who have a demonstrated knowledge of the subject and should be presented in language appropriate for the individuals being trained. Oral quizzes or written tests can ensure that the employees have actually understood the training that they received. An employee's understanding also can be verified by observing the employee at work.

The need to repeat training varies with the circumstances. Retraining should be considered for employees who violate or forget safety measures. Similarly, employees who are transferred to new job assignments or locations may need training even though they may already have received some training in their former position. Establishments with high rates of employee turnover may need to provide training frequently.

21.7.2   TRAINING FOR SUPERVISORS, MANAGERS, AND SECURITY PERSONNEL

To recognize whether employees are following safe practices, management personnel should undergo training comparable to that of the employees and additional training to enable them to recognize, analyze, and establish violence prevention controls. Knowing how to ensure sensitive handling of traumatized employees also is an important skill for management. Training for managers could also address any specific duties and responsibilities they have that could increase their risk of assault. Security personnel need specific training about their roles, including the psychological components of handling aggressive and abusive customers and ways to handle aggression and defuse hostile situations. The team or coordinator responsible for implementation of the program should review and evaluate annually the content, methods, and frequency of training.

Program evaluation can involve interviewing supervisors and employees, testing and observing employees, and reviewing responses of employees to security issues or workplace violence incidents.

Evaluation recordkeeping good records help employers determine the severity of the risks, evaluate the methods of hazard control, and identify training needs. An effective violence prevention program will use records of injuries, illnesses, incidents, hazards, corrective actions, and training to help identify problems and solutions for a safe and healthful workplace.

Employers can tailor their recordkeeping practices to the needs of their violence prevention program. The purpose of maintaining records is to enable the employer to monitor its ongoing efforts, to determine if the violence prevention program is working, and to identify ways to improve it. Employers may find the following types of records useful for this purpose:

- Records of employees and other injuries and illnesses at the establishment.
- Records describing incidents involving violent acts and threats of such acts, even if the incident did not involve an injury or a criminal act. Records of events involving abuse, verbal attacks, or aggressive behavior can help identify patterns and risks that are not evident from the smaller set of cases that actually result in injury or crime.
- Written hazard analyses.
- Recommendations of police advisors, employees, or consultants.
- Up-to-date records of actions taken to deter violence, including work practice controls and other corrective steps.
- Notes of safety meetings and training records.

21.8   PREVENTION PROGRAM EVALUATIONS

Violence prevention programs benefit greatly from periodic evaluation. The evaluation process could involve the following:

- Review the results of periodic safety audits.
- Review post-incident reports. In analyzing incidents, the employer should pay attention not just to what went wrong, but to actions taken by employees that avoided further harm, such as handling a shoplifting incident in such a way as to avoid escalation to violence.
- Examine reports and minutes from staff meetings on safety and security issues.
- Analyze trends and rates in illnesses, injuries, or fatalities caused by violence relative to initial or "baseline" rates.

- Consult with employees before and after making job or worksite changes to determine the effectiveness of the interventions.
- Keep abreast of new strategies to deal with violence in the retail industry.

Management should communicate any lessons learned from evaluating the workplace violence prevention program to all employees. Management could discuss changes in the program during regular meetings of the safety committee, with union representatives, or with other employee groups.

21.9 SUMMARY

Workplace security and violence has emerged as an important occupational safety and health issue in many industries, including the retail trade. These recommendations offer a systematic framework to help an employer protect employees from risks of injury and death from occupationally related violence. By addressing workplace violence as a preventable hazard, employers can develop practical and effective strategies to protect their employees from this serious risk and provide a safe and healthful workplace. The security effort will be improved by addressing workplace violence as an issue.

REFERENCE

National Institute for Occupational Safety and Health (NIOSH), U.S. Department of Health and Human Services. *Violence in the Workplace: Risk Factors and Prevention Strategies (CIB 57)*. Washington, June 1996.

Bureau of Labor Statistics, *Summary of Occupational Injuries and Illnesses in 1994*, Washington, 1996. http://www.bls.gov.

Bureau of Labor Statistics, *Occupational Fatalities in 1996*, Washington, 1997, http://www.bls.gov.

22 Slips and Trips

Slips

Trips

Falls

Accepted warning signs for slips, trips, and falls.

This chapter discusses slips and trips primarily related to falls to the same level. Many slips and trips lead to falls from elevated heights that will be discussed in another chapter. About 7% of all injuries are from slips and trips. In most cases, slips and trips lead to muscle strains and sprains (64%), bruises and contusions (38%), and fractures (20%). Deaths from slips and trips that result in a fall to the same level are normally not fatal, but they have occurred when the worker strikes his/her head or is impaled on an object, or falls on or into an operating piece of equipment.

Slips and trips cause enough disabling injuries to be given their due attention in a safety effort by a company. There are many ways of preventing them. In this chapter, specific attention will be paid to walking–working surfaces and stairs. Although stairs are usually addressed as a form of ladder, they are indeed an integral part of the walking/working surfaces for workers and a critical part of access work areas on different levels.

In most situations, where slip and trip accidents occur, workers tend to lose their footing (50%), lose their balance (13%), or lose their grip (12%). Primarily it has been found that the primary causes are an unsafe mindset (not paying attention to or recognizing the potential hazard), existence of unsafe conditions, or unsafe behavior by the injured party.

Related to stairs the most common hazards occur when descending (83%), carrying an object while descending (57%), not holding onto the handrail (63%), absence of handrail (21%), and unsafe conditions such as ice, snow, object on stairs, or spill. Many times a combination of these hazardous conditions or acts is present.

Some of the hazards that one should look for when inspecting walking/working surfaces are as follows:

- Loose or bent boards or floor tiles
- Unsecured rugs and mats
- Floor surfaces that change elevation
- Broken concrete
- Manholes
- Uncovered drains
- Unsafe stairs
- Slippery surfaces
- Obstructions in walkways
- Improper shoes
- Running or moving too fast for the conditions
- Poor lighting

22.1 PREVENTING SLIPS AND TRIPS

Slips and trips can be prevented by maintaining an alertness and awareness of potential hazards, identifying unsafe conditions and behaviors, selecting the proper tools for the task, and using proper body mechanics. Using good body mechanics entails as follows:

- Not overtilting of the head
- Using all fingers to grip
- Shortening walking stride and pointing feet slightly outward
- Walking with knees slightly bent which help to avoid falling forward
- Balancing all loads that are being carried
- Avoiding overreaching

In using proper body mechanics it is important to maintain a center of balance which requires use of eyes, ears, and muscles; thus, good health is important and anyone under medication, that could affect the sense and body motion, should either not be working or should proceed with great caution. Medication is not the only substance that affects balance. Illegal drugs and alcohol can also have similar effects on the worker's balance. Any vision problems should be corrected. Workers need to be fit and be conditioned for the task and this includes maintaining normal weight. An overweight worker is most likely to lose balance.

In most situation, workers often trip or stumble over unexpected objects in their way. This is why housekeeping is a critical component in preventing slip and trip accidents. Some of the things that should be done regarding housekeeping that will mitigate the potential for slips or trips are as follows:

- Keep everything at work in its proper place.
- Put things away after they are used.
- Have adequate lighting or use a flashlight.
- Walk and change direction slowly, especially when carrying something.
- Make sure the teeth or head on a wrench is in good shape and would not slip when using it.

Some other issues that should not be acceptable are as follows:

- Leaving machines, tools, or other materials on floors
- Blocking walkways or aisles with machines or equipment
- Using a "cheater" on a wrench (get a larger wrench with a longer handle)
- Leaving cord, power cable, or air hoses in walkways
- Placing anything on stairs
- Leaving drawers open
- Carrying or pushing loads that block the vision

Other guidelines in preventing slips are as follows:

- Clean up spills, drips, and leaks immediately.
- Apply sand on icy spots immediately and walk carefully.
- Use slip-resistant floor waxes and polishes in offices and high traffic areas.
- Use steel drain grates and splash guards.
- Use rough or grained steel surfaces in areas where there are often spills.
- Put up signs or barriers to keep people away from temporary slip hazards.
- Wear shoes with antiskid soles and materials that resist oil and acids.
- Avoid turning sharply when walking on slippery surfaces.
- Keep hands at your side not in your pockets.
- Walk slowly and slide your feet on wet, slippery, or uneven surfaces.
- Do not count on other workers to report hazards.

With regards to shoes, no type of shoe soles will prevent slipping on really slippery icy or oily surfaces, but some types of soles are better than others. Some shoe companies actually embed aluminum oxide in the soles to increase the coefficient of friction. Here are some general guidelines to follow:

- Neoprene soles, made with synthetic rubber, can be used safely on most work surfaces, both wet and dry. They are not recommended for oily conditions.
- Crepe soles, rubber with a "crinkled" texture, are best for rough concrete, either dry or wet. They are not suggested for tiles, smooth concrete, or wood surfaces.

- Leather soles can be used for ceramic tiles, wood, concrete surfaces that are wet and greasy, but are not recommended for dry, smooth concrete.
- Soft rubber soles can be used safely when working on most dry surfaces. They are not suggested for wet or greasy concrete.
- Hard rubber soles are best used for greasy concrete and wood. They are not recommended for ceramic tile or dry concrete.

When purchasing antislip floor finishes or covering, the selection is somewhat unlimited including tile, terrazzo, linoleum, and carpeting. It is important to select the right floor treatments or coverings for your given situation. As a general rule, floor treatment materials should have a minimum slip resistance rating of 0.50. This rating is known as the coefficient of friction and you should consult the manufacturers and their specifications to achieve the degree of slip resistance desired. Slip resistant mats or strips should be used in front of sinks, chemical vats, dishwashing areas, walk-in coolers and freezers, ramps, stairs, etc.

22.2   PREVENTING FALLS TO THE SAME LEVEL

Falls to the same level usually occur because of some very simple actions which are highly preventable. These are as follows:

- Running or walking too fast
- Slipping on icy or wet surfaces
- Having poor visibility due to dust, glare, smoke, or carrying a load that blocks one's vision
- Stumbling on loose pant cuffs
- Not wearing appropriate shoes for the job or activity
- Sitting improperly by not keeping all four legs of the chair on the floor

22.3   STAIRWAYS

There is no doubt that stairways are useful for traveling between different levels. Stairways are more convenient than ladders (Figure 22.1). In spite of this, many injuries from slips, trips, and falls transpire on a regular basis in the workplace. Most of the serious injuries and deaths occur while descending a stairway. Going up stairways means that the fall is usually forward and therefore restricted. In most situations, stairway accidents occur from unsafe acts such as the following:

- Climbing or descending stairs without holding onto the handrail
- Carrying a load, especially one that blocks visibility
- Not cleaning known slippery surfaces
- Not concentrating while climbing or descending
- Failing to keep stairs free of clutter
- Forgetting or ignoring safe work practices
- Having slow physical reaction, dizziness, or vision problems

FIGURE 22.1 Set of safe stairs.

The major unsafe conditions that exist regarding stairways that result in accidents are as follows:

- Stairways without handrails
- Tools, equipment, and litter on steps
- Spills left on stairways from failure to immediately clean them up
- Stairs not properly constructed

22.3.1 INSPECTING STAIRWAYS

When inspecting the condition of stairways in the workplace, some of the areas that need to be observed are as follows:

- Handrails and stair rails for placement, smoothness of surfaces, strength, clearance between rail and wall or other objects, and lastly the missing rails where they should be present.

- In looking at stair treads consider their strength, slip resistance, dimensions, evenness of the surface, and visibility of the leading edge.
- Improper/inadequate design, construction, or location of staircases.
- Wet, slippery, or damaged walking or grasping surfaces.
- Improper illumination (there are no general Occupational Safety and Health Administration [OSHA] standards for illumination levels).
- The Illuminating Engineering Society publication should be consulted for recommendations.
- Poor housekeeping.

22.4   PREVENTING SLIPS, TRIPS, AND FALLS IN THE SERVICE INDUSTRY

It seems safe to say that every one of the supersectors of the service industry and each industry sector have their potential hazards that could result in slips, trips, and falls. Wherever there are workers moving around a workplace, the potential for slips, trips, and falls exists. It would be challenging to remove all potential hazards from the workplace. Thus, this particular facet of workplace safety and health requires special attention.

22.5   SUMMARY OF OSHA REGULATIONS

Every flight of stairs having four or more risers is to be provided with a standard railing on all open sides. Handrails are to be provided on at least one side of closed stairways, preferably on the right side descending. Fixed stairways are to have a minimum width of 22 in. Stairs shall be constructed so the riser height and tread width are uniform throughout and do not vary more than one-fourth of an inch. Other general requirements include the following:

- A stairway or ladder must be provided at all worker points of access where there is a break in elevation of 19 in. or more and no ramp, runway, embankment, or personnel hoist are provided.
- When there is only one point of access between levels, it must be kept clear to permit free passage of workers. If free passage becomes restricted, a second point of access must be provided and used.
- Where there are more than two points of access between levels, at least one point of access must be kept clear.
- Stairways must be installed at least 30°—and no more than 50°—from the horizontal.
- Where doors or gates open directly onto a stairway, a platform must be provided that extends at least 20 in. beyond the swing of the door.
- All stairway parts must be free of dangerous projections such as protruding nails.
- Slippery conditions on stairways must be corrected.
- When ascending or descending stairways workers must take extreme care.
- Many serious injuries and even fatalities occur when workers slip and fall on stairways.

22.5.1 Walking/Working Surfaces (29 CFR 1910.21 and .22)

Slips and trips to the same level constitute a major number of general industry accidents. They cause approximately two deaths a year of all accidental deaths. The OSHA standards for walking and working surfaces apply to all permanent places of employment, except where only domestic, mining, or agricultural work is performed.

Working/walking surfaces that are wet need to be covered with nonslip materials. All spilled materials should be cleaned up immediately (Figure 22.2). Any holes in the floor, sidewalk, or other walking surfaces, should be covered or otherwise made safe. All aisles and passageways are to be kept clean and marked as appropriate. There is to be safe clearance for walking in aisles where motorized or mechanical handling equipment is operating.

Materials or equipment should be stored in such a way that sharp projections will not interfere with the walkway. Changes of direction or elevations should be readily identifiable. There should be adequate headroom provided for the entire length of any aisle or walkway.

FIGURE 22.2 Sign indicating slippery conditions.

Some of the most frequently overlooked general requirements involve house-keeping. All places of employment, passageways, storerooms, and service rooms are to be kept clean and orderly and in a sanitary condition. The floor of every workroom is to be maintained in a clean and, so far as possible, dry condition. Where wet processes are used, drainage is to be maintained and gratings, mats, or raised platforms are to be provided. Every floor, working place, and passageway are to be kept free of protruding nails, splinters, holes, or loose boards.

22.5.2 AISLES AND PASSAGEWAYS (29 CFR 1910.17, .22, AND .176)

Aisle and passageways must be free from debris and kept clear and in good repair with no obstruction across or in the aisles that could create a hazard during travel (Figure 22.3). Permanent aisles and passageways must be appropriately marked. Where mechanical material handling equipment is used, sufficient safe clearance must be allowed for aisles, at loading docks, through doorways, and wherever turns or passages must be made. Aisles and passageways used by mechanical equipment are to be kept clean and in good repair with no obstructions across or in aisles that could create a hazards. All permanent aisles and passageways should be appropriately marked. Improper aisle widths coupled with poor housekeeping and vehicle traffic can cause injury to employees, damage the equipment and material, and limit exit space in times of emergencies. Walking areas should be covered and/or guard-rails are to be provided to protect workers from the hazards of open pits, tanks, vats, ditches, etc.

FIGURE 22.3   Aisle that has an even floor and lack of clutter.

22.5.3 FLOORS (GENERAL CONDITIONS) (29 CFR 1910.22 AND .23)

All floor surfaces are to be kept clean, dry, and free from protruding nails, splinters, loose boards, holes, or projections. Where wet processes are used, drainage is to be maintained and false floors, platforms, mats, or other dry standing places are to be provided where practical.

In buildings or other structures, used for mercantile, business, industrial, or storage purposes, the loads approved by the building officials are to be marked on plates securely affixed by the owner of the building, or their duly authorized agent, in a conspicuous place in each space to which they relate. Such plates must not be removed or defaced but, if lost, removed, or defaced, shall be replaced by the owner or his agent. It is unlawful to place on any floor or roof of a building or other structure, a load greater than what is permitted.

22.5.4 GUARDING FLOOR AND WALL OPENINGS (29 CFR 1910.23)

Floor openings and holes, wall openings and holes, and the open sides of platforms may create hazards. People may fall through the openings or over the sides to the level below. Objects, such as tools or parts, may fall through floor or wall openings and holes use the following guidelines:

- A floor hole is an opening measuring less than 12 in. but more than 1 in. in at least one dimension, in any floor, platform, pavement, or yard, through which materials but not persons may fall.
- Floor openings measuring 12 in. or more in at least one dimension, in any floor, platform, pavement, yard, through which a person may fall.
- Platforms are working space for persons, elevated above the surrounding floor or ground.
- A wall hole is an opening of less than 30 in. but more than 1 in., of unrestricted width, in any wall or partition.
- Wall openings are at least 30 in. high and 18 in. wide, in any wall or partition.

22.5.5 PROTECTION OF FLOOR OPENINGS

Standard railings are to be provided on all exposed sides of a stairway opening, hatchway, or chute floor opening, except at the stairway entrance. For infrequently used stairways, where traffic across the opening prevents the use of a fixed standard railing, the guard is to consist of a hinged floor opening cover of standard strength and construction along with removable standard railings and toeboard on all exposed sides, except at the stairways entrance. This is to guard against a person walking directly into the opening.

A standard railing consists of a top rail, mid rail, and post, and is to have a vertical height of 42 in. from the upper surface of the top rail, platform, runway, or ramp level. Mid rails are to be 21 in. A standard toeboard is 4 in. in vertical height with not more than 0.25 in. clearance above the floor level.

Floor openings can be covered rather than guarded with rails. When the floor opening cover is removed a temporary guardrail should be put in its place, or an attendant should be stationed at the opening to warn personnel. Every floor-hole into which persons can accidentally walk into is to be guarded by either a standard railing with standard toeboard on all exposed sides, or a floor-hole cover that should be hinged in place.

Every open-sided floor, platform, or runway 4 ft or more above adjacent floor or ground level is to be guarded by a standard railing with toeboard on all open sides, except where there is entrance to a ramp, stairway, or fixed ladder. Wherever tools, machine parts, or materials are likely to be used on the runway, a toeboard is to be provided on all exposed sides. Runways not less than 18 in. wide used exclusively for special purposes may have the railing on one side omitted where operating conditions necessitate. Regardless of height, open-sided floors, walkways, platforms, or runways above or adjacent to dangerous equipment, pickling, or galvanizing tanks, degreasing units, and similar hazards are to be guarded with a standard railing and toeboard.

22.5.6 RAILINGS (29 CFR 1910.23)

The general requirements apply to all stair rails and handrails for stairways having four or more risers, or rising more than 30 in. in height—whichever is less—must have at least one handrail. Stair width is measured clear of all obstructions except handrails. The following are the guidelines for railings:

- On stairways less than 44 in. wide having both sides enclosed, at least one handrail is to be provided, preferably on the right side descending since most individuals are strong right handed or are accustomed to rails being on the right.
- On stairways less than 44 in. wide with one side open, at least one stair rail must be provided on the open side.
- On stairways less than 44 in. wide having both sides open, two stair rails must be provided, one for each side.
- On stairways that are between 44 and 88 in. wide, one handrail is to be provided on each enclosed side and one stair rail on each open side.
- On stairways 88 in. or more in width, one handrail must be provided on each enclosed side, one stair rail on each open side, and one intermediate stair rail must be placed approximately in the middle of the stairs.

A standard stair rail should be similar to a standard railing, but the vertical height should not be more than 34 in. nor less than 30 in. from the upper surface of the top rail to the surface of the tread in line with the face of the riser at the forward edge of the tread.

A standard handrail consists of a lengthwise member mounted directly on a wall or partition by means of brackets attached to the lower side of the handrail to keep a smooth, unobstructed surface along the top and both sides of the handrail. The brackets should maintain the rail 3 in. from the wall and be no more than 8 ft apart.

The height of handrails should be no more than 34 in. nor less than 30 in. from the upper surface of the handrail to the surface of the tread in line with the face of the riser or to the surface of the ramp.

Winding stairs should have handrails that are offset to prevent people from walking on any portion of the treads where the width is less than 6 in.

A stair rail also must be installed along each unprotected side or edge. When the top edge of a stair rail system also serves as a handrail, the height of the top edge must be no more than 37 in. nor less than 36 in. from the upper surface of the stair rail to the surface of the tread. Stair rails installed after March 15, 1991, must not be less than 36 in. in height.

Mid rails, screens, mesh, intermediate vertical members, or equivalent inter-mediate structural members must be provided between the top rail and stairway steps to the stair rail system. Mid rails, when used, must be located midway between the top of the stair rail system and the stairway steps. Screens or mesh, when used, must extend from the top rail to the stairway step and along the opening between top rail supports.

Intermediate vertical members, such as balusters, when used, must not be more than 19 in. apart. Other intermediate structural members, when used, must be installed so that there are no openings of more than 19 in. wide.

Handrails and the top rails of the stair rail systems must be able to withstand, without failure, at least 200 lb of weight applied within 2 in. of the top edge in any downward or outward direction, at any point along the top edge. The height of handrails must not be more than 37 in. nor less than 30 in. from the upper surface of the handrail to the surface of the tread. The height of the top edge of a stair rail system used as a handrail must not be more than 37 in. nor less than 36 in. from the upper surface of the tread.

Stair rail systems and handrails must be surfaced to prevent injuries such as punctures or lacerations and to keep clothing from snagging. Handrails must provide an adequate handhold for employees to grasp to prevent falls. The ends of stair rail systems and handrails must be built to prevent dangerous projections, such as rails protruding beyond the end posts of the system. Temporary handrails must have a minimum clearance of 3 in. between the handrail and walls, stair rail systems, and other objects. Unprotected sides and edges of stairway landings must be provided with standard 42 in. guardrail systems.

22.5.7   STAIRS, FIXED INDUSTRIAL (29 CFR 1910.23 AND .24)

Fixed stairways should be provided for access from one structure to another where operations necessitate regular travel between levels, and for access to operating platforms at any equipment that requires attention routinely during operations. Fixed stairs should also be provided where access to elevations is daily or at each shift where such work may expose employees to harmful substances, or for which purposes the carrying of tools or equipment by hand is normally required. This includes interior and exterior stairs around machinery, tanks, and other equipment, and stairs leading to and from floors, platforms, and pits. This section of the regulation does not apply to stairs used for fire exit purposes, construction operations, private residences, or to articulated stairs, such as those installed on floating roof tanks, the angle of which changes with the rise and fall of the base support. Spiral stairways are not permitted except for special limited usage and secondary access situation where it is not practical to provide a conventional stairway.

Fixed industrial stairs are to be used to provide access to and from places of work where operations necessitate regular travel between levels. The general requirements for fixed industrial stairs are as follows:

- They must be strong enough to carry five times the normally anticipated live load.
- At the very minimum, any fixed stairway should be able to carry safely a moving concentrated load of 1000 lb.
- All fixed stairways must have a minimum width of 22 in.
- Fixed stairs are to be installed at an angle $30°-50°$ to the horizontal.
- Vertical clearance above any stair tread to the overhead obstruction is to be at least 7 ft measured from the leading edge of the tread.

The length of a staircase is important. Long flights of steps without landing should be avoided whenever possible. The OSHA standard does not specify the exact number or placement of landing. The National Safety Council recommends landings at every 10th or 12th tread. The intermediate landings and platforms on stairways are to be no less than the stair width and a minimum of 30 in. in length measured in the direction of travel.

22.5.8 TOEBOARDS (29 CFR 1910.23)

Toeboards are used to protect workers from being struck by objects falling from elevated areas. Railings protecting floor openings, platforms, and scaffolds are to be equipped with toeboards whenever persons can pass beneath the open side, wherever there is moving machinery, or wherever there is equipment with which falling material could cause a hazard. A standard toeboard is to be at least 4 in. in height with no more than a 0.25 in. clearance above the floor level and may be of any substantial material, either solid or open, with openings not to exceed 1 in. in greatest dimension.

22.6   CHECKLIST FOR WALKING/WORKING SURFACES

This checklist is no guarantee for preventing slips, trips, and falls, but at least the potential hazards will have been addressed. It is important that employers call attention to human potential by assuring that all employees are trained in the prevention of slips, trips, and falls. Figure 22.4 is an example of a checklist for walking and working surfaces.

22.7   SUMMARY

The typical causes of slips and trips are the presence of oil or water on floors and surfaces. Inferior lighting plays a role in these types of accidents. At times stairs and work platforms are improperly constructed. Stairs and walking areas often become

---

Walking–working surfaces checklist

Answer the following either yes or no to determine if your safety program is paying attention to items which could prevent slips, trips, and falls and that you have complied with OSHA regulations.

General work environment

☐ Yes ☐ No Is a documented, functioning housekeeping program in place?
☐ Yes ☐ No Are all worksites clean, sanitary, and orderly?
☐ Yes ☐ No Are work surfaces kept dry or is appropriate means taken to assure the surfaces are slip-resistant?
☐ Yes ☐ No Are all spill hazardous materials or liquids, including blood and other potential infectious materials, cleaned up immediately and according to proper procedures?
☐ Yes ☐ No Is combustible scrap, debris, and waste stored safely and removed from the worksite properly?
☐ Yes ☐ No Are accumulations of combustible dust routinely removed from elevated surfaces including the overhead structure of building, etc.
☐ Yes ☐ No Is combustible dust cleaned up with a vacuum system to prevent the dust from going into suspension?
☐ Yes ☐ No Is metallic or conductive dust prevented from entering or accumulating on or around electrical enclosures or equipment?
☐ Yes ☐ No Are covered metal waste can use for oily and paint-soaked waste?

Walkways

☐ Yes ☐ No Are aisles and passageways kept clear?
☐ Yes ☐ No Are aisles and walkways marked as appropriate?
☐ Yes ☐ No Are wet surfaces covered with nonslip materials?
☐ Yes ☐ No Are hole in floor, sidewalk, or other walking surface repaired properly, covered, or otherwise made safe?
☐ Yes ☐ No Is there safe clearance for walking in aisles where motorized or mechanical handling equipment is operating?
☐ Yes ☐ No Are materials or equipment stored in such a way that sharp projectives will not interfere with the walkway?
☐ Yes ☐ No Are spilled materials cleaned up immediately?
☐ Yes ☐ No Are changes of direction or elevation readily identifiable?
☐ Yes ☐ No Are bridges provided over conveyors or similar hazards?
☐ Yes ☐ No Are aisles or walkways that pass near moving or operating machinery, welding operation, or similar operations arranged so employees will not be subjected to potential hazards?
☐ Yes ☐ No Is adequate headroom provided for the entire length of any aisle or walkway?
☐ Yes ☐ No Are standard guardrails provided wherever aisle or walkway surfaces are elevated more than 30 in. above any adjacent floor or the ground?

Floor and wall openings

☐ Yes ☐ No Are floor openings guarded by a cover, a guardrail, or equivalent on all sides (except at the entrance to stairways or ladder)?
☐ Yes ☐ No Are toeboards installed around the edges of permanent floor openings (where persons may pass below the opening)?
☐ Yes ☐ No Are skylight screens of such construction and mounting that they will withstand a load of at least 200 lb?

FIGURE 22.4   Checklist for walking/working surfaces.

(*continued*)

☐ Yes	☐ No	Is the glass in the windows, doors, glass wall, etc. which are subject to human impact, sufficient thickness and type for the condition of use?
☐ Yes	☐ No	Are grate or similar covers over floor openings such as floor drains of such design that foot traffic or rolling equipment will not be affected by the grate spacing?
☐ Yes	☐ No	Are unused portions of service pits and pits not actually in use either covered or protected by guardrails
☐ Yes	☐ No	Are manhole covers, trench covers, and similar covers, plus their supports designed to carry a truck rear axle load of at least 20,000 lb when located in roadways and subjected to vehicle traffic?
☐ Yes	☐ No	Are floor or wall openings in fire-resistive construction provided with doors or covers compatible with the fire rating of the structure and provided with a self-closing feature when appropriate?

Stairs and stairways

☐ Yes	☐ No	Are standard stair rails or handrails on all stairways having four or more risers?
☐ Yes	☐ No	Are all stairways at least 22 in. wide?
☐ Yes	☐ No	Do stairs have landing platforms not less than 30 in. in the direction of travel and extend 22 in. in width at every 12 ft or less of vertical rise?
☐ Yes	☐ No	Do stairs angle not more than 50° and no less than 30°?
☐ Yes	☐ No	Are step risers on stairs uniform from top to bottom?
☐ Yes	☐ No	Are steps on stairs and stairways designed or provided with a surface that renders them slip resistant?
☐ Yes	☐ No	Are stairway handrails located between 30 and 34 in. above the leading edge of stair treads?
☐ Yes	☐ No	Do stairway handrails have at least 3 in. of clearance between the handrails and the wall or surface they are mounted on?
☐ Yes	☐ No	Where doors or gates open directly on a stairway, is there a platform provided so the swing on the door does not reduce the width of the platform to less than 21 in.?
☐ Yes	☐ No	Where stairs or stairways exit directly into any area where vehicles may be operated, are adequate barriers and warnings provided to prevent employees stepping into the path of traffic?
☐ Yes	☐ No	Do stairway landings have a dimension measured in the direction of travel, at least equal to the width of the stairway?

FIGURE 22.4 (continued)

worn, broken, or present an uneven walking or stepping surface. Poor floor conditions, such as cracks or holes, protruding nails, and improper floor finishes play a role in slips and trips. There is also the danger posed by floor and wall openings that workers could fall through during work activities.

23 Other Hazards

Tire inflation and repair can be deadly. (Courtesy of Mine Safety and Health Administration.)

The retail, wholesale, and warehousing sectors are the sectors that handle, store, and disperse the largest quantities of goods and materials in the United States. Since all the potential hazards faced by these three sectors cannot be addressed in detail in one book it may be necessary to use *Industrial Safety and Health for Infrastructure Services*, *Industrial Safety and Health for Administrative Services*, and *Industrial Safety and Health for People-Oriented Services* to address other hazards. This chapter provides summaries regarding some of the other common hazards that confront workers in these sectors. Some of these hazards are also covered in some detail in the other three books mentioned above.

Some of the hazards that were not covered extensively in this book are as follows:

- Compressed air
- Lockout/tagout
- Dockboards
- Electrical
- Fueling
- Powered tools
- Scaffolds
- Tire inflation
- Workplace violence

23.1   COMPRESSORS AND COMPRESSED AIR (29 CFR 1910.242)

A compressor supplies compressed air. Great care must be taken to ensure such types of equipment are operating safely. Safety devices for a compressed air system should be checked frequently. Compressors should be equipped with pressure relief valves and pressure gauges.

The air intakes must be installed and equipped so as to ensure that only clean uncontaminated air enters the compressor. This is facilitated by the installation of air filters on the compressor intake.

Before any repair work is done on the pressure system of a compressor, the pressure is to be bled off and the system locked out. All compressors must be operated and lubricated in accordance with the manufacturer's recommendations. Signs are to be posted to warn of the automatic starting feature of compressors. The belt drive system is to be totally enclosed to provide protection from any contact.

No worker should direct compressed air toward a person, and employees are prohibited from using highly compressed air for cleaning purposes. If compressed air is used for cleaning clothes, the pressure is to be reduced to less than 10 psi. When using compressed air for cleaning, employees should wear protective chip guarding eyewear and personal protective equipment.

Safety chains or other suitable locking devices are to be used at couplings of high pressure hose lines where a connection failure would create a hazard. Before compressed air is used to empty containers of liquid, the safe working pressure of the container is to be checked.

When compressed air is used with abrasive blast cleaning equipment, the operating valve type must be held open manually. When compressed air is used to inflate auto tires, a clip-on chuck and an inline regulator preset to 40 psi are required. Compressed air should not be used to clean up or move combustible dust because such action could cause the dust to be suspended in the air and cause a fire or explosion hazard.

23.2   CONTROL OF HAZARDOUS ENERGY SOURCES (LOCKOUT/TAGOUT) (29 CFR 1910.147)

Lockout/tagout deals with the preventing of the release of energy from machines, equipment, and electrical circuits which are perceived to be de-energized. The Occupational Safety and Health Administration (OSHA) estimates compliance with the lockout/tagout standard will prevent about 120 fatalities and approximately 28,000 serious and 32,000 minor injuries every year. About 39 million general industry workers will be protected from accidents during maintenance and servicing of equipment under this ruling.

The standard for the control of hazardous energy sources (lockout/tagout) covers servicing and maintenance of machines and equipment in which the unexpected energization or startup of the machines or equipment or release of stored energy could cause injury to employees. The rule generally requires that energy sources for equipment be turned off or disconnected and that the switch either be locked or labeled with a warning tag. About 3 million workers actually servicing equipment face the greatest risk. These include craft workers, machine operators, and laborers.

OSHA's data show that packaging and wrapping equipment, printing presses, and conveyors account for a high proportion of the accidents associated with lock-out/tagout failures.

Typical injuries include fractures, lacerations, contusions, amputations, and puncture wounds with the average lost time for injuries running 24 days. Agriculture, maritime, and construction employers are not covered under standard 29 CFR 1910.147. Also, the generation, transmission, and distribution of electric power by utilities and work on electric conductors and equipment are excluded. The general requirements for the ruling require employers to

- Develop an energy control program.
- Use locks when equipment can be locked out (Figure 23.1).
- Ensure that new equipment or overhauled equipment can accommodate locks.
- Employ additional means to ensure safety, when tags rather than locks are used by using an effective tagout program.
- Identify and implement specific procedures (generally in writing) for the control of hazardous energy including preparation for shutdown, shutdown,

FIGURE 23.1   Locking out a plug on a saw.

equipment isolation, lockout/tagout application, release of stored energy, and verification of isolation.

- Institute procedures for release of lockout/tagout including machine inspection, notification and safe positioning of employees, and removal of the lockout/tagout device.
- Obtain standardized locks and tags which indicate the identity of the employee using them and which are of sufficient quality and durability to ensure their effectiveness.
- Require that each lockout/tagout device be removed by the employee who applied the device.
- Conduct inspections of energy control procedures at least once a year.
- Train employees in the specific energy control procedures with training reminders as part of the annual inspections of the control procedures.
- Adopt procedures to ensure safety when equipment must be tested during servicing, when outside contractors are working at the site, when a multiple lockout is needed for a crew servicing equipment, and when shifts or personnel change.

Excluded from coverage are normal production operations including repetitive, routine minor adjustments which would be covered under OSHA's machine guarding standards. Work on cord and plug connected electric equipment when it is unplugged, and the employee working on the equipment has complete control of the plug. Hot tap operations involving gas, steam, water, or petroleum products when the employer shows that continuity of service is essential, shutdown is impractical, and documented procedures are followed to provide proven effective protection for employees.

In summary all machinery or equipment capable of movement is required to be de-energized or disengaged and locked out during cleaning, servicing, adjusting, or setting up operations, whenever required. Where the power disconnecting means for equipment does not also disconnect the electrical control circuit, the appropriate electrical enclosures must be identified. A means should be provided to assure the control circuit can also be disconnected and locked out. The locking out of control circuits in lieu of locking out main power disconnects must be prohibited. All equipment control valve handles are to be provided with a means for locking out. Lock-out procedures require that stored energy (mechanical, hydraulic, air, etc.) be released or blocked before equipment is locked out for repairs. Appropriate employees must be provided with individually keyed personal safety locks and are expected to keep personal control of their keys while they use safety locks. Only the employee exposed to the hazard should place or remove the safety lock. Employees must check the safety of the lock out by attempting a startup after making sure no one is exposed. Employees need to be instructed to always push the control circuit stop button immediately after checking the safety of the lock out. A means is to be provided to identify any or all employees who are working on locked-out equipment by their locks or accompanying tags. A sufficient number of accident preventive signs or tags and safety padlocks need to be provided for any reasonably foreseeable repair emergency. When machine operations, configuration, or size requires the operator to leave his/her control station to install tools or perform other operations, and that part of the machine could move if accidentally activated, that element is required to be separated, locked, or blocked out. In the event that equipment or lines cannot be shut down, locked out, and tagged, a safe job procedure is to be established and rigidly followed.

23.3   DOCKBOARDS (29 CFR 1910.30)

Dockboards are to be strong enough to carry the load imposed upon them (See Figure 23.2). Portable dockboards are to be anchored or equipped with devices that will prevent their slipping. Handholds should exist on dockboards which provide a safe and effective means of handling. Railroad cars should be provided with a mechanism that would prevent movement while dockboards are being used.

23.4   ELECTRICAL (29 CFR 1910.303, .304, .305, .331, AND .333)

Electricity is accepted as a source of power without much thought to the hazards encountered. Some employees work with electricity directly, as is the case with engineers, electricians, or people who do wiring, such as overhead lines, cable harnesses, or circuit assemblies. Others, such as office workers and salespeople, work with it indirectly. Approximately 700 workers are electrocuted every year with many workers suffering injuries such as burns, cuts, etc. (Figure 23.3).

FIGURE 23.2   Example of dockboards or plates for loading a trailer at a dock.

FIGURE 23.3 Electrical exposure is common in most service sector businesses. (Courtesy of Mine Safety and Health Administration.)

OSHA's electrical standards address the government's concern that electricity has long been recognized as a serious workplace hazard, exposing employees to such dangers as electric shock, electrocution, fires, and explosions. The objective of the standards is to minimize such potential hazards by specifying design characteristics of safety in use of electrical equipment and systems.

Electrical equipment must be free from recognized hazards that are likely to cause death or serious physical harm to employees. Flexible cords and cables (extension cords) should be protected from accidental damage. Unless specifically permitted flexible cords and cables should not be used as a substitute for the fixed wiring of a structure, where attached to building surfaces, where concealed or where they run through holes in walls, ceilings, or floors, or where they run through doorways, windows, or similar openings. Flexible cords are to be connected to devices and fittings so that strain relief is provided that will prevent pull from being directly transmitted to joints or terminal screws.

A grounding electrode conductor is to be used for a grounding system to connect both the equipment grounding conductor and the grounded circuit conductor to the grounding electrode. Both the equipment grounding conductor and the grounding electrode conductor are to be connected to the ground circuit conductor on the supply side of the service disconnecting means or on the supply side of the system disconnecting means or overcurrent devices if the system is separately derived. For ungrounded service-supplied systems, the equipment grounding conductor should be connected to the grounding electrode conductor at the service equipment. The path to ground from circuits, equipment, and enclosures should be permanent and continuous.

Electrical equipment should be free from recognized hazards that are likely to cause death or serious physical harm. Each disconnecting means should be legibly marked to indicate its purpose, unless the purpose is evident. Listed or labeled equipment should be used or installed in accordance with any instructions included in the listing or labeling. Unused openings in cabinets, boxes, and fittings must be effectively closed.

Safety-related work practices are to be employed to prevent electric shock or other related injuries resulting from either direct or indirect electrical contacts, when work is performed near or on equipment of circuits that are or may be energized. Electrical safety-related work practices cover both qualified persons (those who have training in avoiding the electrical hazards of working on or near exposed energized parts) and unqualified persons (those with little or no such training).

There must be written lockout and/or tagout procedures. Overhead power lines must be de-energized and grounded by the owner or operator of the lines, or other protective measures must be provided before starting work. Protective measures, such as guarding or insulating the lines, must be designed to prevent employees from contacting the lines.

Unqualified employees and mechanical equipment must be at least 10 ft away from overhead power lines. If the voltage exceeds 50,000 V, the clearance distance should be increased 4 in. for each 10,000 V.

OSHA requires portable ladders to have nonconductive side rails if used by employees who would be working where they might contact exposed energized circuit parts.

Conductors are to be spliced or joined with devices identified for such use or by brazing, welding, or soldering with a fusible alloy or metal. All splices, joints, and free ends of conductors should be covered with an insulation equivalent to that of the conductor or with an insulating device suitable for the purpose.

All employees should immediately report any obvious hazard to life or property observed in connection with electrical equipment or lines. Employees need to be instructed to make preliminary inspections and/or appropriate tests to determine what conditions exist before starting work on electrical equipment or lines.

All portable electrical tools and equipment are to be grounded or of double insulated type. Electrical appliances such as vacuum cleaners, polishers, and vending machines must be grounded. Extension cords being used are to have a grounding conductor and multiple plug adapters are prohibited.

Ground-fault circuit interrupters should be installed on each temporary 15 or 20 A, 120 V AC circuit at locations where construction, demolition, modifications, alterations, or excavations are being performed. All temporary circuits are to be protected by suitable disconnecting switches or plug connectors at the junction with permanent wiring. If electrical installations in hazardous dust or vapor areas exist, they need to meet the National Electrical Code (NEC) for hazardous locations.

In wet or damp locations, the electrical tools and equipment must be appropriate for this use or all location or otherwise protected. The location of electrical power lines and cables (overhead, underground, under the floor, other side of walls) are to be determined before digging, drilling, or similar work is begun.

All energized parts of electrical circuits and equipment are to be guarded against accidental contact by approved cabinets or enclosures, and sufficient access and working space must be provided and maintained about all electrical equipment to permit ready and safe operations and maintenance.

Low-voltage protection is to be provided in the control device of motors driving machines or equipment which could cause probable injury from inadvertent starting. Each motor disconnecting switch or circuit breaker should be located within sight of the motor control device and each motor located within sight of its controller. Employees who regularly work on or around energized electrical equipment or lines should be instructed in the cardiopulmonary resuscitation (CPR) methods.

23.5 FUELING

An internal combustion engine should not be fueled with a flammable liquid while the engine is running. Fueling operations are to be done in such a manner that likelihood of spillage will be minimal.

When spillage occurs during fueling operations, the spilled fuel is to be washed away completely, evaporated, or other measures taken to control vapors before restarting the engine. Fuel tank caps are to be replaced and secured before starting the engine.

Fueling hoses are to be of a type designed to handle the specific type of fuel. It is prohibited to handle or transfer gasoline in open containers. No open lights, open flames, sparking, or arcing equipment are allowed during fueling or transfer of fuel operations and no smoking should be permitted.

23.6 PORTABLE (POWER-OPERATED) TOOLS AND EQUIPMENT (29 CFR 1910.243)

Tools are such a common part of our lives that it is difficult to remember that they may pose hazards. All tools are manufactured with safety in mind but, tragically, a serious accident often occurs before steps are taken to search out and avoid or eliminate tool-related hazards. In the process of removing or avoiding the hazards, workers must learn to recognize the hazards associated with the different types of tools and the safety precautions necessary to prevent those hazards. All hazards involved in the use of powered tools can be prevented by following five basic safety rules:

- Keep all tools in good condition with regular maintenance.
- Use the right tool for the job.
- Examine each tool for damage before use.
- Operate according to the manufacturer's instructions.
- Provide and use the proper protective equipment.

Employees and employers have a responsibility to work together to establish safe working procedures. If a hazardous situation is encountered, it should be brought to the attention of the proper individual immediately.

Powered tools can be hazardous when improperly used. There are several types of powered tools, based on the power source they use: electric, pneumatic, liquid fuel, hydraulic, and powder-actuated. Employees should be trained in the use of all tools—not just powered tools. They should understand the potential hazards as well as the safety precautions to prevent those hazards from occurring. The following general precautions should be observed by powered tool users:

- Never carry a tool by the cord or hose.
- Never yank the cord or the hose to disconnect it from the receptacle.
- Keep cords and hoses away from heat, oil, and sharp edges.
- Disconnect tools when not in use, before servicing, and when changing accessories such as blades, bits, and cutters.
- All observers should be kept at a safe distance away from the work area.
- Secure work with clamps or a vise, freeing both hands to operate the tool.
- Avoid accidental starting. The worker should not hold a finger on the switch button while carrying a plugged-in tool.
- Tools should be maintained with care. They should be kept sharp and clean for the best performance. Follow instructions in the user's manual for lubricating and changing accessories.
- Be sure to keep good footing and maintain good balance.
- The proper apparel should be worn. Loose clothing, ties, or jewelry can get entangled in moving parts.
- All portable electric tools that are damaged should be removed from use and tagged "Do Not Use."

Hazardous moving parts of a powered tool need to be safeguarded. For example, belts, gears, shafts, pulleys, sprockets, spindles, drums, flywheels, chains, or other reciprocating, rotating, or moving parts of equipment must be guarded if such parts are exposed to contact by employees. Guards, as necessary, should be provided to protect the operator and others from the following:

- Point of operation
- In-running nip points
- Rotating parts
- Flying chips and sparks

Safety guards must never be removed when a tool is being used. For example, portable circular saws must be equipped with guards. An upper guard must cover the entire blade of the saw. A retractable lower guard must cover the teeth of the saw, except when it makes contact with the work material. The lower guard must automatically return to the covering position when the tool is not being used.

The following handheld powered tools must be equipped with a momentary contact "on–off" control switch: drills, tappers, fastener drivers, horizontal, vertical and angle grinders with wheels larger than 2 in. in diameter, disk and belt sanders, reciprocating saws, saber saws, and other similar tools. These tools may also be equipped with a lock-on control provided so that turnoff can be accomplished by a single motion of the same finger or fingers that turn it on.

The following handheld powered tools may be equipped with only a positive "on–off" control switch: platen sanders, disk sanders with disks 2 in. or less in diameter, grinders with wheels 2 in. or less in diameter, routers, planers, laminate trimmers, nibblers, shears, scroll saws, and jigsaws with blade shanks 0.25 in. wide or less.

Other handheld powered tools such as circular saws having a blade diameter greater than 2 in., chain saws, and percussion tools without positive accessory holding means must be equipped with a constant pressure switch that will shut off the power when the pressure is released.

Employees using electric tools must be aware of several dangers; the most serious being the possibility of electrocution. Among the chief hazards of electric-powered tools are burns and slight shocks which can lead to injuries or even heart failure. Under certain conditions, even a small amount of current can result in fibrillation of the heart and eventual death. A shock also can cause the user to fall off a ladder or other elevated work surfaces.

To protect the user from shock, tools must either be grounded, double insu-lated, or powered by a low-voltage isolation transformer. Three-wire cords contain two current-carrying conductors and a grounding conductor. One end of the grounding conductor connects to the tool's metal housing. The other end is grounded through a prong on the plug. Anytime an adapter is used to accommodate a two-hole receptacle, the adapter wire must be attached to a known ground. The third prong should never be removed from the plug. Double insulation is more convenient. The user and the tools are protected in two ways: by normal insulation on the wires inside, and by a housing that cannot conduct electricity to the operator in the event of a malfunction. These general practices should be followed when using electric tools:

- Electric tools should be operated within their design limitations.
- Gloves and safety footwear are recommended during use of electric tools.
- When not in use, tools should be stored in a dry place.
- Electric tools should not be used in damp or wet conditions.
- Work areas should be well lighted.

Powered abrasive grinding, cutting, polishing, and wire buffing wheels create special safety problems because they may throw off flying fragments. Before an abrasive wheel is mounted, it should be inspected closely and sound- or ring-tested to be sure that it is free from cracks or defects. To test, wheels should be tapped gently with a light nonmetallic instrument. If the wheel sounds cracked or dead, it could fly apart in operation and so must not be used. A sound and undamaged wheel will give a clear metallic tone or "ring." To prevent the wheel from cracking, the user should be sure it fits freely on the spindle. The spindle nut must be tight enough to hold the wheel in place, without distorting the flange. Follow the manufacturer's recom-mendations. It must be ensured that the spindle wheel does not exceed the abrasive wheel specifications. Due to the possibility of a wheel disintegrating (exploding) during startup, the employee should never stand directly in front of the wheel as it accelerates to full operating speed.

Portable grinding tools need to be equipped with safety guards to protect workers not only from the moving wheel surface, but also from flying fragments in case of breakage. In addition, when using a powered grinder

- Always use eye protection.
- Turn off the power when not in use.
- Never clamp a handheld grinder in a vise.

23.7 SCAFFOLDS (29 CFR 1910.28)

Analysis of 1986 Bureau of Labor Statistics (BLS) data to support OSHA's scaffolding standard estimates that, of the 500,000 injuries and illnesses that occur in the construction industry annually, 10,000 are related to scaffolds. In addition, of the estimated 900 occupational fatalities occurring annually, at least 80 are associated with work on scaffolds. Seventy-two percent of the workers injured in scaffold accidents covered by the BLS study attributed the accident either to the planking or support giving way, or to the employee slipping or being struck by a falling object. Plank slippage was the most commonly cited cause.

All scaffolds and their supports must be capable of supporting the load they are designed to carry with safety factor of at least four. All planking is to be of scaffold grade, as recognized by grading rules for the type of wood used. The maximum permissible spans for 2 in. by 9 in. or wider planks are shown in Table 23.1.

The maximum permissible span for 1.25 in. by 9-in. or wider plank for full thickness is 4 ft, with medium loading of 50 psf. Scaffold planks should extend over their supports not less than 6 in. nor more than 18 in. Scaffold planking is to overlap a minimum of 12 in. or secured from movement.

23.8 TIRE INFLATION

Because tires have the potential to release a large amount of energy if they explode, and they can cause injury due to poor inflation procedure, care must be taken to follow safety procedures. Tires are mounted and/or inflated on drop center wheels,

TABLE 23.1
Permissible Plank Span for Scaffolds

Maximum Intended Load (psf)	Maximum Permissible Span Using Full Thickness Undressed Lumber (ft)	Maximum Permissible Span Using Normal Thickness Lumber (ft)
25	10	8
50	8	6
75	6	N/A

must have a safe practice procedure posted and enforced. When tires are mounted and/or inflated on wheels with split rims and/or retainer rings, a safe practice procedure is to be posted and enforced. Each tire inflation hose should have a clip-on chuck with at least 24 in. of hose between the chuck and an in-line hand valve and gauge. The tire inflation control valve must be of automatically shutoff type so that the airflow stops when the valve is released. Employees should be strictly forbidden from taking a position directly over or in front of a tire while it is being inflated. A tire restraining device such as a cage, rack, or other effective means must be used while inflating tires mounted on split rims, or rims using retainer rings.

23.9 WORKPLACE VIOLENCE

Workplace violence has emerged as an important safety and health issue in today's workplace. Its most extreme form, homicide, is the second leading cause of fatal occupational injury in the United States.

Nearly 1000 workers are murdered, and 1.5 million are assaulted in the workplace every year. According to the BLS Census of Fatal Occupational Injuries (CFOI), there were 709 workplace homicides in 1998, accounting for 12% of the total 6026 fatal work injuries in the United States. Environmental conditions associated with workplace assaults have been identified and control strategies implemented in a number of work settings. OSHA has developed guidelines and recommendations to reduce worker exposures to this hazard but is not initiating rulemaking at this time.

According to the Department of Justice's National Crime Victimization Survey (NCVS), assaults and threats of violence against Americans at work number almost 2 million a year. The most common type of workplace violent crime was simple assault with an average of 1.5 million a year. There were 396,000 aggravated assaults, 51,000 rapes and sexual assaults, 84,000 robberies, and 1000 homicides. According to the NCVS, retail sales workers were the prime targets, with 330,000 being attacked every year. They were followed by the police, with an average of 234,200 officers victimized. The risk rate for various occupations was as follows (per 1000):

Police officers (306)
Private security guards (218)
Taxi drivers (184)
Prison guards (117)
Bartenders (91)
Mental health professionals (80)
Gas station attendants (79)
Convenience, liquor store clerks (68)
Mental health custodial workers (63)
Junior high/middle school teachers (57)
Bus drivers (45)
Special education teachers (41)

High school teachers (29)
Elementary school teachers (16)
College teachers (3)

Factors which may increase a worker's risk for workplace assault, as identified by the National Institute for Occupational Safety and Health (NIOSH), are as follows:

- Contact with the public
- Exchange of money
- Delivery of passengers, goods, or services
- Having a mobile workplace such as a taxicab or police cruiser
- Working with unstable or volatile persons in health care, social services, or criminal justice settings
- Working alone or in small numbers
- Working late at night or during early morning hours
- Working in high-crime areas
- Guarding valuable property or possessions
- Working in community-based settings

OSHA's response to the problem of workplace violence in certain industries has been the production of OSHA's guidelines and recommendations to those industries for implementing workplace violence prevention programs. In 1996, OSHA published *Guidelines for Preventing Workplace Violence for Health Care and Social Service Workers*. In 1998, OSHA published *Recommendations for Workplace Violence Prevention Programs in Late-Night Retail Establishments*. The guidelines and recommendations are based on OSHA's *Safety and Health Program Management Guidelines* and contain four basic elements:

- Management commitment and employee involvement. May include simply clear goals for worker security in smaller sites or a written program for larger organizations.
- Worksite analysis. Involves identifying high-risk situations through employee surveys, workplace walkthroughs, and reviews of injury/illness data.
- Hazard prevention and control. Calls for designing engineering and administrative and work practice controls to prevent or limit violent incidents.
- Training and education. Ensures that employees know about potential security hazards and ways to protect themselves and their coworkers.

Although not exhaustive, OSHA's guidelines and recommendations include policies, procedures, and corrective methods to help prevent and mitigate the effects of workplace violence. Engineering controls remove the hazard from the workplace or create a barrier between the worker and the hazard. Administrative and work practice controls affect the way jobs or tasks are performed. Some recommended engineering and administrative controls are as follows:

- Physical barriers such as bullet-resistant enclosures, pass-through windows, or deep service counters
- Alarm systems, panic buttons
- Convex mirrors, elevated vantage points, clear visibility of service and cash register areas
- Bright and effective lighting
- Adequate staffing
- Furniture positioning to prevent entrapment
- Cash-handling controls, use of drop safes
- Height markers on exit doors
- Emergency procedures to use in case of robbery
- Training in identifying hazardous situations and appropriate responses in emergencies
- Video surveillance equipment and closed circuit TV (Figure 23.4)
- Liaison with local police

Post-incident response and evaluation are essential to an effective violence prevention program. All workplace violence programs should provide treatment for victimized employees and employees who may be traumatized by witnessing a workplace violence incident. Several types of assistance can be incorporated into the post-incident response including the following:

- Trauma–crisis counseling
- Critical incident stress debriefing
- Employee assistance programs to assist victims

Workplace homicides are the second leading cause of death in the workplace. Homicide is the number one cause of death for women on the job. Although

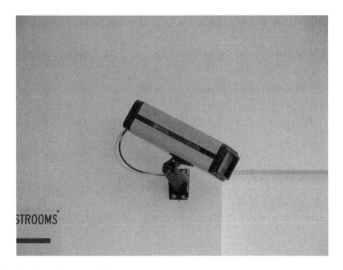

FIGURE 23.4   Security surveillance camera.

workplace murders appear to be declining somewhat, they still represented 15% of all deaths. That is more than 900 workers who went to work but never came home. And 80% of them died at the hands of robbers or other criminals. Almost half of workplace homicides occur in the retail industry, where those working late at night are particularly vulnerable. Employers and employees will want to examine their operations from a variety of perspectives—from work practices to physical barriers to employee training. The options they select to reduce the risk of violence will depend upon their individual circumstances.

For example, a gas station may find pass-through windows with bullet-resistant glass, increased lighting inside the station and over the pumps, and clearing windows of signs to permit an unobstructed view for police officers in the street to be useful measures. A convenience store might use video surveillance equipment, combined with an alarm system, convex mirrors in the store, and drop safes to foil would-be thieves. A liquor store in a high-crime area might increase staffing levels at night and restrict customer access to only one door. A facility that backs up to a wooded area might increase lighting at the rear of the store, lock rear doors at night, and limit deliveries to daytime hours.

All these facilities might find it helpful to train employees in emergency procedures to use in case of potential violence. Employees may also benefit from training in handling aggressive or abusive customers. OSHA's recommendations are not a fixed formula, but rather a listing of commonsensical strategies and practices that can help stop thieves. By making cash more difficult to get, store owners will discourage potential criminals and protect innocent employees.

24 Summary

Providing goods and materials while providing a safe and healthy workplace for workers.

This book is dedicated to those sectors that fulfill the handling, storing, and selling of goods and materials function for the service industry. Thus, the most prominent workplace is the office setting. This book addresses the hazards and safety and health issues that face owners, employers, and workers who work principally in stores and warehousing environments. The industry sectors as defined in this book are those in the retail, wholesale, and warehousing services sectors. All of these service sectors have components that require the storage of goods and materials. The workers in the retail sector interact more with the public than those working in the wholesale and warehousing sectors. To ensure optimum safety and healthful conditions in the retail, wholesale, and warehousing sectors, a comprehensive workplace inspection checklist has been included as a part of the summary of this book.

Use this inspection checklist to evaluate the goods and materials workplaces related to safety and health. Feel free to alter or add to this list to tailor it to your specific workplace. Workplace inspections are a way of identifying hazards in the workplace. Inspections also provide a system to monitor whether hazards have been fixed. All facilities should conduct workplace inspections at least twice a year. If difficulties are noted inspections should be conducted more often.

These checklists are by no means all-inclusive. You should add to them or delete portions or items that do not apply to your operations; however, carefully consider each item as you come to it and then make your decision. You will also need to refer to Occupational Safety and Health Administration (OSHA) standards for complete and specific standards that may apply to your situation. (Note: These checklists are typical for general industry.)

24.1 CHEMICALS

Yes ☐   No ☐   If hazardous substances are used in your processes, do you have a medical or biological monitoring system in operation?

Yes ☐   No ☐   Are you familiar with the threshold limit values (TLVs) or permissible exposure limits (PELs) of airborne contaminants and physical agents used in your workplace?

Yes ☐   No ☐   Have control procedures been instituted for hazardous materials, where appropriate, such as respirators, ventilation systems, and handling practices?

Yes ☐   No ☐   Whenever possible, are hazardous substances handled in properly designed and exhausted booths or similar locations?

Yes ☐   No ☐   Are there written standard operating procedures for the selection and use of respirators where needed?

Yes ☐   No ☐   If you have a respirator protection program, are your employees instructed on the correct usage and limitations of the respirators? Are the respirators National Institute for Occupational Safety and Health (NIOSH) approved for this particular application? Are they regularly inspected and cleaned, sanitized and maintained?

Yes ☐   No ☐   If hazardous substances are used in your processes, do you have a medical or biological monitoring system in operation?

Yes ☐   No ☐   Are you familiar with the TLVs or PELs of airborne contaminants and physical agents used in your workplace?

Yes ☐   No ☐   Have control procedures been instituted for hazardous materials, where appropriate, such as respirators, ventilation systems, and handling practices?

Yes ☐   No ☐   Whenever possible, are hazardous substances handled in properly designed and exhausted booths or similar locations?

Yes ☐   No ☐   Do you use general dilution or local exhaust ventilation systems to control dusts, vapors, gases, fumes, smoke, solvents, or mists which may be generated in your workplace?

Yes ☐   No ☐   Is ventilation equipment provided for removal of contaminants from such operations as production grinding, buffing, spray painting, and/or vapor degreasing, and is it operating properly?

Yes ☐   No ☐   Do employees complain about dizziness, headaches, nausea, irritation, or other factors of discomfort when they use solvents or other chemicals?

Yes ☐   No ☐   Is there a dermatitis problem? Do employees complain about dryness, irritation, or sensitization of the skin? When combustion engines are used, is carbon monoxide kept within acceptable levels?

Yes ☐   No ☐   Is vacuuming used, rather than blowing or sweeping dusts whenever possible for cleanup?

Yes ☐   No ☐   Have you considered the use of an industrial hygienist or environmental health specialist to evaluate your operation?

24.2   COMPRESSED GAS CYLINDERS

Yes ☐   No ☐   Are compressed gas cylinders (CGCs) kept away from radiators and other sources of heat?

Yes ☐   No ☐   Are CGCs stored in well ventilated, dry locations at least 20 ft away from materials such as oil, grease, excelsior, reserve stocks of carbide, acetylene, or other fuels as they are likely to cause acceleration of fires?

Yes ☐   No ☐   Are CGCs stored only in assigned areas?

Yes ☐   No ☐   Are CGCs stored away from elevators, stairs, and gangways?

Yes ☐   No ☐   Are CGCs stored in areas where they will not be dropped, knocked over, or tampered with?

Yes ☐   No ☐   Are CGCs not stored in areas with poor ventilation?

Yes ☐   No ☐   Are storage areas marked with signs such as "Oxygen, No Smoking, or No Open Flames?"

Yes ☐   No ☐   Are CGCs not stored outside generator houses?

Yes ☐   No ☐   Do storage areas have wood and grass cut back within 15 ft?

Yes ☐   No ☐   Are CGCs secured to prevent falling?

Yes ☐   No ☐   Are CGCs stored in a vertical position?

Yes ☐   No ☐   Are protective caps in place at all times except when in use?

Yes ☐   No ☐   Are threads on cap or cylinder not lubricated?

Yes ☐   No ☐   Are all CGCs legibly marked for the purpose of identifying the gas content with the chemical or trade name of the gas?

Yes ☐   No ☐   Are CGCs marked with stencils, stamps, or labels?

Yes ☐   No ☐   Are markings located on the slanted area directly below the cap?

Yes ☐   No ☐   Does each employee determine that CGCs are in a safe condition by means of a visual inspection?

Yes ☐   No ☐   Is each portable tank and all piping, valves, and accessories visually inspected at intervals not to exceed 2.5 years?

Yes ☐   No ☐   Are inspections conducted by the owner, agent, or approved agency?

Yes ☐   No ☐   On insulated tanks, is the insulation not to be removed if, in the opinion of the person performing the visual inspection, external corrosion is likely to be negligible?

Yes ☐   No ☐   If evidence of any unsafe condition is discovered, is the portable tank not to be returned to service until it meets all corrective standards?

24.3   CRANE SAFETY

Yes ☐   No ☐   Is a wind indicator or wind sock placed on all outside cranes and is it visible to the operator?

Yes ☐   No ☐   Is the rated load capacity of the crane on the crane at all times and is it visible from the ground?

Yes ☐   No ☐   Is a fire extinguisher of the appropriate size and type on the crane at all times?

Yes ☐   No ☐   Are all walking surfaces of the non-slip type?

Yes ☐   No ☐   Do all ladders, stairs, and railings comply with requirements of the regulations?

Yes ☐   No ☐   Are all moving parts such as gears, set screws, moving components, or anything dangerous guarded?

Yes ☐   No ☐   Is each independent hoisting unit provided with at least one self-setting brake or holding brake?

Yes ☐   No ☐   Do all electrical equipment and wiring components comply with electrical regulations?

Yes ☐   No ☐   Do all ropes, chains, and cables meet the manufacturer's recommendations?

Yes ☐   No ☐   Is any crane that has a power traveling mechanism equipped with a warning signal to include a visual warning light?

Yes ☐   No ☐   Has the employer ensured all operators of cranes are properly trained?

Yes ☐   No ☐   Are the danger areas under the load and any area below where the load will travel marked and blocked off to prevent other employees from walking under suspended loads?

Yes ☐   No ☐   Are all passages and walkways safe from movement of the crane?

Yes ☐   No ☐   Are cones, warning tapes, or guards erected?

Yes ☐   No ☐   Does the cab allow the operator to see the load at all times?

Yes ☐   No ☐   Is the cab illuminated to allow operator to see sufficiently?

Yes ☐   No ☐   Is there a preventive maintenance program based on manufacturer's recommendations?

Yes ☐   No ☐   Is there a location provided to perform maintenance on cranes where it causes the least interference with surrounding operations?

Yes ☐   No ☐   During maintenance are controls switched off?

Yes ☐   No ☐   Is the main switch locked out and tagged out?

Yes ☐   No ☐   Do the signs posted on the crane, and on the hook where it can be seen from the floor, state "Out of Order?"

Yes ☐   No ☐   Does operating cranes on the same runway as an idle crane have rail stops or suitable means to prevent contact of cranes?

Yes ☐   No ☐   Are all guards in place, safety devices reactivated, and maintenance equipment removed before operating a crane?

Yes ☐   No ☐   Are cranes inspected daily (before every use), monthly, and quarterly?

Yes ☐   No ☐   Are they inspected annually by an outside expert (e.g., manufacturer's representative)?

Yes ☐   No ☐   Is a certificate of the annual inspection retained?

Yes ☐   No ☐   Does the manufacturer's representative inspect cranes annually and retain the certificate?

Before every use, are the following tested:

Yes ☐   No ☐   Hoisting and lowering devices?

Yes ☐   No ☐   Trolley travel?

Yes ☐   No ☐   Bridge travel?

Yes ☐   No ☐   Locking or safety devices?

Yes ☐   No ☐   Inspect all grooves to detect surface defects that may damage ropes?

Yes ☐   No ☐   Inspect all ropes at least once a month?

Yes ☐   No ☐   Inspect rope, cable, or chains for kinks before lifting?

Yes ☐   No ☐   Has a preventative maintenance program based on the manufacturer's recommendations been established?

Yes ☐   No ☐   Are all adjustments or repairs done by a qualified person?

24.4 EMERGENCY RESPONSE AND PLANNING

Yes ☐   No ☐   Is there a written emergency response planning which is available to all employees?

Yes ☐   No ☐   Is there an established procedure specifically outlining the steps to be taken by all employees including route of evacuation, meeting place outside building, and designation of person responsible for verifying that employees are all accounted for?

Yes ☐   No ☐   Have proper evacuation procedures been communicated to everyone before the need for an actual evacuation, and have those procedures been actively practiced in a mock evacuation situation?

Yes ☐   No ☐   Is there an established protocol for determining the need for evacuation?

Yes ☐   No ☐   Is there a designated person responsible for making an evacuation decision?

Yes ☐   No ☐   Is the need for evacuation communicated to employees in such a way that everyone (other than those designated as the initial contacts) receives the same information at the same time?

Yes ☐   No ☐   In the event of electrical failure, is there a backup system for both broadcasting of messages and lighting of escape routes?

Yes ☐   No ☐   Are established escape routes clearly marked, and are maps posted outlining the entire route?

Yes ☐   No ☐   Are escape routes determined to be the shortest safe route possible, allowing adequate room and number of routes for the number of employees?

Yes ☐   No ☐   Are all emergency exits clearly marked and functioning properly?

Yes ☐   No ☐   Are all escape routes free of clutter and tripping hazards?

Yes ☐   No ☐   Is there adequate emergency lighting along the routes?

Yes ☐   No ☐   Are emergency equipment such as fire extinguishers and flashlights located at predetermined sites along escape routes and is this equipment routinely tested for proper operation?

Yes ☐   No ☐   In the event that employees are required to remain within hallways/stairways of escape route for longer than expected, is there adequate ventilation, temperature control, and some type of communication equipment?

Yes ☐   No ☐   Are all established meeting places outside of the building a reasonably safe distance away?

Yes ☐   No ☐   Is there an established method for verification that all employees have left the building, and a way to communicate to emergency personnel the identities and possible locations of those who have not?

24.5   ERGONOMICS

24.5.1   MANUAL MATERIAL HANDLING

Yes ☐   No ☐   Is there lifting of loads, tools, or parts?
Yes ☐   No ☐   Is there lowering of tools, loads, or parts?
Yes ☐   No ☐   Is there overhead reaching for tools, loads, or parts?
Yes ☐   No ☐   Is there bending at the waist to handle tools, loads, or parts?
Yes ☐   No ☐   Is there twisting at the waist to handle tools, loads, or parts?

24.5.2   PHYSICAL ENERGY DEMANDS

Yes ☐   No ☐   Do tools and parts weigh more than 10 lb?
Yes ☐   No ☐   Is reaching greater than 20 in.?
Yes ☐   No ☐   Is bending, stooping, or squatting a primary task activity?
Yes ☐   No ☐   Is lifting or lowering loads a primary task activity?
Yes ☐   No ☐   Is walking or carrying loads a primary task activity?
Yes ☐   No ☐   Is stair or ladder climbing with loads a primary task activity?
Yes ☐   No ☐   Is pushing or pulling loads a primary task activity?
Yes ☐   No ☐   Is reaching overhead a primary task activity?
Yes ☐   No ☐   Do any of the above tasks require five or more complete work cycles to be done within a minute?
Yes ☐   No ☐   Do workers complain that rest breaks and fatigue allowances are insufficient?

24.5.3   OTHER MUSCULOSKELETAL DEMANDS

Yes ☐   No ☐   Do manual jobs require frequent, repetitive motions?
Yes ☐   No ☐   Do work postures require frequent bending of the neck, shoulder, elbow, wrist, or finger joints?

Yes ☐   No ☐   For seated work, do reaches for tools and materials exceed 15 in. from the worker's position?
Yes ☐   No ☐   Is the worker unable to change his/her position often?
Yes ☐   No ☐   Does the work involve forceful, quick, or sudden motions?
Yes ☐   No ☐   Does the work involve shock or rapid buildup of forces?
Yes ☐   No ☐   Is finger-pinch gripping used?
Yes ☐   No ☐   Do job postures involve sustained muscle contraction of any limb?

24.5.4   ENVIRONMENT

Yes ☐   No ☐   Is the temperature too hot or too cold?
Yes ☐   No ☐   Are the worker's hands exposed to temperatures less than 70°F?
Yes ☐   No ☐   Is the workplace poorly lit?
Yes ☐   No ☐   Is there glare?
Yes ☐   No ☐   Is there excessive noise that is annoying, distracting, or producing hearing loss?
Yes ☐   No ☐   Is there upper extremity or whole body vibration?
Yes ☐   No ☐   Is air circulation too high or too low?

24.5.5   GENERAL WORKPLACE

Yes ☐   No ☐   Are walkways uneven, slippery, or obstructed?
Yes ☐   No ☐   Is housekeeping poor?
Yes ☐   No ☐   Is there inadequate clearance or accessibility for performing tasks?
Yes ☐   No ☐   Are stairs cluttered or lacking railings?
Yes ☐   No ☐   Is proper footwear worn?

24.5.6   TOOLS

Yes ☐   No ☐   Is the handle too small or too large?
Yes ☐   No ☐   Does the handle shape require bent wrist to use the tool?
Yes ☐   No ☐   Is the tool hard to access?
Yes ☐   No ☐   Does the tool weigh more than 9 lb?
Yes ☐   No ☐   Does the tool vibrate excessively?
Yes ☐   No ☐   Does the tool cause excessive kickback to the operator?
Yes ☐   No ☐   Does the tool become too hot or too cold?

24.6   FIRE PROTECTION AND PREVENTION

Yes ☐   No ☐   Does the employer provide portable fire extinguishers for small fires?
Yes ☐   No ☐   Are all fire extinguishers clearly marked with symbols that distinctly reflect the type of fire hazard for which they are intended?

Yes ☐ No ☐   Are portable fire extinguishers located where they are readily accessible to employees without subjecting them to possible injury?

Yes ☐ No ☐   Are fire extinguishers fully charged and operable at all times?

Yes ☐ No ☐   Are Class A and D fire extinguishers no more than 75 ft apart?

Yes ☐ No ☐   Are Class B fire extinguishers no more than 50 ft apart?

Yes ☐ No ☐   Are Class C fire extinguishers patterned among class A and B extinguishers where a class C fire hazard exists?

Yes ☐ No ☐   Are all fire extinguishers clearly marked with symbols that distinctly reflect the type of fire hazard for which they are intended?

Yes ☐ No ☐   Are protective clothing, such as respiratory, head, hand, foot, leg, eye, and face guards, worn to protect the entire body?

Yes ☐ No ☐   Are fixed extinguishing systems used on specific fire hazards?

Yes ☐ No ☐   Is an alarm with a delay in place to warn employees before a fixed extinguisher is to be discharged?

Yes ☐ No ☐   Are hazard warning or caution signs posted at the entrance to, and inside, areas protected by systems that use agents known to be hazardous to employees' safety and health?

Yes ☐ No ☐   Are fire detection systems installed and maintained to assure best detection of a fire?

Yes ☐ No ☐   Is an employee alarm system installed that is capable of warning every employee of an emergency?

Yes ☐ No ☐   Is the alarm system such that can be heard above the sound level of the work area?

Yes ☐ No ☐   Are warning lights installed, if there are hearing impaired employees?

Yes ☐ No ☐   Are all firefighting equipment inspected at least annually, and records maintained?

Yes ☐ No ☐   Are portable fire extinguishers inspected at least monthly, and records maintained?

Yes ☐ No ☐   Is any damaged equipment removed immediately from service and replaced?

Yes ☐ No ☐   Is hydrostatic testing done on each extinguisher at least once every 5 years?

Yes ☐ No ☐   Are fixed extinguishing systems inspected annually by a qualified person?

Yes ☐ No ☐   Are fire detection systems tested monthly if they are battery operated?

Yes ☐ No ☐   Is training on the use of portable fire extinguishers, and records of attending employees maintained?

Yes ☐ No ☐   Is training provided to employees designated to inspect, maintain, operate, or repair fixed extinguishing systems?

Yes ☐ No ☐   Is an annual review training required to keep them up to date?

Yes ☐ No ☐   Are all employees trained to recognize the alarm signals for each emergency (fire, tornado, chemical release, etc.)?

Yes ☐ No ☐ Are employees trained for reporting emergencies, locating alarms, and sounding them?

Yes ☐ No ☐ Is training provided on evacuation procedures?

Yes ☐ No ☐ Are drills performed periodically to ensure employees are aware of their duties?

Yes ☐ No ☐ Is all training conducted by a qualified/competent person?

Yes ☐ No ☐ Has the employer established and maintained a written policy that establishes the existence of a fire brigade?

Yes ☐ No ☐ Does the employer use employees who are physically capable of performing the duties as a member of a fire brigade that may be assigned to them during an emergency?

Yes ☐ No ☐ Is the employee provided training by the employer before being assigned any emergency response duties?

Yes ☐ No ☐ Are all fire brigade members trained at least annually, and interior structural firefighters provided with an education session or training at least quarterly?

Yes ☐ No ☐ Did the employer inform the fire brigade members of special hazards, such as storage and use of flammable liquids and/or gases, toxic chemicals, radioactive sources, and water reactive substances that they may encounter during an emergency?

24.7 FORKLIFTS

Yes ☐ No ☐ Do all new forklift meet the American National Standards Institute (ANSI) BS6.1-1969?

Yes ☐ No ☐ Is the ANSI label, load ratings, and/or any plates in place and visible at all times?

Yes ☐ No ☐ Is each forklift examined before each shift and is an operator checklist completed?

Yes ☐ No ☐ Is a defective, unsafe, and out of order forklift removed from service?

Yes ☐ No ☐ Are all repairs done by trained, authorized personnel?

Yes ☐ No ☐ Is a copy of the maintenance report kept on file?

Yes ☐ No ☐ Are lockout/tagout procedures used during maintenance?

Yes ☐ No ☐ Are forklifts operated only by properly licensed operators?

Yes ☐ No ☐ Is refresher training conducted annually?

Yes ☐ No ☐ Are new employees tested despite previous experience?

Yes ☐ No ☐ Are special battery changing areas provided for electric trucks?

Yes ☐ No ☐ Is a hoist or crane provided to lift batteries?

Yes ☐ No ☐ Does proper ventilation exist in areas where exhaust-releasing forklifts are operated?

Yes ☐ No ☐ Are riders not allowed on forklifts?

Yes ☐ No ☐ Are forklifts turned off, controls in neutral, fork lowered, and brakes set when the driver is not in his seat?

Yes ☐ No ☐ Do all forklifts have an overhead guard in place?

Yes ☐   No ☐   Are traffic regulations posted in forklift areas and compliance ensured?

Yes ☐   No ☐   Are only safely arranged loads lifted with a forklift?

Yes ☐   No ☐   Is the forklift operated within its rated capacity?

Yes ☐   No ☐   Are forklifts never fueled while it is running?

Yes ☐   No ☐   Are safety devices never allowed to be removed from the forklift?

Yes ☐   No ☐   Is the forklift maintained clean at all times?

Yes ☐   No ☐   Are operators trained, for the specific machine that the employee will be operating?

Yes ☐   No ☐   Is training repeated annually and are training materials retained?

24.8   HAND AND PORTABLE POWER TOOLS

24.8.1   HAND TOOLS AND EQUIPMENT

Yes ☐   No ☐   Are all tools and equipment (both company's and employees') used by employees at their workplace in good condition?

Yes ☐   No ☐   Are hand tools such as chisels and punches, which develop mushroomed heads during use, reconditioned, or replaced as necessary?

Yes ☐   No ☐   Are broken or fractured handles on hammers, axes, and similar equipment replaced promptly?

Yes ☐   No ☐   Are worn or bent wrenches replaced regularly?

Yes ☐   No ☐   Are appropriate handles used on files and similar tools?

Yes ☐   No ☐   Are employees made aware of the hazards caused by faulty or improperly used hand tools?

Yes ☐   No ☐   Are appropriate safety glasses, face shields, etc. used while using hand tools or equipment that might break or produce sparks?

Yes ☐   No ☐   Are jacks checked periodically to ensure they are in good operating condition?

Yes ☐   No ☐   Are tool handles wedged tightly in the head of all tools?

Yes ☐   No ☐   Are tool cutting edges kept sharp so the tool will move smoothly without binding or skipping?

Yes ☐   No ☐   Are tools stored in dry, secure locations where they would not be tampered with?

Yes ☐   No ☐   Is eye and face protection used when driving hardened or tempered spuds or nails?

24.9   HAZARD COMMUNICATION

Yes ☐   No ☐   Is there a list of hazardous substances used in your workplace?

Yes ☐   No ☐   Is there a written hazard communication program dealing with material safety data sheets (MSDSs), labeling, and employee training?

Yes ☐ No ☐ Is each container for a hazardous substance (i.e., vats, bottles, storage tanks, etc.) labeled with product identity and a hazard warning (communication of the specific health hazards and physical hazards)?

Yes ☐ No ☐ Is there an MSDS readily available for each hazardous substance used?

Yes ☐ No ☐ Is there an employee training program for hazardous substances?

Does this program include the following:

Yes ☐ No ☐ An explanation of what an MSDS is and how to use and obtain one?

Yes ☐ No ☐ MSDS contents for each hazardous substance or class of substances?

Yes ☐ No ☐ Explanation of "Right to Know?"

Yes ☐ No ☐ Identification of where an employee can see the employers written hazard communication program and where hazardous substances are present in their work areas?

Yes ☐ No ☐ The physical and health hazards of substances in the work area, and specific protective measures to be used?

Yes ☐ No ☐ Details of the hazard communication program, including how to use the labeling system and MSDSs?

Are employees trained in the following:

Yes ☐ No ☐ How to recognize tasks that might result in occupational exposure?

Yes ☐ No ☐ How to use work practice and engineering controls and personal protective equipment (PPE) and to know their limitations?

Yes ☐ No ☐ How to obtain information on the type selection, proper use, location removal handling, decontamination, and disposal of PPE?

Yes ☐ No ☐ Who to contact and what to do in an emergency?

24.10 HEALTH HAZARDS

Yes ☐ No ☐ Have any organisms that could cause health problems been identified?

Yes ☐ No ☐ Have any chemicals that could cause health problems been identified?

Yes ☐ No ☐ Has any physical hazard that could cause health problems been identified?

Yes ☐ No ☐ Has any ergonomic hazard that could cause health problems been identified?

Yes ☐ No ☐ Have steps been taken to control potential health hazards?

Yes ☐ No ☐ Have workers been provided with appropriate PPE?

Yes ☐ No ☐ Have employees reported health-related symptoms?

Yes ☐   No ☐   Have the employees undergone a medical examination?
Yes ☐   No ☐   Are inspections conducted on a regular basis?
Yes ☐   No ☐   Is regular environmental and personal monitoring conducted?
Yes ☐   No ☐   Are health concerns or findings addressed promptly?

24.11 LADDERS

Yes ☐   No ☐   Are only Type 1 or Type 1A industrial ladders used?
Yes ☐   No ☐   Do steps on ladders support a minimum load capacity of 250 lb?
Yes ☐   No ☐   Are all ladders inspected for damage before use?
Yes ☐   No ☐   Are ladders not placed against movable objects?
Yes ☐   No ☐   Are ladders placed to prevent movement by lashing or other means?
Yes ☐   No ☐   Are employees shoes free of mud, grease, or other substances that could cause a slip or fall?
Yes ☐   No ☐   Are ladders not placed on unstable bases such as boxes or barrels?
Yes ☐   No ☐   Do employees not stand on the top two steps of a stepladder?
Yes ☐   No ☐   Are ladders used to gain access to a roof that extends at least 3 ft above the point of support, at eave, gutter, or roofline?
Yes ☐   No ☐   Are stepladders fully opened to permit the spreaders to lock?
Yes ☐   No ☐   Are all labels in place and legible on ladders?
Yes ☐   No ☐   Are ladder always moved to prevent and avoid overreaching?
Yes ☐   No ☐   Are single ladders not more than 30 ft?
Yes ☐   No ☐   Do extension ladders up to 36 ft have a 3 ft overlap between sections?
Yes ☐   No ☐   Do extension ladders over 36 ft and up to 48 ft have a 4 ft overlap between sections?
Yes ☐   No ☐   Do extension ladders over 48 ft and up to 60 ft have a 5 ft overlap between sections?
Yes ☐   No ☐   Do two-section extension ladders not exceed 48 ft in total length?
Yes ☐   No ☐   Do ladders over two-sections not exceed 60 ft in total length?
Yes ☐   No ☐   Are adders not used horizontally as scaffolds, runways, or platforms?
Yes ☐   No ☐   Is the area around the top and base of ladders kept free of tripping hazards such as loose materials, trash, cords, hoses, and leaves?
Yes ☐   No ☐   Is the base of a straight or extension ladder set back a safe distance from the vertical or approximately one-fourth of the working length of the ladder?
Yes ☐   No ☐   Are ladders that project into passageways or doorways where they could be struck by personnel, moving equipment, or materials being handled, protected by barricades or guards?
Yes ☐   No ☐   Do employees face the ladder when ascending or descending?
Yes ☐   No ☐   Do employees use both hands when going up or down a ladder?
Yes ☐   No ☐   Are materials or equipment raised or lowered by way of lines?
Yes ☐   No ☐   Are employees trained and educated on the proper use of ladders?

Yes ☐ No ☐ Are repairs done professionally?

Yes ☐ No ☐ Are inspections conducted before every use and defective, broken, or damaged ladders rejected and tagged as "Dangerous. Do Not Use?"

Yes ☐ No ☐ Are the rungs tight in the joint of the side rails?

Yes ☐ No ☐ Do all moving parts operate freely without binding?

Yes ☐ No ☐ Are all pulleys, wheels, and bearings lubricated frequently?

Yes ☐ No ☐ Are rungs kept free of grease and oil?

Yes ☐ No ☐ Are badly worn or frayed ropes replaced immediately?

Yes ☐ No ☐ Are all ladders equipped with slip resistant feet, free of grease, and in good condition?

24.11.1 PORTABLE WOODEN LADDERS

Yes ☐ No ☐ Are all wooden ladders free of splinters, sharp edges, shake, wane, compression failures, decay, and other irregularities?

Yes ☐ No ☐ Are portable stepladders no longer than 20 ft?

Yes ☐ No ☐ Is the step spacing no more than 12 in. apart?

Yes ☐ No ☐ Are stepladders which have a metal spreader or locking device of sufficient strength and size to hold the front and back when open?

24.11.2 PORTABLE METAL LADDERS

Yes ☐ No ☐ Are ladders inspected immediately when dropped or tipped over?

Yes ☐ No ☐ Are the step spacing no more than 12 in. apart?

Yes ☐ No ☐ Are metal ladders used for electrical work or near energized conductors?

24.11.3 FIXED LADDERS

Yes ☐ No ☐ Are the steps no more than 12 in. apart?

Yes ☐ No ☐ Are job made ladders constructed to conform with the established OSHA standards?

Yes ☐ No ☐ Are all fixed ladders painted or treated to prevent rusting?

Yes ☐ No ☐ Do fixed ladders 20 ft or higher have a landing every 20 ft if there is no surrounding cage?

Yes ☐ No ☐ If it has a cage or safety device, is there a landing every 30 ft?

24.12 LIFTING SAFETY

Yes ☐ No ☐ Have all workers been trained on proper lifting techniques?

Yes ☐ No ☐ Was the object inspected to decide on the best grip?

Yes ☐ No ☐ Has the load been sized up to insure it can be lifted?

Yes ☐ No ☐ Are loads kept small to prevent heavy lifts?

Yes ☐   No ☐   Are the feet placed close to the object?
Yes ☐   No ☐   Can the load be gripped firmly?
Yes ☐   No ☐   Are the knees bent while keeping the back straight?
Yes ☐   No ☐   Is the load held close to the worker's body?
Yes ☐   No ☐   Can the worker see past the load?
Yes ☐   No ☐   Did the worker get help for large or heavy objects?
Yes ☐   No ☐   Is the lift more appropriate for a team lift?
Yes ☐   No ☐   Are harsh jerking movements when pushing, pulling, or lifting a load avoided?

Yes ☐   No ☐   Are materials stacked between the knees and waist?
Yes ☐   No ☐   Are gloves worn when handling sharp or rough objects?
Yes ☐   No ☐   Is power lifting equipment used, instead of manual lifting, when possible to prevent injuries?
Yes ☐   No ☐   Is care taken not to drop materials that might hit someone?
Yes ☐   No ☐   Is equipment, carts, and/or table kept at a proper height to prevent back injuries?

24.13   MACHINE GUARDING AND SAFETY

Yes ☐   No ☐   Do the safeguards provided meet the minimum OSHA requirements?
Yes ☐   No ☐   Do the safeguards prevent workers' hands, arms, and other body parts from making contact with dangerous moving parts?
Yes ☐   No ☐   Are the safeguards firmly secured and not easily removable?
Yes ☐   No ☐   Do the safeguards ensure that no objects will fall into the moving parts?
Yes ☐   No ☐   Do the safeguards permit safe, comfortable, and relatively easy operation of the machine?
Yes ☐   No ☐   Can the machine be oiled without removing the safeguard?
Yes ☐   No ☐   Is there a system for shutting down the machinery before safeguards are removed?
Yes ☐   No ☐   Can the existing safeguards be improved?
Yes ☐   No ☐   Is there a point-of-operation safeguard provided for the machine?
Yes ☐   No ☐   Does it keep the operator's hands, fingers, and body out of the danger area?
Yes ☐   No ☐   Is there evidence that the safeguards have been tampered with or removed?
Yes ☐   No ☐   Could you suggest a more practical, effective safeguard?
Yes ☐   No ☐   Could changes be made on the machine to eliminate the point-of-operation hazard entirely?
Yes ☐   No ☐   Are there any unguarded gears, sprockets, pulleys, or flywheels on the apparatus?
Yes ☐   No ☐   Are there any exposed belts or chain drives?

Yes ☐ No ☐   Are there any exposed set screws, key ways, collars, etc.?

Yes ☐ No ☐   Are starting and stopping controls within easy reach of the operator?

Yes ☐ No ☐   If there is more than one operator, are separate controls provided?

Yes ☐ No ☐   Are safeguards provided for all hazardous moving parts of the machine including auxiliary parts?

Yes ☐ No ☐   Have appropriate measures been taken to safeguard workers against noise hazards?

Yes ☐ No ☐   Have special guards, enclosures, or PPE been provided, where necessary, to protect workers from exposure to harmful substances used in machine operation?

Yes ☐ No ☐   Is the machine installed in accordance with National Fire Protection Association and National Electrical Code requirements?

Yes ☐ No ☐   Are there loose conduit fittings?

Yes ☐ No ☐   Is the machine properly grounded?

Yes ☐ No ☐   Is the power supply correctly fused and protected?

Yes ☐ No ☐   Do workers occasionally receive minor shocks while operating any of the machines?

Yes ☐ No ☐   Do operators and maintenance workers have the necessary training in how and why to use the safeguards?

Yes ☐ No ☐   Have operators and maintenance workers been trained in locating and understanding the functioning and use of the safeguards?

Yes ☐ No ☐   Have operators and maintenance workers been trained in how and under what circumstances guards can be removed?

Yes ☐ No ☐   Have workers been trained to act in cases of damaged. missing, or inadequate guards?

Yes ☐ No ☐   Is protective equipment required?

Yes ☐ No ☐   If protective equipment is required, is it appropriate for the job, in good condition, kept clean and sanitary, and stored carefully when not in use?

Yes ☐ No ☐   Is the operator dressed safely for the job (i.e., no loose-fitting clothing or jewelry)?

Yes ☐ No ☐   Have maintenance workers received up-to-date instruction on the machines they service?

Yes ☐ No ☐   Do maintenance workers lock out the machine from its power sources before beginning repairs?

Yes ☐ No ☐   Where several maintenance persons work on the same machine, are multiple lockout devices used?

Yes ☐ No ☐   Do maintenance persons use appropriate and safe equipment in their repair work?

Yes ☐ No ☐   Is the maintenance equipment itself property guarded?

Yes ☐ No ☐   Are maintenance and servicing workers trained in the requirements of 29 CFR 1910.147, lockout/tagout hazard, and do the procedures for lockout/tagout exist before they attempt their tasks?

24.14 MATERIAL HANDLING

24.14.1 MATERIAL HANDLING EQUIPMENT

Yes ☐   No ☐   Are all operators of material handling equipment trained (includes hand trucks, cranes, hoists, fork trucks, or any motorized equipment)?

Yes ☐   No ☐   Are all operators of forklifts trained by a certified instructor?

Yes ☐   No ☐   Are all material handling equipment kept in good repair, and maintained by trained personnel?

Yes ☐   No ☐   Are all material handling equipment inspected before use, daily, monthly, and annually as required?

Yes ☐   No ☐   Are all material handling equipment properly marked with load ratings?

Yes ☐   No ☐   Are forklifts marked "Flammable," if they use propane, or any other compressed gas source?

Yes ☐   No ☐   Are railroad cars, heavy equipment, and rolling hoists or cranes choked or blocked to prevent rolling?

Yes ☐   No ☐   Are grading or ramps installed between two working levels for safe vehicle movement?

Yes ☐   No ☐   Are material handling equipment that pose a danger to equipment or personnel guarded to prevent access within a safe distance?

24.14.2 STORAGE AREAS

Yes ☐   No ☐   Are maximum safe load limits observed?

Yes ☐   No ☐   Are load limits posted for platforms and floors?

Yes ☐   No ☐   Are storage racks stable and secure?

Yes ☐   No ☐   Are stored material neatly stacked, racked, blocked, or interlocked?

Yes ☐   No ☐   Are height limits set and posted to insure stability of stacked material?

Yes ☐   No ☐   Do all aisles, loading docks, doorways, turns, and passages have safe clearances for equipment and material?

Yes ☐   No ☐   Are clearance signs posted in a visible place to warn employees of clearance limits?

Yes ☐   No ☐   Are all ramps, open pits, tanks, vats, ditches, and elevated surfaces 4 ft or more guarded?

24.14.3 HOUSEKEEPING

Yes ☐   No ☐   Are storage areas kept clean, dry, and in good condition?

Yes ☐   No ☐   Are storage areas kept free of tripping and slipping hazards?

Yes ☐ No ☐ Are storage areas kept free of fire hazards (trash, paper, oily rags, or empty flammable liquid containers)?

Yes ☐ No ☐ Are storage areas kept free of explosion hazards (unsecured compressed gas cylinders, flammable vapors, or dusts)?

Yes ☐ No ☐ Are storage areas kept free of pests such as rats, mice, roaches, and other vermin?

24.15 MEANS OF EXIT

Yes ☐ No ☐ Do all exits have an illuminated sign above stating, "Exit"?

Yes ☐ No ☐ Are there signs that state, "Not an exit," placed over doors if there is the possibility that it could be mistaken for an exit, e.g., closets, stairways, and doors?

Yes ☐ No ☐ Are under no circumstances exits locked while the building is occupied?

Yes ☐ No ☐ Are all emergency exit doors equipped with panic bars?

Yes ☐ No ☐ Do all emergency exit doors designated for fire escape lead to a safe area of refuge?

Yes ☐ No ☐ Do all emergency exit doors or passageways have emergency illumination, in case of power failure?

Yes ☐ No ☐ Is there access to exits that are unobstructed at all times?

Yes ☐ No ☐ Are all floor areas around exits clean and dry at all times?

Yes ☐ No ☐ Is an inspection from a fire marshal done at least once a year?

Yes ☐ No ☐ Is a general inspection of exit signs, exit doors, exit accesses, and alarm systems conducted by a trained person who has the authority to rectify any problems?

Yes ☐ No ☐ Is training done on the identification of all exits and their locations?

24.16 MEDICAL SERVICES AND FIRST AID

Yes ☐ No ☐ Are medical facilities and medically trained personnel on-site if possible?

Yes ☐ No ☐ In the absence of any nearby medical facility, have personnel been adequately trained to render first aid?

Yes ☐ No ☐ Are physician-approved first-aid supplies readily available?

Yes ☐ No ☐ Are there facilities for quick drenching or flushing in work areas where the eyes or body may be exposed to injurious corrosive materials or chemicals?

Yes ☐ No ☐ Is a first-aid log kept on employees?

Yes ☐ No ☐ Is an inventory checklist kept of all first-aid supplies?

Yes ☐ No ☐ Are all employees trained on basic first-aid techniques and procedures?

Yes ☐ No ☐ Are all employees trained on usage of PPE while first aid is being performed?

24.17 RIGGING

Yes ☐ No ☐ Is ANSI approved equipment used?

Yes ☐ No ☐ Are daily inspections conducted before use by the user/operator?

Yes ☐ No ☐ Are monthly inspections done by a person trained to recognize defects and authorized to remove equipment from service?

Yes ☐ No ☐ Are annual inspections by the manufacturer or outside contractor done but not required?

Yes ☐ No ☐ Is only the manufacturer allowed to repair these devices?

Yes ☐ No ☐ Do all chains, slings, and cables have an identification tag attached that shows their load rating, limitations, etc.?

Yes ☐ No ☐ Is the load rating never exceeded for chains/slings/cables?

Yes ☐ No ☐ Are only alloy steel chains used?

Yes ☐ No ☐ Are chains inspected before use for wear, abrasions, collapse, visible damage, or any damage no matter how insignificant?

Yes ☐ No ☐ Are damaged chains removed from service?

Yes ☐ No ☐ Do hooks, rings, links, or any coupling device have the same or higher rating as the chain to which it is affixed?

Yes ☐ No ☐ Are wire rope slings and cables inspected before use?

Yes ☐ No ☐ Do all attachments meet the same load standards as the wire rope sling they are attached to?

Yes ☐ No ☐ Are wire rope slings and fiber-core wire ropes operated at temperatures below 200°F?

Yes ☐ No ☐ Are nonfiber core, wire rope slings only used at temperatures below 400°F and above 60°F?

Yes ☐ No ☐ Do the handles of metal mesh slings meet the minimum requirements of the sling?

Yes ☐ No ☐ Are metal mesh slings not impregnated with elastomers used in temperatures not exceeding 500°F or below 20°F?

Yes ☐ No ☐ Are metal mesh slings impregnated with polyvinyl chloride or neoprene, used in the temperature range from 0°F to 200°F?

Yes ☐ No ☐ Are natural or synthetic fiber rope slings only used in temperature ranges of above 20°F–180°F, unless they are wet or frozen?

Yes ☐ No ☐ Are metal mesh slings never spliced except that manufacturers make alterations to slings?

Yes ☐ No ☐ Are natural or synthetic fiber slings removed from service if wear is abnormal, powdered fibers appear between strands, fibers are broken or cut, there is variation in size or roundness of strands, and discoloration, rotting, or distortion of hardware is detected?

Yes ☐ No ☐ Are synthetic web slings uniform in thickness?

Yes ☐ No ☐ Are polyester and nylon webs not used where fumes, vapors, sprays, mists, liquids of acids, phonetics, or caustics are present?

Yes ☐ No ☐ Are synthetic fiber slings removed from service when the following conditions are present: acid or caustic, burns, melting or charring of any part of the sling, snags, punctures, tears or cuts, broken or worn stitches, and distortion of any fitting?

24.18 SLIPS, TRIPS, AND FALLS

Yes ☐ No ☐ Do employees have to walk or work on greasy, oily, or wet floors that are not adequately slip resistant?

Yes ☐ No ☐ Do loads that are carried or pushed interfere with forward vision?

Yes ☐ No ☐ Are the loads to be carried excessive or likely to upset a person's balance?

Yes ☐ No ☐ Do heavy carts have to be pushed up ramps?

Yes ☐ No ☐ Are employees hurried due to time constraints?

Yes ☐ No ☐ Does water puddle on smooth floors on rainy days?

Yes ☐ No ☐ Are there any hard, smooth floors in wet or oily areas?

Yes ☐ No ☐ Are there any leaks of fluids onto the floor from processes or machines?

Yes ☐ No ☐ Is poor drainage causing pooling of fluids?

Yes ☐ No ☐ Is the floor surface uneven and not easily noticed?

Yes ☐ No ☐ Is the floor slippery when wet?

Yes ☐ No ☐ Is any antislip paint, coating profiles or tapes worn smooth or damaged?

Yes ☐ No ☐ Are there any isolated low steps (commonly at doorways)?

Yes ☐ No ☐ Are there any trip hazards due to equipment and other objects left on the floor?

Yes ☐ No ☐ Are there any raised carpet edges or holes worn in carpets?

Yes ☐ No ☐ Are there any tiles becoming unstuck or curling at the edges?

Yes ☐ No ☐ Are there any holes or unevenness in the floor surface?

Yes ☐ No ☐ Do the employees' safety shoes lack grip?

Yes ☐ No ☐ Are the tread patterns on safety footwear too worn?

Yes ☐ No ☐ Are the tread patterns clogged with dirt?

Yes ☐ No ☐ Is there a buildup of polish on floors?

Yes ☐ No ☐ Are wet floor signs not available or not used correctly?

Yes ☐ No ☐ Do you need to provide information/training/advice to contractors regarding slipping and tripping hazards?

Yes ☐ No ☐ Are paper, rubbish, dirt, spills, etc., left on the floor?

Yes ☐ No ☐ Are aisles poorly marked and cluttered?

Yes ☐ No ☐ Are there any trip hazards due to equipment and other movable objects left lying on the ground?

Yes ☐ No ☐ Do spills (wet or dry) occur regularly during work processes?

Yes ☐   No ☐   Is the lighting insufficient for ramps or steps to be seen clearly and without glare?

Yes ☐   No ☐   Do any steps have too small a rise or tread or an excessive nosing?

Yes ☐   No ☐   Are any step edges (nosings) slippery or hard to see?

Yes ☐   No ☐   Are the steps uneven or are there excessive variations in step dimensions?

Yes ☐   No ☐   Are handrails inadequate on stairs?

Yes ☐   No ☐   Are ramps too steep, or too slippery?

Yes ☐   No ☐   Is there insufficient lighting in passageways and at flooring transitions, ramps, or stairs?

Yes ☐   No ☐   Does the lighting throw distracting shadows or produce excessive glare?

Yes ☐   No ☐   Is there a buildup of moss or other vegetation on pathways?

Yes ☐   No ☐   Are there any surface transitions not easily noticed (any ridge that is as high as a footwear sole or higher)?

Yes ☐   No ☐   Are there potholes in footpaths or walkways?

Appendix A   Common Exposures or Accident Types

The common exposures or accident types help standardize the review of hazards. There are 11 basic types of accidents:

- Struck-against
- Struck-by
- Contact-with
- Contacted-by
- Caught-in
- Caught-on
- Caught-between
- Fall-to-same-level
- Fall-to-below
- Overexertion
- Exposure

Hazards should be looked with these common accident types in mind to identify procedures, processes, occupations, and tasks, which present a hazard that could cause one of the accident types in the following section.

A.1   ACCIDENT TYPES

A.1.1   STRUCK-AGAINST TYPES OF ACCIDENTS

Look at these first four basic accident types—struck-against, struck-by, contact-with, and contacted-by—in more detail, with the job step walk-round inspection in mind. Can the worker strike against anything while doing the job step? Think of the worker moving and contacting something forcefully and unexpectedly—an object capable of causing injury. Can he/she forcefully contact anything that will cause injury? This forceful contact may be with machinery, timber or bolts, protruding objects or sharp, jagged edges. Identify not only what the worker can strike against, but also how the contact can come about. This does not mean that every object around the worker must be listed.

A.1.2   STRUCK-BY TYPES OF ACCIDENTS

Can the worker be struck by anything while doing the job step? The phrase "struck by" means that something moves and strikes the worker abruptly with force. Study the work environment for what is moving in the vicinity of the worker, what is about to move, or what will move as a result of what the worker does. Is unexpected movement possible from normally stationary objects? Examples are ladders, tools, containers, supplies, and so on.

A.1.3 Contact-by and Contact-With Types of Accidents

The subtle difference between contact with and contact-by injuries is that in the first, the agent moves to the victim, while in the second, the victim moves to the agent. Can the worker be contacted by anything while doing the job step? The contacted by accident is one in which the worker could be contacted by some object or agent. This object or agent is capable of injuring by nonforceful contact. Examples of items capable of causing injury are chemicals, hot solutions, fire, electrical flashes, and steam.

Can the worker come in contact with some agent that will injure without forceful contact? Any type of work that involves materials or equipment, which may be harmful without forceful contact, is a source of contact with accidents. There are two kinds of work situations, which account for most of the contact with accidents. One situation is working on or near electrically charged equipment, and the other is working with chemicals or handling chemical containers.

A.1.4 Caught-In and Caught-On Types of Accidents

The next three accident types involve "caught" accidents. Can the person be caught in, caught on, or caught between objects? A caught in-accident is one in which the person, or some part of his/her body, is caught-in an enclosure or opening of some kind. Can the worker be caught on anything while doing the job step? Most caught-on accidents involve worker's clothing being caught on some projection of a moving object. This moving object pulls the worker into an injury contact. Or, the worker may be caught on a stationary protruding object, causing a fall.

A.1.5 Caught-Between Types of Accidents

Can the worker be caught between any objects while doing the job step? Caught-between accidents involve having a part of the body caught between something moving and something stationary, or between two moving objects. Always look for pinch points.

A.1.6 Fall-to-Same-Level and Fall-to-Below Types of Accidents

Slip, trip, and fall accident types are one of the most common accidents occurring in the workplace. Can the worker fall while doing a job step? Falls are such frequent accidents that we need to look thoroughly for slip, trip, and fall hazards. Consider whether the worker can fall from something above ground level, or whether the worker can fall to the same level. Two hazards account for most fall-to-same-level accidents: slipping hazards and tripping hazards. The fall-to-below accidents occur in situations where employees work above ground or above floor level, and the results are usually more severe.

A.1.7 OVEREXERTION AND EXPOSURE TYPES OF ACCIDENTS

The next two accident types are overexertion and exposure. Can the worker be injured by overexertion; that is, can he/she be injured while lifting, pulling, or pushing? Can awkward body positioning while doing a job step cause a sprain or strain? Can the repetitive nature of a task cause injury to the body? An example of this is excessive flexing of the wrist, which can cause carpal tunnel syndrome (which is abnormal pressure on the tendons and nerves in the wrist).

Finally, can exposure to the work environment cause injury to the worker? Environmental conditions such as noise, extreme temperatures, poor air, toxic gases and chemicals, or harmful fumes from work operations should also be listed as hazards.

Appendix B   Sample and Blank Material Safety Data Sheets

 | *Material Safety Data Sheet*

24 Hour Emergency Telephone: 918-450-2551
CHEMTREC: 1-800-424-9300

National Response in Canada
CANUTEC: 613-996-6666

Outside U. S. and Canada
Chemtrec: 703-527-3887

From: CDR Chemicals, Inc.

3 Reactor Drive
Solution, ZT 98765

All non-emergency questions should be
direct to Customer Service (1-866-999-1100)
for assistance

ACETONE

MSDS Number: A0446 --- *Effective Date: 04/10/01*

1. Product Identification

Synonyms: Dimethylketone; 2-propanone; dimethylketal
CAS No.: 67-64-1
Molecular Weight: 58.08
Chemical Formula: (CH3)2CO
Product Codes:
J.T. Baker: 5356, 5580, 5805, 9001, 9002, 9003, 9004, 9005, 9006, 9007, 9008, 9009, 9010, 9015, 9036, 9125, 9254, 9271, A134, V655
Mallinckrodt: 0018, 2432, 2435, 2437, 2438, 2440, 2443, 2445, 2850, H451, H580, H981

2. Composition/Information on Ingredients

```
Ingredient                                CAS No         Percent    Hazardous
----------------------------------------  -----------    -------    ---------

Acetone                                   67-64-1        99 - 100%    Yes
```

3. Hazards Identification

Emergency Overview

DANGER! EXTREMELY FLAMMABLE LIQUID AND VAPOR. VAPOR MAY CAUSE FLASH FIRE. HARMFUL IF SWALLOWED OR INHALED. CAUSES IRRITATION TO SKIN, EYES AND RESPIRATORY TRACT. AFFECTS CENTRAL NERVOUS SYSTEM.

J.T. Baker SAF-T-DATA^(tm) Ratings (Provided here for your convenience)

--

Health Rating: 1 - Slight
Flammability Rating: 4 - Extreme (Flammable)
Reactivity Rating: 2 - Moderate
Contact Rating: 1 - Slight
Lab Protective Equip: GOGGLES; LAB COAT; VENT HOOD; PROPER GLOVES; CLASS B EXTINGUISHER
Storage Color Code: Red (Flammable)

--

Potential Health Effects

Inhalation:
Inhalation of vapors irritates the respiratory tract. May cause coughing, dizziness, dullness, and headache. Higher concentrations can produce central nervous system depression, narcosis, and unconsciousness.
Ingestion:
Swallowing small amounts is not likely to produce harmful effects. Ingestion of larger amounts may produce abdominal pain, nausea and vomiting. Aspiration into lungs can produce severe lung damage and is a medical emergency. Other symptoms are expected to parallel inhalation.
Skin Contact:
Irritating due to defatting action on skin. Causes redness, pain, drying and cracking of the skin.
Eye Contact:
Vapors are irritating to the eyes. Splashes may cause severe irritation, with stinging, tearing, redness and pain.
Chronic Exposure:
Prolonged or repeated skin contact may produce severe irritation or dermatitis.
Aggravation of Pre-existing Conditions:
Use of alcoholic beverages enhances toxic effects. Exposure may increase the toxic potential of chlorinated hydrocarbons, such as chloroform, trichloroethane.

4. First Aid Measures

Inhalation:
Remove to fresh air. If not breathing, give artificial respiration. If breathing is difficult, give oxygen. Get medical attention.

Ingestion:
Aspiration hazard. If swallowed, vomiting may occur spontaneously, but DO NOT INDUCE. If vomiting occurs, keep head below hips to prevent aspiration into lungs. Never give anything by mouth to an unconscious person. Call a physician immediately.

Skin Contact:
Immediately flush skin with plenty of water for at least 15 minutes. Remove contaminated clothing and shoes. Get medical attention. Wash clothing before reuse. Thoroughly clean shoes before reuse.

Eye Contact:
Immediately flush eyes with plenty of water for at least 15 minutes, lifting upper and lower eyelids occasionally. Get medical attention.

5. Fire Fighting Measures

Fire:
Flash point: -20C (-4F) CC
Autoignition temperature: 465C (869F)
Flammable limits in air % by volume:
lel: 2.5; uel: 12.8
Extremely Flammable Liquid and Vapor! Vapor may cause flash fire.

Explosion:
Above flash point, vapor-air mixtures are explosive within flammable limits noted above. Vapors can flow along surfaces to distant ignition source and flash back. Contact with strong oxidizers may cause fire. Sealed containers may rupture when heated. This material may produce a floating fire hazard. Sensitive to static discharge.

Fire Extinguishing Media:
Dry chemical, alcohol foam or carbon dioxide. Water may be ineffective. Water spray may be used to keep fire exposed containers cool, dilute spills to nonflammable mixtures, protect personnel attempting to stop leak and disperse vapors.

Special Information:
In the event of a fire, wear full protective clothing and NIOSH-approved self-contained breathing apparatus with full facepiece operated in the pressure demand or other positive pressure mode.

6. Accidental Release Measures

Ventilate area of leak or spill. Remove all sources of ignition. Wear appropriate personal protective equipment as specified in Section 8. Isolate hazard area. Keep unnecessary and unprotected personnel from entering. Contain and recover liquid when possible. Use non-sparking tools and equipment. Collect liquid in an appropriate container or absorb with an inert material (e.g., vermiculite, dry sand, earth), and place in a chemical waste container. Do not use combustible materials, such as saw dust. Do not flush to sewer! If a leak or spill has not ignited, use water spray to disperse the vapors, to protect personnel attempting to stop leak, and to flush spills away from exposures. U.S. Regulations (CERCLA) require reporting spills and releases to soil, water and air in excess of reportable quantities. The toll free number for the U.S. Coast Guard National Response Center is (800) 424-8802.

J.T. Baker SOLUSORB(R) solvent adsorbent is recommended for spills of this product.

7. Handling and Storage

Protect against physical damage. Store in a cool, dry well-ventilated location, away from any area where the fire hazard may be acute. Outside or detached storage is preferred. Separate from incompatibles. Containers should be bonded and grounded for transfers to avoid static sparks. Storage and use areas should be No Smoking areas. Use non-sparking type tools and equipment, including explosion proof ventilation. Containers of this material may be hazardous when empty since they retain product residues (vapors, liquid); observe all warnings and precautions listed for the product.

8. Exposure Controls/Personal Protection

Airborne Exposure Limits:
Acetone:
-OSHA Permissible Exposure Limit (PEL):
1000 ppm (TWA)

-ACGIH Threshold Limit Value (TLV):
500 ppm (TWA), 750 ppm (STEL) A4 - not classifiable as a human carcinogen
Ventilation System:
A system of local and/or general exhaust is recommended to keep employee exposures below the Airborne Exposure Limits. Local exhaust ventilation is generally preferred because it can control the emissions of the contaminant at its source, preventing dispersion of it into the general work area. Please refer to the ACGIH document, *Industrial Ventilation, A Manual of Recommended Practices*, most recent edition, for details.
Personal Respirators (NIOSH Approved):
If the exposure limit is exceeded, a half-face organic vapor respirator may be worn for up to ten times the exposure limit or the maximum use concentration specified by the appropriate regulatory agency or respirator supplier, whichever is lowest. A full-face piece organic vapor respirator may be worn up to 50 times the exposure limit or the maximum use concentration specified by the appropriate regulatory agency or respirator supplier, whichever is lowest. For emergencies or instances where the exposure levels are not known, use a full-face piece positive-pressure, air-supplied respirator. WARNING: Air-purifying respirators do not protect workers in oxygen-deficient atmospheres.
Skin Protection:
Wear impervious protective clothing, including boots, gloves, lab coat, apron or coveralls, as appropriate, to prevent skin contact.
Eye Protection:
Use chemical safety goggles and/or a full face shield where splashing is possible. Maintain eye wash fountain and quick-drench facilities in work area.

9. Physical and Chemical Properties

Appearance:
Clear, colorless, volatile liquid.
Odor:
Fragrant, mint-like
Solubility:
Miscible in all proportions in water.
Specific Gravity:
0.79 @ 20C/4C
pH:
No information found.
% Volatiles by volume @ 21C (70F):
100
Boiling Point:
56.5C (133F) @ 760 mm Hg
Melting Point:
-95C (-139F)
Vapor Density (Air=1):
2.0
Vapor Pressure (mm Hg):
400 @ 39.5C (104F)
Evaporation Rate (BuAc=1):
ca. 7.7

10. Stability and Reactivity

Stability:
Stable under ordinary conditions of use and storage.
Hazardous Decomposition Products:
Carbon dioxide and carbon monoxide may form when heated to decomposition.
Hazardous Polymerization:
Will not occur.
Incompatibilities:
Concentrated nitric and sulfuric acid mixtures, oxidizing materials, chloroform, alkalis, chlorine compounds, acids, potassium t-butoxide.
Conditions to Avoid:
Heat, flames, ignition sources and incompatibles.

11. Toxicological Information

Oral rat LD50: 5800 mg/kg; Inhalation rat LC50: 50,100mg/m3; Irritation eye rabbit, Standard Draize, 20 mg severe; investigated as a tumorigen, mutagen, reproductive effector.

```
--------\Cancer Lists\---------------------------------------------------
                                    ---NTP Carcinogen---
Ingredient                          Known    Anticipated    IARC Category
------------------------------      -----    -----------    -------------
Acetone (67-64-1)                   No          No            None
```

12. Ecological Information

Environmental Fate:
When released into the soil, this material is expected to readily biodegrade. When released into the soil, this material is expected to leach into groundwater. When released into the soil, this material is expected to quickly evaporate. When released into water, this material is expected to readily biodegrade. When released to water, this material is expected to quickly evaporate. This material has a log octanol-water partition coefficient of less than 3.0. This material is not expected to significantly bioaccumulate. When released into the air, this material may be moderately degraded by reaction with photochemically produced hydroxyl radicals. When released into the air, this material may be moderately degraded by photolysis. When released into the air, this material is expected to be readily removed from the atmosphere by wet deposition.

Environmental Toxicity:
This material is not expected to be toxic to aquatic life. The LC50/96-hour values for fish are over 100 mg/l.

13. Disposal Considerations

Whatever cannot be saved for recovery or recycling should be handled as hazardous waste and sent to a RCRA approved incinerator or disposed in a RCRA approved waste facility. Processing, use or contamination of this product may change the waste management options. State and local disposal regulations may differ from federal disposal regulations. Dispose of container and unused contents in accordance with federal, state and local requirements.

14. Transport Information

Domestic (Land, D.O.T.)

Proper Shipping Name: ACETONE
Hazard Class: 3
UN/NA: UN1090
Packing Group: II
Information reported for product/size: 350LB

International (Water, I.M.O.)

Proper Shipping Name: ACETONE
Hazard Class: 3.1
UN/NA: UN1090
Packing Group: II
Information reported for product/size: 350LB

15. Regulatory Information

```
--------\Chemical Inventory Status - Part 1\-------------------------------
Ingredient                                        TSCA  EC   Japan  Australia
--------------------------------------------------  ----  ---  -----  ---------
Acetone (67-64-1)                                 Yes   Yes  Yes    Yes

--------\Chemical Inventory Status - Part 2\-------------------------------
                                                        --Canada--
Ingredient                                        Korea  DSL  NDSL  Phil.
--------------------------------------------------  -----  ---  ----  -----
Acetone (67-64-1)                                 Yes   Yes  No    Yes

--------\Federal, State & International Regulations - Part 1\----------------
                                                  -SARA 302-    ------SARA 313------
Ingredient                                        RQ   TPQ     List  Chemical Catg.
------------------------------------------  ---  -----   ----  --------------
Acetone (67-64-1)                                 No   No      Yes   No

--------\Federal, State & International Regulations - Part 2\----------------
                                                       -RCRA-    -TSCA-
Ingredient                                        CERCLA  261.33    8(d)
------------------------------------------  ------  ------   ------
Acetone (67-64-1)                                 5000    U002      No
```

```
Chemical Weapons Convention:  No     TSCA 12(b):  Yes    CDTA:  Yes
SARA 311/312:  Acute: Yes       Chronic: No    Fire: Yes Pressure: No
Reactivity: No              (Pure / Liquid)
```

Australian Hazchem Code: 2[Y]E
Poison Schedule: No information found.
WHMIS:
This MSDS has been prepared according to the hazard criteria of the Controlled Products
Regulations (CPR) and the MSDS contains all of the information required by the CPR.

---

16. Other Information

NFPA Ratings: Health: 1 Flammability: 3 Reactivity: 0
Label Hazard Warning:
DANGER! EXTREMELY FLAMMABLE LIQUID AND VAPOR. VAPOR MAY
CAUSE FLASH FIRE. HARMFUL IF SWALLOWED OR INHALED. CAUSES
IRRITATION TO SKIN, EYES AND RESPIRATORY TRACT. AFFECTS CENTRAL
NERVOUS SYSTEM.
Label Precautions:
Keep away from heat, sparks and flame.
Keep container closed.
Use only with adequate ventilation.
Wash thoroughly after handling.
Avoid breathing vapor.
Avoid contact with eyes, skin and clothing.
Label First Aid:
Aspiration hazard. If swallowed, vomiting may occur spontaneously, but DO NOT
INDUCE. If vomiting occurs, keep head below hips to prevent aspiration into lungs.
Never give anything by mouth to an unconscious person. Call a physician immediately. If
inhaled, remove to fresh air. If not breathing, give artificial respiration. If breathing is
difficult, give oxygen. In case of contact, immediately flush eyes or skin with plenty of
water for at least 15 minutes. Remove contaminated clothing and shoes. Wash clothing
before reuse. In all cases, get medical attention.
Product Use:
Laboratory Reagent.
Revision Information:
No changes.

Material Safety Data Sheet

U.S. Department of Labor

May be used to comply with

OSHA's Hazard Communication Standard,
29 CFR 1910.1200. Standard must be
consulted for specific requirements.

Occupational Safety and Health
Administration
(Non-Mandatory Form)
Form Approved
OMB No. 1218-0072

IDENTITY *(As Used on Label and List)*	Note: Blank spaces are not permitted. If any item is not applicable, or no information is available, the space must be marked to indicate that.

Section I

Manufacturer's Name	Emergency Telephone Number
Address *(Number, Street, City, State, and ZIP Code)*	Telephone Number for Information
	Date Prepared
	Signature of Preparer *(optional)*

Section II - Hazard Ingredients/Identity Information

Hazardous Components (Specific Chemical Identity; Common Name(s))	OSHA PEL	ACGIH TLV	Other Limits Recommended	%*(optional)*

Section III - Physical/Chemical Characteristics

Boiling Point		Specific Gravity ($H_2O = 1$)	
Vapor Pressure (mm Hg.)		Melting Point	
Vapor Density (AIR = 1)		Evaporation Rate (Butyl Acetate = 1)	
Solubility in Water			
Appearance and Odor			

Section IV - Fire and Explosion Hazard Data

Flash Point (Method Used)	Flammable Limits	LEL	UEL
Extinguishing Media			
Special Fire Fighting Procedures			
Unusual Fire and Explosion Hazards			

(Reproduce locally)   OSHA 174, Sept. 1985

Section V - Reactivity Data

Stability	Unstable		Conditions to Avoid
	Stable		
Incompatibility *(Materials to Avoid)*			
Hazardous Decomposition or Byproducts			
Hazardous Polymerization	May Occur		Conditions to Avoid
	Will Not Occur		

Section VI - Health Hazard Data

Route(s) of Entry:	Inhalation?	Skin?	Ingestion?

Health Hazards *(Acute and Chronic)*			

Carcinogenicity:	NTP?	IARC Monographs?	OSHA Regulated?

Signs and Symptoms of Exposure

Medical Conditions Generally Aggravated by Exposure

Emergency and First Aid Procedures

Section VII - Precautions for Safe Handling and Use

Steps to Be Taken in Case Material is Released or Spilled

Waste Disposal Method

Precautions to Be taken in Handling and Storing

Other Precautions

Section VIII - Control Measures

Respiratory Proctection *(Specify Type)*		
Ventilation	Local Exhaust	Special
	Mechanical *(General)*	Other
Protective Gloves		Eye Protection
Other Protective Clothing or Equipment		
Work/Hygienic Practices		

U.S.G.P.O.: 1986 - 491 - 529/45775

Appendix C   Personal Protective Equipment Hazard Assessment

Date:	Location:
Assessment Conducted By:	
Specific Tasks Performed at this Location:	

Hazard Assessment and Selection of Personal Protective Equipment

I. Overhead Hazards -
- Hazards to consider include:
- Suspended loads that could fall
- Overhead beams or loads that could be hit against
- Energized wires or equipment that could be hit against
- Employees work at elevated site who could drop objects on others below
- Sharp objects or corners at head level

Hazards Identified:

Head Protection

Hard Hat:	Yes	No
If yes, type:		

    ☐ **Type A** (impact and penetration resistance, plus low-voltage electrical insulation)

    ☐ **Type B** (impact and penetration resistance, plus high-voltage electrical insulation)

    ☐ **Type C** (impact and penetration resistance)

II. Eye and Face Hazards -
- Hazards to consider include:
- Chemical splashes
- Dust
- Smoke and fumes
- Welding operations
- Lasers/optical radiation
- Projectiles

Hazards Identified:

Eye Protection

Safety glasses or goggles	Yes	No
Face shield	Yes	No

III. Hand Hazards -
- Hazards to consider include:
- Chemicals
- Sharp edges, splinters, etc.
- Temperature extremes
- Biological agents
- Exposed electrical wires
- Sharp tools, machine parts, etc.
- Material handling

Hazards Identified:

Hand Protection

Gloves	Yes	No
☐ Chemical resistant ☐ Temperature resistant ☐ Abrasion resistant ☐ Other (Explain)		

IV. Foot Hazards -
- Hazards to consider include:
- Heavy materials handled by employees
- Sharp edges or points (puncture risk)
- Exposed electrical wires
- Unusually slippery conditions
- Wet conditions
- Construction/demolition

Hazards Identified:

Foot Protection

Safety shoes	Yes	No
Types:		
☐ Toe protection		
☐ Metatarsal protection		
☐ Puncture resistant		
☐ Electrical insulation		
☐ Other (Explain)		

V. Other Identified Safety and/or Health Hazards:

Hazard	Recommended Protection

I certify that the above inspection was performed to the best of my knowledge and ability, based on the hazards present on _____

_____

(Signature)

Appendix D   Assault Incident Report Form

Note: This type of form can be used to report any threatening remark or act of physical violence against a person or property, whether experienced or observed. Individuals may be more forthcoming with information if the form is understood to be voluntary and confidential. The form also needs to identify where it should be sent after completion (for example, workplace violence prevention group or safety committee representative).

Date of incident	Year	Month	Day of week
Location of incident (Map, sketch on reverse side)			

Name of victim	Gender Male___Female___
Victim description _____Employee job title_____ _____Client _____Visitor	Member of labor organization? Yes_____No_____

Assigned work location (if employee)

Supervisor	Has supervisor been notified? Yes_____ No_____

Describe the assault incident.

List any witnesses to the incident (name and phone).

Did the assault involve a firearm? If so, describe.

Did the assault involve another weapon (not a firearm)? If so, describe.

Was the victim injured? If yes, please describe.

Who was responsible for assault?

_____Stranger   _____Coworker   If other, describe.
_____Personal relation   _____Supervisor
_____Client/patient/patron/customer   _____Other

What was the gender of the person(s)   _____Male
who committed the assault?   _____Female

Please check any risk factors applicable to this incident:

Note: Each company should develop and include a list of potential risk factors that may apply in its worksite. For example:

_____working with money

_____working in a high-crime area

_____working with drugs

What steps could be taken to avoid a similar incident in the future?
(To avoid recreating trauma, sound judgment should be exercised in deciding when to request this information.)

Bibliography

American Nation Standards Institute. *Combustion Engine-Powered Industrial Trucks*, No. 558 (ANSI B56.4-1980). New York, 1980.

Blosser, F. *Primer on Occupational Safety and Health*. Washington DC: The Bureau of National Affairs, Inc., 1992.

Bureau of Labor Statistics. *Fatal Workplace Injuries in 1992, A Collection of Data and Analysis*. Report 870. Washington, DC, April 1994.

Bureau of Labor Statistics, U.S. Department of Labor. *National Census of Fatal Occupational Injuries*. Washington, DC, 1999.

Bureau of Labor Statistics, U.S. Department of Labor. *Career Guide to Industries, 2006–07 Edition*. Utilities. Available at http://www.bls.gov/oco/cg/cgs018.htm, September 20, Washington, DC, 2006.

Bureau of Labor Statistics, U.S. Department of Labor. Available at http://www.bls.gov, Washington DC, 2007.

Bureau of Labor Statistics, U.S. Department of Labor. *National Census of Fatal Occupational Injuries in 2005*. Washington, DC, Available at http://bls.gov.

Bureau of Labor Statistics, U.S. Department of Labor. *Workplace Injuries and Illnesses in 2004*. Washington, DC, Available at http://bls.gov.

Cailliet, R. *Low Back Pain Syndrome*, 2nd ed. Philadelphia: F.A. Davis Co., 1968.

California Department of Industrial Relations (Cal/OSHA). *Easy Ergonomics: A Practical Approach for Improving the Workplace*. Sacramento, CA, 1999.

California Department of Labor. *Guidelines for Security and Safety of Health Care and Community Service Workers*. Sacramento, March 1998. Available at http://www.ca.gov.

Centers for Disease Control and Prevention, U.S. Department of Health and Human Services. *Occupational Health and Safety Manual*. Atlanta, GA, 2002.

Chaffin, D.B. and Andersson, G.B.J. *Occupational Biomechanics*. New York: John Wiley & Sons, 1984.

Cohen, A.L. et al. *Elements of Ergonomics Programs*. Cincinnati, OH: U.S. Department of Health and Human Services/NIOSH, 1997.

Eastman Kodak Company, *Ergonomic Design for People at Work*, Vol. 1 and 2. Belmont, CA: Lifetime Learning Publications, 1983.

Jensen, R. Lift truck training: It's here and it works. *Modern Materials Handling*. pp. 72–78. September 1989.

Lifting Safety: Tip to Help Prevent Back Injuries. Available at http://familydoctor.org/174.xml, 2006.

Mine Safety and Health Administration, U.S. Department of Labor. *Accident Prevention, Safety Manual No. 4*. Beckley, WV. Revised in 1990.

Mine Safety and Health Administration, U.S. Department of Labor. *Job Safety Analysis: A Practical Approach, Instruction Guide No. 83*. Beckley, WV, 1990.

Mine Safety and Health Administration, U.S. Department of Labor. *Job Safety Analysis, Safety Manual No. 5*. Beckley, WV. Revised in 1990.

Mine Safety and Health Administration, U.S. Department of Labor. *Safety Observation, MSHA IG 84*. Beckley, WV. Revised in 1991.

Mital, A., Nicholson, A.S., and Ayoub, M.M. *A Guide to Manual Materials Handling*. New York: Taylor & Francis, 1993.

Moran, M.M. *Construction Safety Handbook*. Rockville, MD: Government Institutes Inc., 1996.

National Institute for Occupational Safety and Health (NIOSH), DHHS. *Work Practices Guide for Manual Lifting.* 1981 and 1991.

National Institute for Occupational Safety and Health (NIOSH), U.S. Department of Health and Human Services. *Violence in the Workplace: Risk Factors and Prevention Strategies (CIB 57).* Washington DC, June 1996.

National Institute for Occupational Safety and Health, U.S. Department of Health and Human Services. *Elements of Ergonomics Programs (DHHS 97-117).* 1997.

National Mine Health and Safety Academy, U.S. Department of Labor. *Accident Prevention Techniques.* Beckley, WV, 1984.

Occupational Safety and Health Administration, U.S. Department of Labor. *Office of Construction and Maritime Compliance Assistance: OSHA Instruction STD 3-1.1.* Washington, DC, June 22, 1987.

Occupational Safety and Health Administration, U.S. Department of Labor. *Federal Register: Safety and Health Program Management Guidelines.* Vol. 54, No. 16, pp. 3904–3916, Washington, DC, January 26, 1989.

Occupational Safety and Health Administration, U.S. Department of Labor. *OSHA 10 and 30 Hour Construction Safety and Health Outreach Training Manual.* Washington DC, 1991.

Occupational Safety and Health Administration, U.S. Department of Labor. *Walking–Working Surfaces (Slips, Trips, and Falls).* Washington DC, 1992.

Occupational Safety and Health Administration, Office of Training and Education, U.S. Department of Labor. *OSHA Voluntary Compliance Outreach Program: Instructors Reference Manual.* Des Plaines, IL, 1993.

Occupational Safety and Health Administration, U.S. Department of Labor. *Field Inspection Reference Manual (FIRM): OSHA Instruction CPL 2.103.* Washington DC, September 26, 1994.

Occupational Safety and Health Administration, U.S. Department of Labor. *Citation Policy for Paperwork and Written Program Requirement Violations: OSHA Instruction CPL 2.111.* Washington DC, November 27, 1995.

Occupational Safety and Health Administration, U.S. Department of Labor. *General Industry Digest (OSHA 2201).* Washington DC: GPO, 1995.

Occupational Safety and Health Administration, U.S. Department of Labor. *Voluntary Compliance Outreach Program, Walking and Working Surfaces—Subpart D.* Washington DC, 1995.

Occupational Safety and Health Administration, U.S. Department of Labor, OSHA Web site for Small Business. *Flammable and Combustible Liquids—1910.106.* Washington DC, 1997.

Occupational Safety and Health Administration, U.S. Department of Labor. *Stairways and Ladders (OSHA 3124).* Washington DC, 1997.

Occupational Safety and Health Administration, Construction, U.S. Department of Labor. Code of Federal Regulations. Title 29, Part 1926. Washington DC: GPO, 1998.

Occupational Safety and Health Administration, General Industry, U.S. Department of Labor. Code of Federal Regulations. Title 29, Part 1910. Washington DC: GPO, 1998.

Occupational Safety and Health Administration, U.S. Department of Labor. Subject Index, "Internet." April, 1999. Available at http://www.osha.gov.

Occupational Safety and Health Administration, U.S. Department of Labor. *29* Code of Federal Regulations *1910.* Washington DC: GPO, 1999.

Occupational Safety and Health Administration, U.S. Department of Labor. *29* Code of Federal Regulations *1926.* Washington DC, 1999.

Occupational Safety and Health Administration, Office of Training and Education, U.S. Department of Labor. *Manual for Trainer Course in OSHA Standards for the General Industry.* Des Plaines, IL, 2001.

Occupational Safety and Health Administration, U.S. Department of Labor. *How to Plan for Workplace Emergencies and Evacuations (OSHA 3088)*. Washington DC, 2001.

Occupational Safety and Health Administration, U.S. Department of Labor. *Job Hazard Analysis and Control (1910.917–922)*, Subject Index, "Internet." 2001. Available at http://www.osha.gov.

Occupational Safety and Health Administration, U.S. Department of Labor. *Material Handling and Storage (OSHA 2236)*. Washington DC, 2002.

Occupational Safety and Health Administration, U.S. Department of Labor. *Hand and Power Tools (OSHA 3080)*. Washington DC, 2002.

Occupational Safety and Health Administration, U.S. Department of Labor. *Personal Protective Equipment (OSHA 3141-12R)*. Washington DC, 2003.

Occupational Safety and Health Administration. Available at http://www.osha.gov. Washington DC, 2006.

Occupational Safety and Health Administration, General Industry, U.S. Department of Labor. Code of Federal Regulations. Title 29, Part 1910. Washington DC: GPO, 2006.

Occupational Safety and Health Administration, U.S. Department of Labor. *OSHA Handbook for Small Businesses (OSHA 2209)*. Washington DC, 2006.

Occupational Safety and Health Administration. *OSHA Technical Manual*. Washington DC, 2007.

Occupational Safety and Health Administration, U.S. Department of Labor. Available at http://www.osha.gov, 2007.

Office of Nuclear Energy, U.S. Department of Energy. *Root Cause Analysis Guidance Document*. Washington DC, February, 1992.

Operator training: More than just driving a lift truck. *Materials Handling Engineering*. pp. 33–46. June 1990.

Petersen, D. *Techniques of Safety Management: A Systems Approach*, 3rd ed. Goshen, NY: Aloray Inc., 1989.

Reese, C.D. *Mine Safety and Health for Small Surface Sand/Gravel/Stone Operations: A Guide for Operators and Miners*. Storrs, CT: University of Connecticut Press, 1997.

Reese, C.D. *Material Handling Systems: Designing for Safety and Health*. New York: Taylor & Francis, 2000.

Reese, C.D. *Accident/Incident Prevention Techniques*. New York: Taylor & Francis, 2001.

Reese, C.D. *Annotated Dictionary of Construction Safety and Health*. Boca Raton, FL: CRC/Lewis Publishers, 2000.

Reese, C.D. *Occupational Health and Safety Management: A Practical Approach*. Boca Raton, FL: CRC/Lewis Publisher, 2003.

Reese, C.D. *Office Building Safety and Health*. Boca Raton, FL: CRC Press, 2004.

Reese, C.D. and Eidson, J.V. *Handbook of OSHA Construction Safety & Health*, 2nd ed. Boca Raton, FL: CRC/Lewis Publishers, 1999.

Schwind, G.F. Forklift training. *Industrial Maintenance and Plant Operation*. p. 34. January 1990.

U.S. Department of Energy. *Hoisting and Rigging Manual*. Washington DC, 1991.

U.S. Department of Energy. *OSH Technical Reference Manual*. Washington DC, 1993.

U.S. Department of Labor. *Occupational Safety and Health Standards for General Industry (29 CFR 1910)*. Washington DC, 2003.

U.S. Office of Personnel Management. *Dealing with Workplace Violence: A Guide for Agency Planners*. Washington DC, February 1998.

Werner Ladder Company, *Climbing Pro Training Manual*. Greenville, PA, 1996.

Index

A

Accidents
  causes of, 88
    direct, 89–90
    environmental factors, 94
    indirect, 90
    personal factors, 93–94
    policies and decisions, 93
    unsafe acts and conditions, 92
  compressed gases and
    cylinder, 140–148
    general precautions of, 152
    hazards of, 154–156
    storage, 152–154
  definition of, 6
Advanced notice of proposed rulemaking
      (ANPRM), 55
Air-purifying respirators, 178
American Conference of Governmental Industrial
      Hygienist (ACGIH), 106, 139
American National Standards Institute (ANSI),
      55, 301
American Society of Mechanical Engineers
      (ASME), 301, 306
ANSI Z87.1 standard, 223
Approved industrial trucks, 291–292
Asphyxiation, 155; *see also* Chemical hazards
Axes, 228–229; *see also* Hand tools

B

Back injuries and disorders, 252–253
Box and socket wrenches, 229; *see also* Hand tools
Bureau of Labor Statistics (BLS) data, 13, 97, 259,
      311, 351

C

Carcinogens, 108–109; *see also* Health hazards
Cardiopulmonary resuscitation (CPR), 348;
      see also Chemical hazards
Carpenter's/claw hammer, 229; *see also* Hand
      tools
Census of Fatal Occupational Injuries
      (CFOI), 352
CFR numbering system, 59–60; *see also*
      Occupational Safety and Health
      Administration (OSHA)

Chemical Abstracts Service (CAS), 108
Chemical hazards
  cancer-causing chemicals, 133–134
  monitoring surveys for, 132
    airborne contaminants, types
      of, 127–128
  chemicals, 106–109
    acids and bases, 130–131
    adhesives and sealants, 131
    cleaners, 129–130
    fuels, 131–132
    paints, 131
    solvents, 129
  entry routes and action modes
    eyes, 119–121
    ingestion, 124
    lungs and inhalation, 121–122
    respiration, 122–123
    skin absorption, 123–124
    exposure guidelines, 125–127
Chisels, 229; *see also* Hand tools
Clothing, accessory, and general merchandise
      stores, 27–28, 35–37
Clothing and accessory store managers, 37;
      see also Retail trade
Code of Federal Regulations (CFR), 57
Compressed Gas Association (CGA), 140
Compressed gases, 137–138
  corrosive and toxic gases, 138–140
  cryogenic accidents
    prevention of, 157
    safety of, 151–156
  cylinder accidents prevention
    compressed air safety guidelines,
      150–151
    cryogenic safety for, 151–156
    cylinders usage, 140–141
    handling of, 141–143
    identification and colour coding of,
      147–148
    movement and transportation,
      146
    OSHA compressed gas regulations,
      157–158
    safety usage, 140–141
    storage of, 143–146
  and cylinder checklist, 159–161
  hoses and regulators of, 148–150
  OSHA regulations, 157–159
  safety guidelines of, 150–151
  service industry, 157

403